SPIDERS
OF LEICESTERSHIRE & RUTLAND

JOHN CROCKER • JONATHAN DAWS

Illustrations by M J Roberts

LOUGHBOROUGH NATURALISTS' CLUB

in association with
KAIROS PRESS
Newtown Linford
Leicestershire

1996

Copyright © John Crocker & Jonathan Daws, 1996
ISBN 1-871344-09-3
First Edition 1996

Printed in Great Britain, by AplhaGraphics, Leicester, using DocuTech.
Body text in Century Schoolbook BT, 10.5pt. and Minister Book, 10.5pt.

Cover picture by M J Roberts — *Mastigusa macrophthalma* m. (x10)

Cover design by David Rowell, The Really Useful Company.

Spider illustrations by M J Roberts.

LOUGHBOROUGH NATURALISTS' CLUB

in association with

Kairos Press
552 Bradgate Road
Newtown Linford
Leicester LE6 0HB

SPIDERS
OF LEICESTERSHIRE & RUTLAND

Publication of this book has been made
possible by sponsorship grants from

In addition, there has been generous
grant aid from the following:–

British Arachnological Society
Leicestershire Entomological Society
Leicestershire Museums Service
British Gas plc
Charnwood Borough Council
Charnwood Wildlife Project
Humphries Rowell Associates Ltd
Leicestershire & Rutland Trust
for Nature Conservation
Mrs A. Marmont
Bardon Roadstone plc

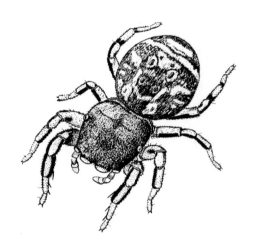

Ballus chalybeius f. (x8.5)

TABLE OF CONTENTS

Foreword 7
Acknowledgements 8

CHAPTER ONE
Introduction to Leicestershire 9
    The Regions 10
    Geology 13
    Climate and Weather 15
    Woodland 16
    Wetland 17
    Grassland, Moorland and Heath 20

CHAPTER TWO
Check List of the Spiders of Leicestershire and Rutland — VC55 22

CHAPTER THREE
An Historical Review of Arachnology in Leicestershire and Rutland, Watsonian Vice-county 55 30
    Introduction 30
        PART I 1795-1955
    The Cradle of Natural Philosophy 31
    Early Stirrings of Interest 32
    William Agar 33
    George B. Chalcraft 34
    The Victoria County History 36
    The Early Twentieth Century 37
    H. St.J. K. Donisthorpe 39
    The Comity of Spiders (W.S. Bristowe, 1939-41) 40
    Leicestershire Fauna Survey (1947-1953) 42
        PART II 1955-1995
    The Post-1959 Period 44
    Literature – Local Publications 46
    Literature – the National Scene 47
    Rutland 50
    Important Identification Aids 51
    Garden and House Spiders 52
    Revival 54
    First Spider Records for VC55 55
    Detailed Field Records for VC55 57

CHAPTER FOUR
Material and Methods, Voucher Specimens and Red Data Book Species 61
    Material 61
    Participation 62
    Methods 63
    Collecting 63
    Spirit Collections 65
    Recording 66
    Voucher Material 68
    Living Fossils 70
    Mastigusa in Baltic Amber 71
    Lepthyphantes midas 72
    Lepthyphantes beckeri 74

CHAPTER FIVE
Atlas of Leicestershire and Rutland Spiders — 76

CHAPTER SIX
Habitat Evaluation — 188
   The Natural Areas Concept — 188
   Leicestershire Natural Areas — 189
   The Lincolnshire Limestone (NA20) — 190
   Charnwood Forest (NA19) — 190
   Trent Valley and Levels (NA18) — 193
   Factors Affecting the Distribution of Spiders — 194
   Habitat Diversity — 195
   Site Evaluation — 196
   Woodland — 197
   The Forest Vision – a New National Forest — 199

APPENDIX I
Site Statistics and Gazeteer — 200
   List of 100 Sites with Spider Statistics and Scores — 200
   Gazetteer of Main Sites — 202

APPENDIX II
Forty-nine Key Sites in the County, and their Spiders — 213
   Schedule of the 49 sites — 214
   Table a) Calcareous and heath grassland — 215
   Table b) Coal Measures heath grassland — 219
   Table c) Moorland and parkland — 222
   Table d) Riparian habitats — 226
   Table e) Reservoirs and open water — 229
   Table f) Woodland — 232
   Table g) House and garden spiders — 235

APPENDIX III
Natural History Societies — 238
   The British Arachnological Society — 238
   The Spider Recording Scheme — 238
   Leicestershire Entomological Society — 238
   Natural History Section of the
       Leicester Literary and Philosophical Society — 239
   Loughborough Naturalists' Club — 238

REFERENCES — 240

INDEX — 243

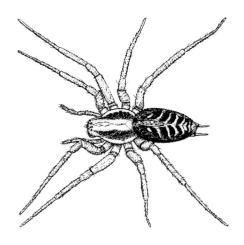

Agelena labyrinthica f. (x3)

FOREWORD
by David R. Nellist
President of the British Arachnological Society

Many years ago, in his foreword to a book on Northaw Great Wood in Hertfordshire, the naturalist and author Maxwell Knight wrote that although he was honoured to have been asked to write it he did not feel qualified for the task. I echo his sentiments, and yet I have great pleasure in contributing the foreword to a remarkable book and in congratulating the authors on their achievement.

I have known John Crocker now for nearly thirty years, indeed it was John who first persuaded me to join the Council of the British Arachnological Society, almost as many years ago. I clearly remember the day we met – on a Sunday at a field meeting at Box Hill, Surrey. As a relative newcomer to arachnology I was overwhelmed by his enthusiasm, his knowledge and his attention to detail, and by the fact that his species list, together with notes on some of the rarer species he had collected, dropped through my letter box on the following Tuesday morning! These aspects of his personality have been amply confirmed during the intervening years, particularly in his contributions and service to the British Arachnological Society, of which he is an Honorary Member, and by his devotion to the study of the spiders of Leicestershire. I know Jon Daws less well, but we have worked together out in the field, and I certainly remember his contributions to a survey of the shingle ridges on Orford Ness off the Suffolk coast, organised by the Spider Recording Scheme back in 1994. His innovative approach to collecting impressed me and ensured that he sampled habitats overlooked by other members of the survey team.

The collaborative effort of these two authors has produced an excellent and timely book. It contains, amongst other topics, a check-list of the spiders of Leicestershire and Rutland (VC55), information on the diversity of the habitats within the area, accounts of key sites and their spider fauna, and an atlas showing the distribution of all the recorded species within the vice-county. Not least it is published at a time when the need for such information by, for example, site managers, planners and others, has never been greater. The data is needed in the ongoing struggle to slow and then halt the accelerating pace of destruction, change and isolation of invertebrate sites and the loss of the species they contain. The recently published report of the Biodiversity Steering Group, established by the Government in the wake of the Biodiversity Convention signed by the Prime Minister in Rio de Janeiro in 1992, contains proposals for the preparation of action plans for the conservation of a wide range of habitats and species (including spiders) within the United Kingdom. The criteria for the selection of those species for which action plans will be written include: species whose numbers and range have declined substantially in recent years, are endemic, or under threat. It is acknowledged that the gathering and interpretation of the data necessary for the application of these criteria, and the selection of key sites, will only be possible with the cooperation of enthusiastic amateurs who have the appropriate expertise and the specialised knowledge to survey and then monitor the fauna of their local areas.

The authors of this book are two such enthusiastic and expert amateurs. They have assembled over 19 000 individual records on the occurrence and distribution of spiders within Leicestershire and Rutland, have listed 326 species including one new to Britain and several included in the Red Data Book or regarded as Nationally Notable, and then assigned a score to one hundred sites based on the number and status of local and rare species. They have undoubtedly placed VC55 in the vanguard of such studies. Of course, in so doing they have highlighted the very unsatisfactory situation in many of the other vice-counties in the United Kingdom where we have a paucity of information on the occurrence and distribution of the spider fauna. However this book will, I am sure, be read with great interest and studied in detail, and will now serve as the model for future work in these under-recorded vice-counties. Again I congratulate the authors on their achievement.

David R. Nellist
St. Albans, Hertfordshire
March 1996

ACKNOWLEDGEMENTS

Throughout this book, the authors have endeavoured to acknowledge sources of their information, and to credit all who have submitted records or collected specimens for identification. Many people have contributed one or more records and these are acknowledged elsewhere - in particular on pages 44-45, 51, 54-56 and 61-62; these embrace a large number of individuals to whom the authors are indebted for their help and co-operation, and we apologise to any whose names have been inadvertently omitted.

We are grateful to the many authorities, landowners and tenants who have allowed us access to their property to look for spiders. We also express our gratitude to the following institutions: British Museum (Natural History), Leicestershire Museums Service, National Museums and Galleries on Merseyside, and Manchester University Museum, for the loan of specimens. We owe a special debt of gratitude to Dr Eric Duffey, Peter Harvey, Ian Evans and Derek Lott for inspiration, support and encouragement over many years; also to John Mousley, Jan Dawson, Steven Grover, Jane McPhail, Anona Finch, Harry Ball and Deborah Proctor for their practical help and advice. Ian Evans has allowed us to draw heavily on his notes in our preparation for the historical section of this book; Tony Russell-Smith, Vlastimil Ruzicka and Søren Langemark have not only lent specimens of *Lepthyphantes midas*, but have also allowed us to use their notes, for which we are most grateful. Dr Peter Merrett has provided us with useful information on points of ecology, distribution and taxonomy, and has rendered much assistance by confirming the identities of certain critical species. We would also like to thank Mr John Martin for his assistance in the interpretation of the geology of the county. We express our gratitude to David Nellist for consenting to write the Forward, to David Rowell for his distinctive cover design and careful attention in processing his initial sketches into the finalised covers, and to Mike Roberts for the meticulous way in which the drawings of spiders were prepared at short notice. Also our thanks go to Fred Wanless for his illustrations of *Lepthyphantes beckeri* genitalia, prepared over twenty years ago! To Peter Gamble and Ian Butterfield we are beholden for constructive advice and contributions to the content of this book, and to the Leicestershire County Council for permission to use information from their 1993 population survey in the preparation of our map on page 11. Our thanks also go to Tony Squires, Ian Evans, Peter Merrett, Marcene Crocker and Judy Johnson for reading preliminary drafts of the manuscript and making useful suggestions, and particularly to Elizabeth Orr and John Ward for their painstaking reading and correction of the final manuscript.

Without the availability of a suitable computer mapping program, it is unlikely that this publication would have been undertaken. The authors are indebted to Dr Alan Morton, author of 'DMAP for Windows', for advice and practical help in this particular application, and especially for his willingness to enhance the program to ensure we could obtain the highest quality output. Likewise, our thanks are also expressed to Stanley Dobson for the program 'SPIREC', written primarily at our request as a means of computerising the 19 000 field records of Leicestershire spiders, which are the basis of this book. Stan was also most helpful in writing protocols for transferring field records to the mapping program.

The Loughborough Naturalists' Club and the authors express their gratitude to the sponsors and those who have generously supported this publication with grant-aid, which has enabled this specialist work to be published at a much lower price than would otherwise have been possible. The authors would like to thank the Loughborough Naturalists' Club and Kairos Press for undertaking publication of this work. We are extremely grateful to Robin Stevenson of Kairos Press for the great care he has taken in the production of this book, and for his co-operation in bringing the whole work together and handling the practical details of publication

Overseeing design and publication has been the responsibility of John Crocker, Chairman of the L.N.C. Publications sub-committee. He would like to express his personal thanks to members of the sub-committee, to Judy Johnson for her efficient attention to detail in providing secretarial services, to John Ward for keeping financial matters under control, and to Helen Ikin for carrying out research for the gazeteer; and finally to his wife Marcene who has helped him in the ways only a wife can - as a constant and enthusiastic companion during forty years of arachnological activities, for encouragement when things have been difficult and a great practical help in everything he has done in the field of natural history since 1956.

CHAPTER ONE

Introduction to Leicestershire

Leicestershire and Rutland together make up the Watsonian Vice-county 55 (Dandy, 1969), collectively known as Leicestershire. Situated close to the centre of England it is entirely landlocked and as far from the sea as anywhere in Britain. It covers an area of 255 372 hectares (986 square miles). The highest point is the summit of Bardon Hill, 278 metres (912 feet) above sea level on the south-west edge of Charnwood Forest, with the lowest point at Lockington Grounds on the northern boundary with Nottinghamshire, lying at 27 metres (90 feet) where the River Soar joins the River Trent. The majority of the county lies below 122 metres (400 feet) with four isolated upland areas – Charnwood Forest in the north-west, the Southern Heights in the extreme south, the Eastern Wolds in central-east Leicestershire, and the Northern Wolds in the north-east bordering the Vale of Belvoir (see map 1).

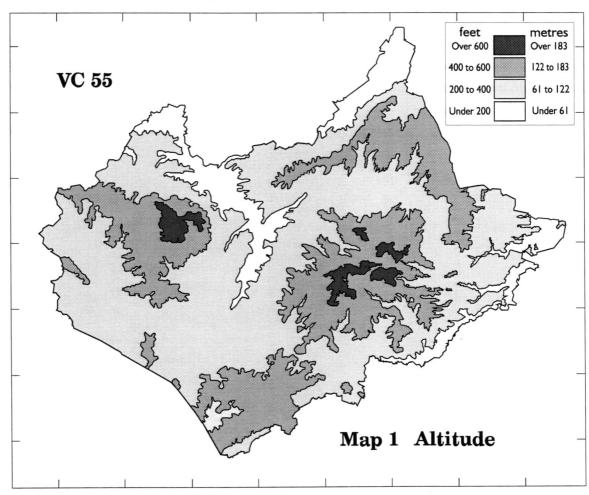

Map 1 Altitude

Leicestershire is bounded by six counties, Staffordshire, Derbyshire, Nottinghamshire, Lincolnshire, Northamptonshire and Warwickshire, and stands as a geographical stepping stone between the high ground of northern and western Britain and the English lowlands to the south and east. More so in the past than today, Leicestershire has been influenced by the topography and ecosystems of its surrounding counties and still retains pockets of residual habitat reminiscent of grander examples elsewhere. An intensively agriculturalised county, overworked by extractive industries and under pressure from a growing population and expanding urban and rural infrastructure, it nevertheless preserves an interesting flora and fauna which is evident from the material presented in this publication.

The extensive Staffordshire moorland beyond the western boundary of Leicestershire merges northwards into the high ground of Derbyshire's Peak District, which marks the southern limit of the true uplands of northern Britain. On the eastern flank of the High Peak Nottinghamshire's gentle lowland landscape is typified by the large tract of parkland and woodland around Sherwood, formerly containing the estates of five dukes and including the distinctive Budby heathland and Birklands ancient forest. A dominant feature of the landscape is the River Trent which drains virtually the whole of Nottinghamshire and has considerable influence on its tributaries as a wetland corridor, notably the River Soar in Leicestershire. Lincolnshire's coast, with its extensive foreshore and expansive intertidal zone, sand dunes and salt marshes, is close enough to have a considerable influence on distant Leicestershire, and despite the loss of traditional fenland, carr woodland and inland marshes, which were once a major feature of this low-lying coastal county, this influence on Leicestershire is still evident. Again, we must consider the historical implications of the proximity of Northamptonshire's finest and richest ancient woodlands, remnants of the great forests of Rockingham, Salcey and Whittlebury on adjoining south-eastern Leicestershire; and prime calcareous grassland to match anything in the Cotswolds, with remnants such as that at Barnack Hills and Holes. As with the River Trent in the north, so also in the south-east, the River Welland at one time provided a linear highway along its flood-plain between inland sites and the maritime wetlands around the coast. Despite river dredging, straightening and deepening, remnant wetland habitats survive. Warwickshire's greatest influence on Leicestershire has been the atmospheric pollution brought in from the industrial heartland since the eighteenth century.

Many people know Leicestershire only as a county they have passed through by means of its large network of major roads, with the city of Leicester now a traffic island since the creation of its new inner and outer ring roads. But despite the self-evident unattractiveness of its urban sprawl and agrarian monotony, Leicestershire has some pleasant countryside and an interesting flora and fauna influenced by both its geographical position and its rich history. As with most of Britain, the Leicestershire countryside is man-made. It is characterised by a rolling landscape of predominantly agricultural use, scattered with small spinneys and farming villages centred around their parish church. There are small remnants of semi-natural habitats that have survived the plough and occupy land that is unprofitable to improve, together with sites of former industries such as quarries and railways which provide sanctuary for a hard-pressed flora and fauna that have all but disappeared from ancient strongholds. On a miniature scale, Charnwood Forest represents the upland habitats of northern Britain, the county's river valleys contain remnant wetland faunas and the pockets of calcareous grassland of east Leicestershire and Rutland provide a foothold for a number of rare plants and invertebrates not found elsewhere in the county. Although Leicestershire is poorly wooded, some of the prime sites that have survived are of extremely high quality and strenuous efforts are being made to safeguard these.

Today Leicestershire is divided into 9 administrative districts. This total includes Rutland which, before 1974, was a county in its own right and may well be again in the future. The county is further divided into 287 civil parishes which have traditionally been the basic recording unit for wildlife since natural history surveys began in the eighteenth century. Neither of these systems is convenient nor yet desirable for modern flora and fauna recording due, in part, to frequently changing boundaries. Nationally, the established Watsonian Vice-county system is adopted as the basis for county recording and this has the advantage of roughly equally sized units. The larger counties are divided whilst smaller ones are integrated, as in the case of the former Leicestershire and Rutland. Modern mapping is now based on the Ordnance Survey national grid and biological recording has followed this principle which enables much greater accuracy and ease of data manipulation. Overlying, and independent from the Watsonian Vice-county network, a national grid of 10 kilometre squares has been adopted for recording flora and fauna over the British Isles. In line with the Leicestershire and Rutland floras, the 10 km squares for the spider atlas in this publication are further divided into 25 tetrads, each 2 km square, for plotting purposes. Individual records are established by six-figure map references.

The Regions

For an overview, the county is divided into seven visually distinct regions, the boundaries of which are arbitrary, except that in the case of Rutland the district administrative boundary is adopted.

The City: Leicester sits in the centre of the county, with the major arteries of commerce radiating from it. With the post-war explosion of new housing developments on its periphery it has grown to merge with its expanding satellite towns and villages to form a conglomerate known as Greater

Leicester. The city has, in the past, been unusual in that it has developed available building land within its confines wherever possible, so that the large areas of derelict land one associates with big cities such as Birmingham and Greater London are at a premium. The city sits astride the River Soar, which to north and south has associated green wedges (*e.g.* the Riverside Park) extending into its centre.

The area incorporated in Greater Leicester holds just over half of the county's population (see Map 2). The other main towns are Loughborough (in the north), Coalville and Ashby-de-la-Zouch (in the north-west), Hinckley (in the south-west), Lutterworth and Market Harborough (in the south), Melton Mowbray (in the north-east) and Oakham (in Rutland). In the county as a whole, the area east of the Soar Valley is the least populated part, since it is predominantly rural in nature and lacks the labour intensive industries of west Leicestershire. Industrial growth in the west of the county has given rise to rapid expansion of settlements, such as Coalville and Whitwick where new light industry is replacing coal mining.

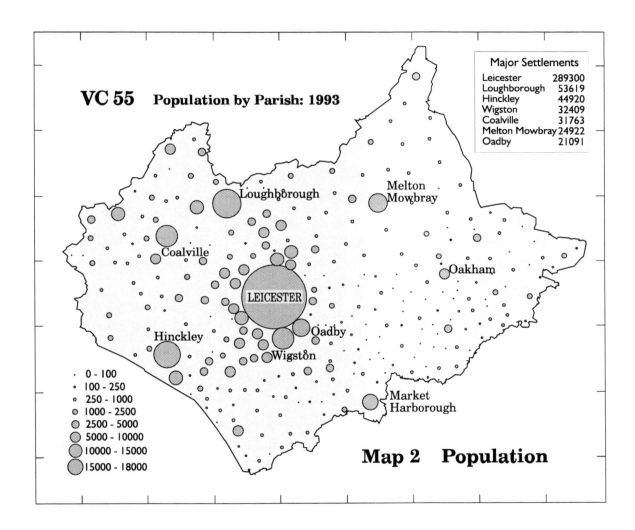

Map 2   Population

The Soar Valley: The Soar Valley divides the county in two, separating the coalfields and high ground of Charnwood in the west from the rolling hills of east Leicestershire. Along with its tributaries it drains over half the county. The valley floor is mainly improved inundation meadow which, despite huge flood-alleviation schemes, can still be under water for several weeks during the winter period on account of the river's large catchment area and low drainage gradient as it meanders northwards toward the Trent. Heavy rainstorms result in a rapid rise in water levels of the River Soar. The River Wreake which enters from the east, a few miles north of Leicester, is the main tributary of the River Soar. It flows from the north-east of the county draining the land around Melton Mowbray. Both the Soar and Wreake valleys have active gravel extraction projects along parts of their lengths. Several abandoned workings have been made into fishing lakes, incorporated into country parks or are used for water sports.

Charnwood Forest: To the north-west of the City and south-west of Loughborough lies the high ground of Charnwood Forest. Charnwood (the name means woodland with piles of rock) is characterised by rocky outcrops, stone walls and bracken-dominated moorland and the relic landscape of former deer parks. Today, the remnants of this once widespread moorland are separated by large tracts of arable fields, improved pasture and leys. Although this area looks naturally well-wooded, much of the Charnwood woodlands are in fact nineteenth and twentieth century plantings. The whole district of Charnwood has been extensively studied in a field-by-field survey by the Loughborough Naturalists' Club (Crocker, 1981).

Leicestershire Coalfield: To the west of Charnwood, centred on Ashby-de-la-Zouch, are the Leicestershire and South Derbyshire coalfields. The soils are acidic and would at one time have held reasonably large areas of lowland heath such as existed at Nailstone Wiggs. Coal mining has been a feature of the landscape of the area for around ten centuries, with the production of deep-mine coal ceasing in the late 1980s. More recently open-casting around Ashby and Coleorton has taken place. The whole coalfield area is prone to subsidence as old mines collapse, occasionally producing some interesting habitats including the lakes that have appeared between Measham and Donisthorpe on the Saltersford Brook. The coalfield countryside is quite bleak, with large sections of the landscape having been fashioned by the mining industry. In places remnant tips are still being restored to agricultural use or as public amenity land. Many of the mining towns have a prevailing air of decay with former sites of coal production made into either museums, parks or new industrial estates.

The north-west corner beyond the Thringstone Fault, which marks the northern boundary of the coalfield, is well wooded and rural in character with small mixed farms. A notable feature is the occurrence of several small outcrops of limestone. The best known of these is Breedon Hill. Here the Norman church stands on the remaining half of the hill, close to the precipice that marks the edge of quarrying activity. East Midlands Airport at Castle Donington has the biggest industrial impact on the area now that the Donington coal-fired power station has closed down. As this airport is an expanding enterprise, the future take-up of land and associated atmospheric pollution from increased flights are factors which could have further deleterious impact on sensitive ecosystems such as Donington Park. These comments also apply to the Donington Park racing circuit which attracts huge numbers of spectators at sporting events and pop festivals.

South-west Leicestershire: Although mainly agricultural, this region is less intensively managed than the east of the county, having smaller fields and more permanent grassland. The landscape is flatter and the Ashby Canal meanders through it, lacking the need for any locks. The Rivers Sence and Tweed flow gently through this quiet agrarian corner of the county.

East Leicestershire: To the east of the River Soar the landscape is mainly undulating and dotted with spinneys and copses. The two highest points are Whatborough Hill at 230 metres (754 Feet) and Robin-a-Tiptoe at 221 metres (724 feet), both of which give magnificent views over the surrounding countryside. The soft rocks of this area have produced some fine escarpments, as at Tilton-on-the-Hill and Burrough Hill. North of these is the Melton Ridge, which marks the western edge of the ironstone in Leicestershire. This ridge, extending from Melton Mowbray north-eastwards to Belvoir Castle, rises over 100 metres in places and is well-wooded along much of its length. The topography of the area produces well-drained land with fertile soils and rich grassland which, along with similar grassland further east in the district of Rutland, are used for the fattening of sheep before market. This side of the county has suffered the worst impact of prairie farming. This is noticeable around Houghton-on-the-Hill, for example, and is even more evident further east. Scattered throughout the region is evidence of medieval open-field systems in the ridge and furrow of now permanent pasture.

Rutland: Except for the interruption of the Vale of Catmose, the District of Rutland slopes gently from the high ground of the north and west into the Welland Valley. Much of the land is given over to arable farming but there are some leys and improved grassland. At its centre is Britain's largest man-made reservoir – Rutland Water – which incorporates several nature reserves and allows public access to much of its 40 km (25 miles) of shore-line. The eastern part of Rutland is predominantly limestone, the soils of which contrast with the neutral clays of the west. Today species that once inhabited floristically-diverse limestone grassland are restricted mainly to disused quarries and roadside verges. Limestone and ironstone have also made a visual impact on the landscape with many of the older houses and barns being made from these materials, resulting in a Cotswold appearance in some areas.

Geology

Solid Geology: In simple terms the solid geology of Leicestershire forms a bridge between the West Midlands and eastern England, being composed of elements of both. The bands of rock cross the county in a north to south direction, with the oldest rocks being found in the west. These dip eastwards and underlie successive strata of younger Jurassic rocks in the east of the county and Lincolnshire. The River Soar runs along the geological boundary between the Triassic rocks of the west and the Jurassic rocks of the east (see Map 3).

West Leicestershire is predominantly Triassic mudstone (Keuper marl) and sandstone. These rocks extend from North Derbyshire through the west of the county and into Warwickshire. In the Charnwood Forest area there are outcrops of Pre-Cambrian rocks which contain granites, slate and volcanic rocks. These Pre-Cambrian outcrops contain some of the oldest rocks in Britain (700 million years old) with Leicestershire lying at the south-eastern edge of their occurrence in England and having their only outcrop in the Midlands. Such rocks are more reminiscent of those found in the Cairngorms of Scotland and the heights of Snowdonia. Amongst these outcrops to south and east of Charnwood are diorite and granite formations which are, or have been, quarried for roadstone and which have supplied most of the constructional and road maintenance needs of south-east England. There is a further line of several smaller outcrops of igneous rock south-west of Leicester, the largest of which is at Croft Hill.

The coalfield area of north-west Leicestershire is centred on the market town of Ashby-de-la-Zouch which stands on the anticline between the Leicestershire and South Derbyshire coalfields. A large area of coal measures (with outcropping seams of coal) is exposed between Coleorton and Swannington, and between Heather, Measham and the Derbyshire border. On the exposed coal measures the coal, shale, mudstone and sandstone make for a naturally acidic substrate, and these conditions are extended into the vicinity of deep mine operations as far south as Bagworth and Merry Lees, where extensive weathered spoil tips are a feature of the coal mining process.

Another feature of north-west Leicestershire is the carboniferous limestone which outcrops in several small exposures. These are similar in age to the limestone of the Peak District that was laid down around 380 million years ago, but differ in that the calcium carbonate constituent in the rock has been replaced by magnesium carbonate.

The east of the county is made up of Jurassic limestone, ironstone and clays, which are representative of the southern East Midlands, Lincolnshire and Cambridgeshire. In the past both ironstone and limestone were mined in east Leicestershire and their abandoned workings are scattered across eastern Rutland and north-east Leicestershire. The ironstone workings have provided local stone for many of the older buildings in the east of the county. Today there are still several limestone quarries producing limestone for cement and building stone, the most well-known of which is at Ketton.

Nearly all the rock types found within Leicestershire have been quarried at one time or another. Historically there has been quarrying activity since around 4000 BC, when stone axes were being made from Charnwood Forest stone. The Romans are known to have quarried slate at Swithland and smelted locally quarried ironstone in Rutland. Today, we are left with a complex legacy of physical features which derive from past mining and quarrying activities. Now weathered and blended into the landscape, they provide havens for a flora and fauna under threat of extinction in the county. In contrast, current massive quarrying and open-cast operations are being carried out at prime sites, such as Buddon Wood and Bardon Hill, so that the long-term losses, not only of the flora and fauna but also amenity and landscape, outweigh the short-term financial gains.

Drift Geology: Much of Leicestershire's solid geology is hidden beneath drift material brought from further north by glaciers during the last ice age. The commonest drift material is boulder clay, of which there are two main types in Leicestershire formed from different parent materials; red boulder clay derived from Triassic rock debris originating in the north-west of England, and chalky boulder clay with a Jurassic clay base and including chalk 'boulders' originating in the north-east. Red boulder clay is generally found overlying the Trias to the west of the River Soar, with chalky boulder clay to the east over Lias, from which its main constituent derives. These boulder clays are a mixture of debris derived from all the rock types over which the ice has advanced

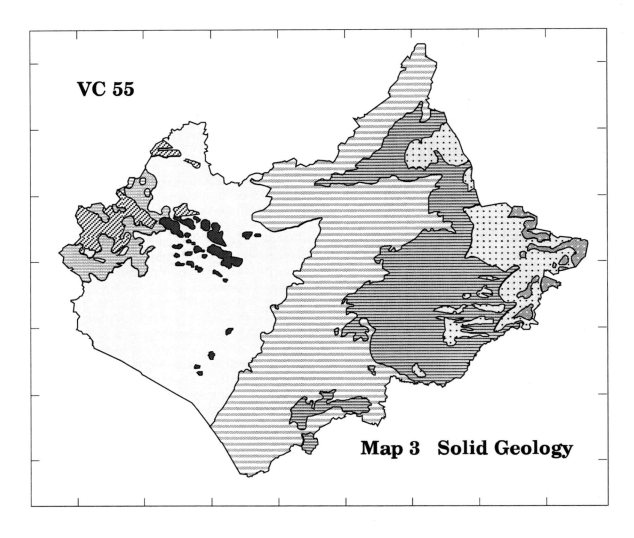

Map 3 Solid Geology

Jurassic Rocks: *Younger estuarine rocks in the east of the county overlie the older Triassic mudstones which dip to the east under the Jurassic clays, ironstones and limestones.*

Great Oolite: limestone, 150 m.y. old. Small deposits in the extreme east of the county. These, the youngest rocks in Leicestershire's solid geology were laid down in coastal swamps subject to intermittent marine incursions. They are part of the upper estuarine series in Rutland which rest directly on the Lincolnshire limestone.

Inferior Oolite: limestone etc. 175 m.y. old, comprising clays, silts, ferruginous sandstones and oolitic limestones.

Upper & Middle Lias: ironstone and clay, 185 m.y. old. The upper Lias outcrops in the south of the county around Uppingham. Low grade iron ore is exposed in the marlstone rock bed of the Middle Lias in north Leicestershire, the upper part of which has workable ironstone, extensively exploited north of Melton Mowbray.

Lower Lias: clay, mudstone & limestone, 200 m.y. old. Extensive deposits of fossiliferous clays laid down in a quiet water environment. These outcrops in Leicestershire are known as the White Lias.

Triassic Rocks: *new red sandstone, 225 m.y. old. Exposed in west Leicestershire overlying the older coal measures and Pre-Cambrian rocks.*

Keuper Marl : clay and mudstones.

Keuper Sandstone.

Lower and Middle Coal Measures: 290 m.y. old. An area of exposed coal seams, shale, mudstone and sandstone centred on Ashby-de-la-Zouch, where the lower coal measures outcrop.

Carboniferous Limestone: 340 m.y. old (Breedon and Grace Dieu). Small deposits in north-west Leicestershire, extensively quarried for lime-making, roadstone and aggregate.

Pre-Cambrian: 700 m.y. old Charnian agglomerates and Pre-Cambrian intrusive volcanic plugs, vents and larvas (Peldar Tor, High Sharpley); also Pre-Cambrian intrusive Syenite (Markfield, Groby, Bradgate, Hammercliffe, Copt Oak, Bawdon Castle), with younger (400 m.y. old) Granodiorite (Mountsorrel), and outlying masses of igneous rock of uncertain age at Enderby, Croft, Narborough and Sapcote.

and can be over 30 metres deep in places. They may also contain erratics from as far away as Scandinavia. The main deposits of boulder clay form a wide but broken band running through the county from north-east to south-west on land below 122 m (400 ft). There are large areas having no drift cover at all in the west, north-east and south-east of the county and here the solid geology has weathered to form the soils.

Along the major valleys of the Rivers Soar, Wreake and Welland, and in parts of south-west Leicestershire, there are discontinuous sand and gravel deposits left behind by the melt waters of retreating glaciers. All the river valleys also contain more recent alluvial deposits, with the largest deposit at Lockington Marsh in north-west Leicestershire, where there are several square miles of flood meadow on this substrate.

The soils of Leicestershire that we see today have been derived from a number of constituent components and modified by many other factors. During the Pleistocene and Holocene periods – over the last 250 000 years – drift material laid down by the glaciers and their melt waters has been subjected to the effects of weathering, temperature, drainage, biological activity and the influence of man. The results have been to evolve a range of soil types far more complex than that presented by the underlying solid or drift geology, such as pockets of acidic soil in calcareous districts and lime-rich soil on acidic or neutral land.

For a more detailed description of the structure, climate and habitats of the pre-1974 counties of Leicestershire and Rutland, the reader is directed to the introductory chapters of the two floras published by Leicestershire Museums Service:

>Messenger, Guy (1971) *Flora of Rutland*
>Primavesi, A.L. & Evans, P.A. (1988) *Flora of Leicestershire*

Climate and Weather

To a large extent the climate of Leicestershire is influenced by its position in central England and the county's own topography. East Leicestershire is open to the easterly and north-easterly winds coming in over the low-lying fens of Lincolnshire and Cambridgeshire, though to some extent the high ground of east Leicestershire takes the brunt of any winter storms from the north-east. The west of the county is sheltered from the westerly winds by the high ground of Derbyshire and the West Midlands, whilst the high ground of Charnwood further protects the Soar Valley from cold winds and rain coming from this direction.

Wind: The predominant winds across the county are westerlies, between south-west and north-west, but south-westerlies are more frequent. These have, in the past, brought pollution from the Birmingham and Wolverhampton conurbations; however, there is strong evidence with the reappearance of lichens in the county that this problem is decreasing. The summer months often have north-easterlies, bringing clear skies and good weather.

Rainfall: Leicestershire is one of the driest counties in Britain, with a mean annual rainfall of 700 mm in the north, falling to 560 mm in the south-east. This compares with a national average of 1100 mm for the British Isles and 940 mm for England and Wales. The wettest months are August and November to January, whilst the driest months are from February to June.

Temperature: The warmest months in Leicestershire are July and August, whilst the coldest is January except that the higher ground in the north-west usually reaches lower mean temperatures in February. Most frosts occur between December and February, with July and August being frost free. There is an increasing incidence of frost across the county from north-west to south-east. Mean annual temperatures are about average for the British Isles, around 9.2 degrees C. On average, snow or sleet falls on less than 20 days per year. Snow usually lies for less than 16 days, even on the higher ground of the Charnwood hills. The growing season in Leicestershire is generally from the beginning of April to the middle of November when mean temperatures do not fall below 6 degrees C.

Sunshine: Leicestershire is one of the least sunny counties in England with an average of only 40 days per year when there are more than 9 hours of sunshine. This compares with 70 days for counties situated on the south coast. Across the county the south and east get slightly more sunshine, but the difference is negligible.

Fogs and Mists: Over the last couple of decades there has been a marked reduction in the incidence of fog and smog with the improvement in air quality due to the introduction of efficient methods of power generation, the establishment of smokeless fuel zones in urban areas and the demise of some of the largest industries. This has helped to reduce the incidence of fog and smog in the county, which now occurs on fewer than 10 days a year. However, during the spring and autumn there are often mists in the river valleys and on the hill tops.

Woodland

Leicestershire today has just three percent of its surface covered by woodland, with only three other counties in England having less. The poorly wooded nature of the county has a long history and the shortage of woodland, first detectable in the Iron Age, has endured to the present day. The Domesday Survey (1086) records around 13 000 hectares of woodland, mostly from west Leicestershire and Rutland, which is a mere five percent of the county's surface area (255 372 ha). In 1982 a census of woodland trees in the county revealed a total area of only 7 700 hectares, including 700 ha of coppice-with-standards, 800 ha of coppice and 1 258 ha of scrub woodland (Squires & Jeeves, 1994). Much of the rest was plantation (conifer and broadleaf), the predominant conifer being Norway spruce. Many spinneys and small woods have been established over the last two centuries as fox coverts for hunting and for preserving game species such as pheasants. These woodlands now form a significant landscape feature of the Leicestershire countryside and provide valuable habitats for wildlife. Overall, the chief broadleaf tree in the county is still oak.

Ancient Woodland: There is no primary woodland in Leicestershire and only about 130 ancient woodland sites, that is semi-natural sites that have been continuously wooded since before 1640 A.D. The rest is secondary woodland, growing on land once used for other purposes. The distribution of ancient woodland in the county lies in three main areas with a little in the south-west around Hinckley. These are in the north-west centred on Charnwood and extending up to the Derbyshire border; in the south-east stretching in a broad band from Tilton and Owston to Great Easton; and in Rutland in another wide band from the central area to the north-eastern county boundary with Lincolnshire. Some of this ancient woodland occurs on ridge and furrow, pointing to the fact that at some time the land has been clear-felled and ploughed. Fortunately this occurred at a time when there were local reservoirs of woodland species preserved in linear woodland relics along green lanes, stream valleys and adjacent scrubland. These species were able to re-colonise sites when the land returned to woodland.

The best examples of ancient woodland are Burley Wood in Rutland, Swithland Wood on Charnwood Forest and several of the woodlands in central-east Leicestershire such as Owston Wood, Skeffington Wood and Prior's Coppice. Although half these woods contain varying amounts of plantation (conifer and broad-leaved), they still hold large areas of semi-natural habitat that contain a diverse and interesting flora and fauna. Burley Wood is the largest remaining piece of ancient woodland in Leicestershire; it is privately owned with restricted access and incorporates a former deer park. It is an oak-ash wood with a number of rare tree and shrub species. In contrast, Swithland Wood is a country park with open public access and rates very highly for its floristic diversity. It is predominantly sessile oak on an acid soil but with good stands of small-leaved lime and pedunculate oak. Despite intense visitor pressure the wood still retains its distinguished ecological status and has the largest list of higher plants for any wood in the county. Buddon Wood, on the north-eastern edge of Charnwood Forest in north-west Leicestershire was, before clear felling, burning and quarrying, one of the most important woods in England. External political pressures overruled the ecological arguments for preservation and powerful commercial interests have exploited this unique habitat to the point of near total destruction. The ancient woodlands in the central eastern part of the county are mostly ash-dominated with stands of planted oak, lying chiefly on heavy clay and are scheduled Sites of Special Scientific Interest. They are privately owned but some have public footpaths running through them.

In the past many woodlands were managed on a coppice-with-standards system, with oak being prized for timber trees and the coppice consisting of ash, field maple, hazel and (locally) small-leaved lime. This method of management had drastically declined by the early twentieth century with the onset of the easy availability of cheaper alternatives for products previously made from locally grown timber, as well as the decline in the use of wood for fuel. The majority of native woodlands are dominated by either oak or ash, with both species of native oak growing naturally in parts of Charnwood and Rutland, pedunculate oak having been planted widely in the other parts of the county. The remainder of the county's ancient woodlands are dominated by ash and field maple, which allow more light through to the ground layer. Where oak is present in these woods it is believed to be introduced, since natural regeneration from seed is seldom found.

Conifer Plantations: The presence of conifer plantations across the county can be attributed mostly to the policies of the Forestry Commission. This national body, which was set up after the end of the First World War, promoted these fast growing trees in a bid to replenish timber resources exploited during the war. This policy has also been carried out on sites of ancient woodland, which have been wholly or partly felled, such as Owston Wood, Pickworth Great Wood and Martinshaw Wood.

Pasture Woodland: Pasture woodland is a very important habitat in Leicestershire, as it contains most of the county's oldest trees, some of which have been pollarded for centuries. These old trees often have large cavities, and may be completely hollow. This type of habitat is found predominantly in old deer parks, such as Bradgate Park and Donington Park, and around the remnants of great estates and large country houses like Croxton Park.

Wetland

The major wetland areas in Leicestershire today are the river valleys and their network of tributary streams and ditches (see Map 4) which contain a combination of flood meadows, ox-bows, canals and flooded gravel pits. Gravel extraction began in earnest in 1940 and this industry has left a few groups of lakes along the Rivers Soar and Wreake, some of which have been incorporated into country parks. Others, in the south-west of the county have been made into fishing lakes. The rest of the abandoned gravel workings have been landfilled and returned to their previous land use.

Rivers: The dominant river of the county is the River Soar, which is formed by the convergence of the Soar Brook with several other streams near Sharnford in south-west Leicestershire. It then flows northwards and is joined by the Grand Union Canal south of Leicester. Between Leicester and the River Trent sections of the river have been canalised and this stretch is known as the Leicester Navigation. The River Soar flows into the Trent at the Lockington Grounds, on the border with Nottinghamshire; where the two rivers meet there is a substantial area of marsh and flood meadow (SK4830) known as Lockington Marsh, part of which is a Site of Special Scientific Interest.

The major tributary of the River Soar is the River Wreake which drains a large proportion of north-east Leicestershire. This river rises at Bescaby under the name of the River Eye and flows in a south-westerly direction, through Melton Mowbray to the Eye Kettleby Mill (SK737181), from which point it is known as the River Wreake. It then continues to meander on a westerly course until it joins the River Soar at Cossington (SK595127).

For a short distance (10 km), between Donington Park and the Lockington Grounds in the north-west of the county, the River Trent marks the boundary between Leicestershire and Derbyshire. This great river drains about three-quarters of Leicestershire through its tributaries, mainly the River Soar. It is also joined by the Rivers Devon and Smite from the Vale of Belvoir, the Mease from the west, and the Anker with its tributaries the Sence and Tweed from the south-west. After incorporating the River Soar the Trent journeys some 120 kilometres northwards until it flows into the Humber Estuary.

The Rivers Avon and Swift drain a small area of south-west Leicestershire westward via the River Severn into the Bristol Channel. In south-east Leicestershire the county boundary runs along the River Welland between Husbands Bosworth in the south and Duddington in the east. This river drains most of south-east Leicestershire including nearly all of the district of Rutland. The Welland rises near Sibbertoft in Northamptonshire (SP6783) and, leaving Leicestershire, journeys a further 50 kilometres in a north-easterly direction to enter the Wash in its western corner. For approximately the last 25 kilometres it is tidal. Within Leicestershire, the Welland has several tributaries including the Eye Brook, Langton Brook, Saddington Brook and the Rivers Chater and Gwash.

The River Witham rises in east Leicestershire just inside the county border with Lincolnshire. It drains a negligible area of east Leicestershire, through Lincolnshire, before flowing into the Wash at Boston a few kilometres north of the River Welland.

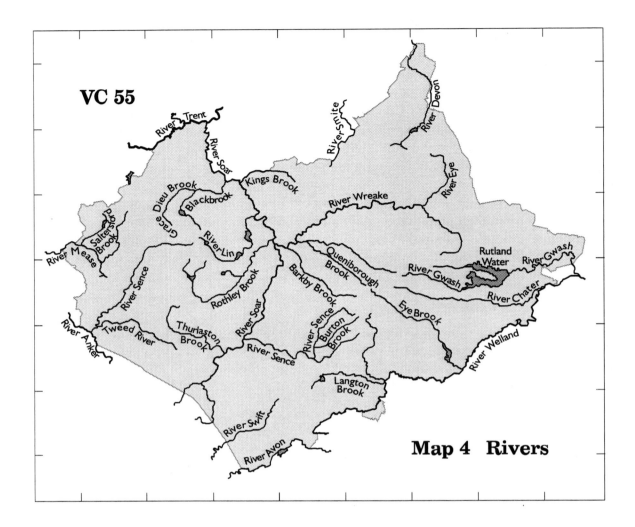

Map 4 Rivers

Many of the county's smaller water courses have been graded, straightened and their banks landscaped. This treatment has had disastrous consequences on the wildlife of these waterways with the destruction of traditional micro-habitats. The River Eye, between Hambridge and Melton Mowbray, is virtually the only piece of river to escape the excesses of the county's drainage schemes. Part of this river is a Site of Special Scientific Interest and also a Grade 2 Nature Conservation Review site.

Pollution from sewage and industrial effluent first began to seriously affect the county's river system during the early decades of this century. This problem was later compounded by agricultural chemicals which were carried into the water courses from surrounding farmland. The situation is slowly improving due to legislation and close monitoring by statutory bodies such as the Environment Agency, and today the main threat to the county's river systems is from the flood prevention schemes. These channelise the waterways and drain their associated marshes and flood meadows which rivers need to store excess water in times of flood. If this drained land is then built on, further flooding and further alleviation schemes inevitably result.

Canals: To promote greater trade with the rest of Britain, six canals were built and two of the county's major rivers were made navigable between the years 1770-1820. The River Soar was the first river to be made navigable and the work was completed in two stages. The first of these, from the River Trent to Loughborough was completed in 1778, with the second stage from Loughborough to Leicester being completed in 1794. Both schemes were achieved by cutting through several of the larger meanders and by building locks. The Union Canal, opened in 1809, was promoted to link Leicester and the Trent with the Grand Junction Canal and London, but was not completed. This task was taken up by the Grand Union Company who opened the final link in 1814.

The River Wreake was also made navigable as far as Melton Mowbray by 1797, again by cutting through its larger meanders, but was abandoned for navigation in 1877.

The six canals had to be specially constructed along with feeder reservoirs where local water sources were inadequate to maintain constant water levels. The canals were:

	Completed	Abandoned
Charnwood Canal	1794	1804
Grantham Canal	1797	1960s
Oakham Canal	1803	1846
Ashby Canal	1804	still open in part
Leics & Northants Union Canal	1809	still open
Grand Union Canal	1814	still open

The Leicestershire and Northamptonshire Union Canal joined the River Soar navigation just south of Leicester, and following amalgamation with other canals eventually connected London with the River Trent as the Grand Union Canal. In Leicestershire the Grand Union Canal enters the county at North Kilworth and mostly follows the contours northwards, except to pass through the Saddington and Husbands Bosworth tunnels. Its most spectacular point is at Foxton, where a rise of 23 metres (75 feet) was achieved by the construction of ten consecutive locks, known as Foxton Locks. Here also was the famous 'inclined plane', one of the wonders of the canal age, where counter-balanced tanks containing barges were wound up and down the inclined haulage by steam power, from bottom level to top level.

Feeder Reservoirs: The construction of canals during the eighteenth century brought the need for a reliable and sustained water supply to maintain navigable levels within the canal system. This requirement for water to make up losses and maintain navigation levels was particularly demanding where gradients had to be negotiated by means of locks, which use a lot of water each time they are operated. Many reservoirs were constructed to supply this demand, often several being required to supply different levels of the same canal.

The Grand Union Canal has two feeder reservoirs within the county, one at Saddington and the other on the Northamptonshire border at Welford. The stretch of canal from Kilby Bridge southward to Foxton Locks is a Site of Special Scientific Interest and a local Trust nature reserve.

The Charnwood Canal was situated in north-west Leicestershire. It was only 13 kilometres (8 miles) long and ran from Osgathorpe to Nanpantan, from where a horse-drawn tramway linked it to the Grand Union Canal at Loughborough. The Charnwood Canal had a very short life. Intended for coal transportation from the Coleorton and Swannington mines to the Soar navigation, it failed after its feeder reservoir – the Blackbrook – burst its dam in 1799. Although the dam was repaired, the canal was abandoned in 1804 through lack of trade. The coal mines closed because they could not compete with cheap Derbyshire coal and the unprofitable limestone freight alone was not enough to justify maintenance costs. Today the canal survives mainly as an earth work, with occasional ponding in wet periods.

The Grantham Canal in north-east Leicestershire was built to join the town of Grantham to the River Trent at Nottingham. It was finally abandoned for navigation in the early 1960s, but the whole length within the county and its feeder reservoir at Knipton, still retain their water levels. There are plans to make this canal navigable once more, but this would involve the construction of new bridges where roads have been built across it in the last thirty years, and a massive dredging operation. The length of canal between Harby and Redmile (SK747317 – SK790359) is a Site of Special Scientific Interest and a local Trust nature reserve.

The Oakham Canal was built to link Oakham to the canal network at Melton Mowbray, but was bought and closed by the Midland Railway Company in 1846. The railway line follows a similar course to that of the former canal, over which it was built in many places. There are remnants of the former canal surviving as a few linear pools, but most of its length is now dry.

The Ashby Canal was built to connect Ashby-de-la-Zouch to the network of canals of the West Midlands and joins the Coventry Canal at Bedworth, in Warwickshire (SP3688). Due to the effects of subsidence from coal mining, the northern 10 kilometres (6 miles) of the canal have been drained and filled in with pulverised fuel ash. Since the 1960s the Ashby Canal has terminated at the Snarestone Tunnel, with the whole length south of that point still navigable today. The Ashby Canal feeder reservoir was situated at its northern tip, near Moira, but became redundant in the early 1940s due to the effects on the canal of mining subsidence. The reservoir was finally drained and the site open-cast mined in the 1960s.

Open Water: Before the beginning of the nineteenth century the county's largest area of open water was Groby Pool, which is situated 8 kilometres (5 miles) north-west of Leicester. Sediment stratigraphy and the pollen record show the lake to be of relatively recent origin, created possibly some time during the twelfth century A.D. (David, 1989). Today, this large lake covers an area of some 13 hectares (32 acres), but may have been more than twice this size in the past. It has some excellent reedswamp and wet carr.

Reservoirs – domestic & industrial water supplies: The growing population and industrial development of Leicester in the nineteenth century produced a need for more clean water. To meet this demand a programme of reservoir construction began on the southern margins of the Charnwood Forest in the 1850s. These reservoirs were built to contain local precipitation but unfortunately demand quickly outstripped supply and eventually water had to be piped from larger reservoirs in Derbyshire and North Wales. Not all the reservoirs built within the county supply its residents. Water from Eyebrook Reservoir supplies the Corby district and Rutland Water supplies other towns in Northamptonshire.

The major reservoirs wholly or partly within the county are:

	Completed	Million Litres	Million Gallons
Thornton Reservoir	1854	1 472	324
Cropston Reservoir	1870	2 528	556
Swithland Reservoir	1894	2 228	490
Blackbrook Reservoir	1906	2 300	506
Stanford Reservoir	1928	1 340	295
Eyebrook Reservoir	1940	8 000	1 781
Staunton Harold Reservoir	1965	6 655	1 464
Rutland Water	1977	124 000	27 000

Today, despite the great loss in the number of field ponds over previous decades, with entire parishes now devoid of such water bodies, there is a larger area of freshwater in the county than ever before.

Marsh, bog and reedswamp: What little marshland there is in the county is associated mostly with the reservoirs and river systems. These include the cut-offs along the Rivers Soar and Wreake, many with good stands of reedswamp and willow-carr; inundation meadows such as Seaton on the Welland and Loughborough Big Meadow on the Soar; and places like Narborough Bog which have fen and reedswamp communities. The largest wetland area is at Lockington Marsh, at the confluence of the River Soar and River Trent, where there is a wide range of wetland habitats including inundation meadow, reedswamp, willow-carr and marsh. Away from these riparian habitats the best marshy areas – all SSSIs – are Great Bowden Pit (bog and reedswamp), Botcheston Bog (marsh), Newton Burgoland (marsh) and Misterton Marshes (fen and reedswamp). Great Bowden Pit is a hundred-year-old borrow-pit, formed when the railway line was built and contains a large area of *sphagnum* and cotton grass with invading reedmace. Botcheston Bog comprises mostly herb-rich hay meadow, with ancient damp grassland/marsh in one corner containing a number of rare marsh plants. Newton Burgoland is a grazed marsh/sedge bed periodically flooded by the adjacent stream. Misterton Marsh has a large area of reedswamp with smaller areas of fen and contains plants such as tussock sedge and water avens.

Grassland, Moorland and Heath

Grassland: There is a wide range of grassland types within the county from acidic in the Charnwood uplands to alkaline in the east, with occasional patches of each occurring outside their usual range. Neutral grassland is the predominant natural grassland of Leicestershire and is the vegetation type for which the county was once best known. Not only has the majority of permanent grassland been ploughed, but much of that which remains has been modified by the use of agricultural chemicals. Some herb-rich meadows have survived, but the remaining unimproved grasslands in the county are confined to sites that are physically hard to work or unprofitable to improve. These areas include a few flood meadows along the major river valleys, the rocky acidic grasslands of Charnwood Forest where hard rocks lie just below the surface and disused quarries.

During the Second World War large areas of permanent grassland were converted to arable as part of the war effort towards self-sufficiency, with some of the grassland being ploughed for the first time since the Napoleonic Wars. In just four years (1939-1943) the acreage under arable within the county rose from 11% to 50%. Following this extension of arable farming, subsequent decades saw the average field size increase dramatically, with the inevitable loss of miles of hedgerow and redundancy of field ponds. East Leicestershire and its traditional sheep farming suffered most from this arable revolution as the visual character of the countryside changed from rolling green sward to vast prairies of monoculture.

Limestone grassland has been confined to small areas in the north-east and north-west and has suffered similarly from modified land-utilisation, with only remnants surviving in isolated locations. Natural acid grassland, once an extensive feature of Charnwood Forest and areas with thin soils on acidic rocks, has suffered less but nevertheless has also been destroyed by changing land-use and improvement since the Second World War. Much also has been lost due to the cessation of low intensity grazing and the subsequent invasion of bracken, also by chemical improvement and ploughing. Some small areas of heath grassland survive but these are always under threat from bracken which is now a dominant characteristic of much of the Charnwood Forest landscape.

Heath & Moorland: There is very little heath left in Leicestershire, most of it being either grass or lichen, the former often dominated by bracken. What little moorland habitat remains is concentrated on the tops of the Charnwood Forest area, as at High Sharpley and Charnwood Lodge Nature Reserve where also, on Timberwood Hill, is the best example of *Vaccinium/Calluna* heath in the county. Lowland heath is restricted to a few small pockets in the north-west of the county and an isolated fragment at Luffenham Heath Golf Course, in Rutland.

On urban industrial waste ground – where the soil is often rocky and extremely poor – a lichen heath community usually associated with heathlands just after they have been burnt, can appear. This lichen heath flora, along with highly mobile beetles that are also associated with recently burnt heathlands, has been found on the site of the former Aylestone power station (demolished 1986), close to the centre of Leicester. Unfortunately this interesting habitat is considered an eyesore rather than a natural asset and is liable to 'cosmetic' improvement.

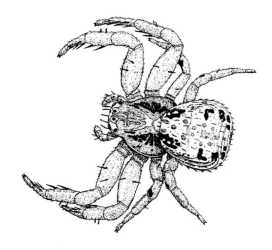

Ozyptila trux f. (x7)

CHAPTER TWO

Check list of the spiders of Leicestershire and Rutland — VC55

This list of **326** species is based on Merrett, Locket & Millidge (1985) *A check list of British spiders*, with amendments listed in Merrett & Millidge (1992) Amendments to the check list of British spiders, as reproduced in the British Arachnological Society's *Members' Handbook* Section 2.3, March 1992.

The species status is established from current knowledge of the British spider fauna. Rare or threatened species are indicated in two categories, Notable and Red Data Book species: thirteen Notable species in the category **Nb** (recorded from fewer than 16 counties) are listed, and one in category **Na** (recorded from fewer than 5 counties), as defined by Merrett, P. (1990) *A Review of the Nationally Notable Spiders of Great Britain*.

There are no Endangered species (**RDB 1**) known to occur in VC55, but there is one species *Lepthyphantes midas* which is considered Vulnerable (**RDB 2**), i.e. taxa believed likely to move into the Endangered category in the near future, and one *Mastigusa macrophthalma* in the category Rare (**RDB 3**), i.e. taxa with small populations which are not at present Endangered or Vulnerable, but are at risk. *Lepthyphantes beckeri* Wunderlich (1973), is recorded here for the first time in Britain from a single specimen. The species is insufficiently known and therefore falls into the category **RDB K**, as defined in Bratton, J.H. (1991) *British Red Data Books - 3, Invertebrates other than Insects*.

There are no published revisions to the status of British spiders since Merrett (1990) and Bratton (1991). It is not the purpose of this current work to present new data for Britain, as it is felt that this matter will be better addressed when the draft results of the ten year national spider survey are available in 1997. However, following discussions with Dr Merrett (pers.comm. 3.2.96), it is considered appropriate to down-grade two previously listed nationally notable species on the Leicestershire list (*Philodromus praedatus* (Nb) and *Pityohyphantes phrygianus* (Na)), to 'local' status, as these now appear to be well established and apparently extending their range. The national status of a further eight 'common' or 'local' species included in the following check list is subject to review, but is unchanged here, and should be viewed in a regional rather than national context. These are: *Harpactea hombergi, Zelotes latreillei, Clubiona lutescens, Clubiona diversa, Xysticus erraticus, Cnephalocotes obscurus, Micrargus apertus* and *Lepthyphantes leprosus*. Except as discussed above, status shown as 'common' or 'local' is a national definition, and though generally applicable to Leicestershire should be compared with the regional status discussed in chapter five. Synanthropic species are those associated with man and are distributed by human agency. The first number after Status is the Biological Records Centre species identity number, and the Index number is the systematic (check list) number, used for data manipulation.

	Status	BRC	Index
Family AMAUROBIIDAE			
Amaurobius fenestralis (Stroem, 1768)	Common	301	30
.... *similis* (Blackwall, 1861)	Common	303	40
.... *ferox* (Walckenaer, 1830)	Common	302	50
Family DICTYNIDAE			
Dictyna arundinacea (Linnaeus, 1758)	Common	401	60
.... *uncinata* Thorell, 1856	Common	405	90
.... *latens* (Fabricius, 1775)	Local	402	100
Lathys humilis (Blackwall, 1855)	Local	601	140
Family OONOPIDAE			
Oonops pulcher Templeton, 1835	Common	1102	220
.... *domesticus* de Dalmas, 1916	Synanthropic	1101	230

	Status	BRC	Index
Family DYSDERIDAE			
Dysdera erythrina (Walckenaer, 1802)	Common	1202	240
.... *crocata* C.L.Koch, 1838	Common	1201	250
Harpactea hombergi (Scopoli, 1763)	Local	1301	260
Family SEGESTRIIDAE			
Segestria senoculata (Linnaeus, 1758)	Common	1403	280
Family PHOLCIDAE			
Pholcus phalangioides (Fuesslin, 1775)	Synanthropic	1601	320
Psilochorus simoni (Berland, 1911)	Synanthropic	1701	330
Family GNAPHOSIDAE			
Drassodes lapidosus (Walckenaer, 1802)	Common	1802	360
.... *cupreus* (Blackwall, 1834)	Common	1801	370
Haplodrassus signifer (C.L.Koch, 1839)	Local	1903	390
.... *silvestris* (Blackwall, 1833)	Notable: Nb	1904	430
Scotophaeus blackwalli (Thorell, 1871)	Synanthropic	2001	450
Zelotes latreillei (Simon, 1878)	Local	2203	490
.... *apricorum* (L.Koch, 1876)	Local	2201	500
Urozelotes rusticus (L.Koch, 1872)	Synanthropic	2401	550
Drassyllus pusillus (C.L.Koch, 1833)	Local	2503	570
Micaria pulicaria (Sundevall, 1832)	Common	2802	630
Family CLUBIONIDAE			
Clubiona corticalis (Walckenaer, 1802)	Common	2904	680
.... *reclusa* O.P. Cambridge, 1863	Common	2913	690
.... *stagnatilis* Kulczynski, 1897	Common	2916	710
.... *pallidula* (Clerck, 1757)	Common	2911	750
.... *phragmitis* C.L.Koch, 1843	Local	2912	760
.... *terrestris* Westring, 1851	Common	2919	770
.... *neglecta* O.P. Cambridge, 1862	Local	2909	780
.... *lutescens* Westring, 1851	Common	2908	800
.... *comta* C.L.Koch, 1839	Common	2903	810
.... *brevipes* Blackwall, 1841	Common	2901	820
.... *trivialis* C.L.Koch, 1843	Common	2920	830
.... *diversa* O.P. Cambridge, 1862	Local	2905	860
.... *subtilis* L.Koch, 1867	Local	2918	870
Cheiracanthium virescens (Sundevall, 1833)	Local	3003	900
Family LIOCRANIDAE			
Agroeca brunnea (Blackwall, 1833)	Local	3201	910
.... *proxima* (O.P. Cambridge, 1871)	Common	3205	920
Phrurolithus festivus (C.L.Koch, 1835)	Common	3101	1020
Family ZORIDAE			
Zora spinimana (Sundevall, 1833)	Common	3604	1040
Family ANYPHAENIDAE			
Anyphaena accentuata (Walckenaer, 1802)	Common	3701	1080
Family HETEROPODIDAE			
Micrommata virescens (Clerck, 1757) †	Local	3801	1090
Family THOMISIDAE			
Diaea dorsata (Fabricius, 1777)	Local	4001	1110
Misumena vatia (Clerck, 1757)	Common	4101	1120
Xysticus cristatus (Clerck, 1757)	Common	4304	1140
.... *audax* (Schrank, 1803)	Common	4302	1150
.... *erraticus* (Blackwall, 1834)	Local	4305	1170

	Status	BRC	Index
Xysticus lanio C.L.Koch, 1835	Common	4307	1180
.... *ulmi* (Hahn, 1831)	Local	4312	1190
Ozyptila scabricula (Westring, 1851)	Notable: Nb	4407	1270
.... *sanctuaria* (O.P. Cambridge, 1871)	Local	4406	1290
.... *praticola* (C.L.Koch, 1837)	Local	4405	1300
.... *trux* (Blackwall, 1846)	Common	4409	1310
.... *atomaria* (Panzer, 1801)	Common	4401	1330
.... *brevipes* (Hahn, 1826)	Local	4403	1340

Family PHILODROMIDAE

	Status	BRC	Index
Philodromus dispar Walckenaer, 1826	Common	4505	1350
.... *aureolus* (Clerck, 1757)	Common	4501	1360
.... *praedatus* O.P. Cambridge, 1871	Local	4510	1370
.... *cespitum* (Walckenaer, 1802)	Common	4503	1380
.... *collinus* C.L.Koch, 1835	Notable: Nb	4504	1400
.... *albidus* Kulczynski, 1911	Notable: Nb		1450
Tibellus oblongus (Walckenaer, 1802)	Common	4702	1500

Family SALTICIDAE

	Status	BRC	Index
Salticus scenicus (Clerck, 1757)	Common	4803	1510
.... *cingulatus* (Panzer, 1797)	Common	4801	1520
Heliophanus cupreus (Walckenaer, 1802)	Common	4902	1540
.... *flavipes* (Hahn, 1832)	Common	4903	1550
Ballus chalybeius (Walckenaer, 1802)	Local	5201	1620
Neon reticulatus (Blackwall, 1853)	Common	5301	1630
Euophrys frontalis (Walckenaer, 1802)	Common	5404	1650
.... *erratica* (Walckenaer, 1826)	Local	5403	1680
.... *aequipes* (O.P. Cambridge, 1871)	Local	5401	1690
.... *lanigera* (Simon, 1871)	Synanthropic	5406	1710
Sitticus pubescens (Fabricius, 1775)	Local	5503	1730
Evarcha falcata (Clerck, 1757)	Common	5702	1780

Family LYCOSIDAE

	Status	BRC	Index
Pardosa agrestis (Westring, 1861)	Notable: Nb	6401	1880
.... *monticola* (Clerck, 1757)	Common	6407	1900
.... *palustris* (Linnaeus, 1758)	Common	6410	1910
.... *pullata* (Clerck, 1757)	Common	6413	1920
.... *prativaga* (L.Koch, 1870)	Common	6411	1930
.... *amentata* (Clerck, 1757)	Common	6403	1940
.... *nigriceps* (Thorell, 1856)	Common	6408	1950
.... *lugubris* (Walckenaer, 1802)	Common	6406	1960
Alopecosa pulverulenta (Clerck, 1757)	Common	6704	2040
.... *barbipes* (Sundevall, 1833)	Common	6701	2060
Trochosa ruricola (Degeer, 1778)	Common	6802	2080
.... *robusta* (Simon, 1876)	Notable: Nb	6801	2090
.... *terricola* Thorell, 1856	Common	6804	2100
Arctosa perita (Latreille, 1799)	Local	6904	2130
Pirata piraticus (Clerck, 1757)	Common	7103	2170
.... *hygrophilus* Thorell, 1872	Common	7101	2190
.... *uliginosus* (Thorell, 1856)	Local	7106	2200
.... *latitans* (Blackwall, 1841)	Local	7102	2210

Family PISAURIDAE

	Status	BRC	Index
Pisaura mirabilis (Clerck, 1757)	Common	7301	2240

Family ARGYRONETIDAE

	Status	BRC	Index
Argyroneta aquatica (Clerck, 1757)	Local	7501	2270

		Status	BRC	Index
Family AGELENIDAE				
Agelena labyrinthica (Clerck, 1757)		Common	7601	2280
Textrix denticulata (Olivier, 1789)		Common	7701	2290
Tegenaria gigantea Chamberlin & Ivie, 1935	†	Synanthropic	7804	2300
.... *saeva* Blackwall, 1844	†	Synanthropic	7807	2310
.... *parietina* (Fourcroy, 1785)	†	Synanthropic	7805	2330
.... *agrestis* (Walckenaer, 1802)		Local	7801	2340
.... *domestica* (Clerck, 1757)		Synanthropic	7803	2350
.... *silvestris* L.Koch, 1872		Local	7808	2360
Coelotes atropos (Walckenaer, 1830)		Common	7901	2380
Cicurina cicur (Fabricius, 1793)		Local	8001	2400
Cryphoeca silvicola (C.L.Koch, 1834)		Common	8101	2410
Mastigusa macrophthalma (Kulczynski, 1897)		RDB 3	8202	2430
Family HAHNIIDAE				
Antistea elegans (Blackwall, 1841)		Local	8401	2450
Hahnia montana (Blackwall, 1841)		Common	8504	2460
.... *nava* (Blackwall, 1841)		Local	8505	2490
.... *helveola* Simon, 1875		Local	8502	2500
Family MIMETIDAE				
Ero cambridgei Kulczynski, 1911		Common	8602	2520
.... *furcata* (Villers, 1789)		Common	8603	2530
Family THERIDIIDAE				
Episinus angulatus (Blackwall, 1836)		Local	8701	2560
Euryopis flavomaculata (C.L.Koch, 1836)		Local	8801	2590
Crustulina guttata (Wider, 1834)		Local	9001	2670
Steatoda bipunctata (Linnaeus, 1758)		Common	9102	2710
Anelosimus vittatus (C.L.Koch, 1836)		Common	9203	2740
Achaearanea lunata (Clerck, 1757)		Local	9301	2760
.... *simulans* (Thorell, 1875)		Notable: Nb	9303	2790
Theridion sisyphium (Clerck, 1757)		Common	9413	2810
.... *impressum* L.Koch, 1881		Local	9405	2820
.... *pictum* (Walckenaer, 1802)		Local	9410	2830
.... *varians* Hahn, 1833		Common	9415	2840
.... *simile* C.L.Koch, 1836		Common	9412	2860
.... *melanurum* Hahn, 1831		Synanthropic	9407	2880
.... *mystaceum* L.Koch, 1870		Common	9408	2890
.... *blackwalli* O.P. Cambridge, 1871		Local	9403	2900
.... *tinctum* (Walckenaer, 1802)		Local	9414	2910
.... *bimaculatum* (Linnaeus, 1767)		Common	9402	2920
.... *pallens* Blackwall, 1834		Common	9409	2930
Enoplognatha ovata (Clerck, 1757)		Common	9504	2960
.... *thoracica* (Hahn, 1833)		Local	9506	2980
Robertus lividus (Blackwall, 1836)		Common	9603	3020
.... *neglectus* (O.P. Cambridge, 1871)		Local	9604	3040
Pholcomma gibbum (Westring, 1851)		Common	9701	3070
Theonoe minutissima (O.P. Cambridge, 1879)		Local	9801	3080
Family NESTICIDAE				
Nesticus cellulanus (Clerck, 1757)		Local	9901	3090
Family TETRAGNATHIDAE				
Tetragnatha extensa (Linnaeus, 1758)		Common	10001	3100
.... *pinicola* L.Koch, 1870		Notable: Nb	10005	3110
.... *montana* Simon, 1874		Common	10002	3120
.... *obtusa* C.L.Koch, 1837		Local	10004	3130
.... *nigrita* Lendl, 1886		Local	10003	3140

	Status	BRC	Index
Tetragnatha striata L.Koch, 1862	Notable: Nb	10006	3150
Pachygnatha clercki Sundevall, 1823	Common	10101	3160
.... *listeri* Sundevall, 1830	Local	10103	3170
.... *degeeri* Sundevall, 1830	Common	10102	3180

Family METIDAE

	Status	BRC	Index
Metellina segmentata (Clerck, 1757)	Common	10203	3190
.... *mengei* (Blackwall, 1869)	Common	10201	3200
.... *merianae* (Scopoli, 1763)	Common	10202	3210
Meta menardi (Latreille, 1804)	Local	10302	3220
Zygiella x-notata (Clerck, 1757)	Common	10403	3240
.... *atrica* (C.L.Koch, 1845)	Common	10401	3250

Family ARANEIDAE

		Status	BRC	Index
Gibbaranea gibbosa (Walckenaer, 1802)		Local	10602	3280
Araneus diadematus Clerck, 1757		Common	10703	3300
.... *quadratus* Clerck, 1757		Common	10705	3310
.... *marmoreus* Clerck, 1757	†	Local	10704	3320
Larinioides cornutus (Clerck, 1757)		Common	10801	3340
.... *sclopetarius* (Clerck, 1757)		Local	10803	3350
.... *patagiatus* (Clerck, 1757)		Local	10802	3360
Nuctenea umbratica (Clerck, 1757)		Common	10901	3370
Agalenatea redii (Scopoli, 1763)		Local	11001	3380
Atea sturmi (Hahn, 1831)		Local	11201	3400
Araniella cucurbitina (Clerck, 1757)		Common	11302	3420
.... *opistographa* (Kulczynski, 1905)		Local	11305	3430
Hypsosinga pygmaea (Sundevall, 1832)		Local	11503	3490
Cercidia prominens (Westring, 1851)		Local	11701	3530
Cyclosa conica (Pallas, 1772)		Local	11901	3550

Family LINYPHIIDAE

	Status	BRC	Index
Ceratinella brevipes (Westring, 1851)	Common	12201	3580
.... *brevis* (Wider, 1834)	Common	12202	3590
.... *scabrosa* (O.P. Cambridge, 1871)	Local	12203	3600
Walckenaeria acuminata Blackwall, 1833	Common	12301	3610
.... *antica* (Wider, 1834)	Common	12303	3630
.... *cucullata* (C.L.Koch, 1836)	Common	12308	3650
.... *atrotibialis* (O.P. Cambridge, 1878)	Local	12304	3670
.... *capito* (Westring, 1861)	Local	12305	3680
.... *incisa* (O.P. Cambridge, 1871)	Notable: Nb	12312	3690
.... *dysderoides* (Wider, 1834)	Local	12310	3700
.... *nudipalpis* (Westring, 1851)	Common	12317	3720
.... *furcillata* (Menge, 1869)	Local	12311	3760
.... *unicornis* O.P. Cambridge, 1861	Common	12320	3770
.... *cuspidata* Blackwall, 1833	Common	12309	3800
.... *vigilax* (Blackwall, 1853)	Local	12321	3810
Dicymbium nigrum (Blackwall, 1834)	Common	12402	3820
.... *tibiale* (Blackwall, 1836)	Local	12403	3840
Entelecara acuminata (Wider, 1834)	Common	12501	3850
.... *erythropus* (Westring, 1851)	Local	12504	3870
Moebelia penicillata (Westring, 1851)	Common	12601	3910
Hylyphantes graminicola (Sundevall, 1830)	Local	12701	3920
Gnathonarium dentatum (Wider, 1834)	Common	12801	3930
Tmeticus affinis (Blackwall, 1855)	Local	13001	3950
Gongylidium rufipes (Linnaeus, 1758)	Common	13101	3960
Dismodicus bifrons (Blackwall, 1841)	Common	13201	3970
Hypomma bituberculatum (Wider, 1834)	Common	13301	3990
.... *cornutum* (Blackwall, 1833)	Common	13302	4010

	Status	BRC	Index
Metopobactrus prominulus (O.P. Cambridge, 1872)	Local	13401	4020
Baryphyma pratense (Blackwall, 1861)	Local	13604	4040
.... *trifrons* (O.P. Cambridge, 1863)	Local	13605	4070
Gonatium rubens (Blackwall, 1833)	Common	13703	4090
.... *rubellum* (Blackwall, 1841)	Common	13702	4100
Maso sundevalli (Westring, 1851)	Common	13802	4120
.... *gallicus* Simon, 1894	Notable: Na	13801	4130
Peponocranium ludicrum (O.P. Cambridge, 1861)	Common	13901	4150
Pocadicnemis pumila (Blackwall, 1841)	Common	14002	4160
.... *juncea* Locket & Millidge, 1953	Common	14001	4170
Oedothorax gibbosus (Blackwall, 1841)	Common	14204	4190
.... *f. tuberosus* (Blackwall, 1841) †	Common	14206	4191
.... *fuscus* (Blackwall, 1834)	Common	14203	4200
.... *agrestis* (Blackwall, 1853)	Local	14201	4210
.... *retusus* (Westring, 1851)	Common	14205	4220
.... *apicatus* (Blackwall, 1850)	Local	14202	4230
Pelecopsis parallela (Wider, 1834)	Local	14406	4270
.... *nemoralis* (Blackwall, 1841)	Local	14405	4280
Silometopus elegans (O.P. Cambridge, 1872)	Local	14502	4320
.... *reussi* (Thorell, 1871)	Local	14504	4340
Cnephalocotes obscurus (Blackwall, 1834)	Local	14701	4370
Ceratinopsis stativa (Simon, 1881)	Local	15002	4430
Evansia merens O.P. Cambridge, 1900	Local	15101	4440
Tiso vagans (Blackwall, 1834)	Common	15202	4450
Troxochrus scabriculus (Westring, 1851)	Local	15301	4470
Minyriolus pusillus (Wider, 1834)	Common	15401	4480
Tapinocyba praecox (O.P. Cambridge, 1873)	Local	15504	4490
.... *pallens* (O.P. Cambridge, 1872)	Local	15503	4500
.... *insecta* (L.Koch, 1869)	Local	15501	4510
Microctenonyx subitaneus (O.P. Cambridge, 1875)	Local	15701	4540
Thyreosthenius parasiticus (Westring, 1851)	Local	15902	4560
.... *biovatus* (O.P.-Cambridge, 1875) †	Local	15901	4570
Monocephalus fuscipes (Blackwall, 1836)	Common	16002	4580
.... *castaneipes* (Simon, 1884)	Local	16001	4590
Lophomma punctatum (Blackwall, 1841)	Local	16101	4600
Saloca diceros (O.P. Cambridge, 1871)	Notable: Nb	16201	4610
Gongylidiellum vivum (O.P. Cambridge, 1875)	Common	16303	4620
.... *latebricola* (O.P. Cambridge, 1871)	Local	16301	4630
Micrargus herbigradus (Blackwall, 1854)	Common	16402	4650
.... *apertus* (O.P. Cambridge, 1871)	Local	16401	4660
.... *subaequalis* (Westring, 1851)	Local	16404	4670
Erigonella hiemalis (Blackwall, 1841)	Common	16701	4720
.... *ignobilis* (O.P. Cambridge, 1871	Local	16702	4730
Savignia frontata (Blackwall, 1833)	Common	16801	4740
Diplocephalus cristatus (Blackwall, 1833)	Common	16902	4750
.... *permixtus* (O.P. Cambridge, 1871)	Common	16905	4760
.... *latifrons* (O.P. Cambridge, 1863)	Common	16904	4770
.... *picinus* (Blackwall, 1841)	Common	16906	4790
Araeoncus humilis (Blackwall, 1841)	Local	17002	4810
Panamomops sulcifrons (Wider, 1834)	Local	17101	4830
Lessertia dentichelis (Simon, 1884)	Local	17201	4840
Typhochrestus digitatus (O.P. Cambridge, 1872)	Local	17401	4860
Milleriana inerrans (O.P. Cambridge, 1885)	Local	17501	4880
Erigone dentipalpis (Wider, 1834)	Common	17705	4900
.... *atra* Blackwall, 1833	Common	17703	4910
.... *arctica* (White, 1852)	Local	17702	4930
.... *longipalpis* (Sundevall, 1830)	Local	17706	4940

	Status	BRC	Index
Prinerigone vagans (Audouin, 1826)	Local	17801	5000
Leptorhoptrum robustum (Westring, 1851)	Local	18301	5060
Drepanotylus uncatus (O.P. Cambridge, 1873)	Local	18401	5070
Halorates distinctus (Simon, 1884)	Local	18701	5140
Asthenargus paganus (Simon, 1884)	Local	19201	5210
Ostearius melanopygius (O.P. Cambridge, 1879)	Naturalised	9501	5240
Porrhomma pygmaeum (Blackwall, 1834)	Common	19709	5260
.... *convexum* (Westring, 1851)	Local	19702	5270
.... *pallidum* Jackson, 1913	Local	19708	5290
.... *campbelli* F.O.P. Cambridge, 1894	Local	19701	5300
.... *microphthalmum* (O.P. Cambridge, 1871)	Local	19705	5310
.... *errans* (Blackwall, 1841)	Notable: Nb	19704	5320
.... *egeria* Simon, 1884	Local	19703	5330
Agyneta subtilis (O.P. Cambridge, 1863)	Common	19806	5370
.... *conigera* (O.P. Cambridge, 1863)	Common	19802	5380
.... *decora* (O.P. Cambridge, 1871)	Local	19803	5390
.... *cauta* (O.P. Cambridge, 1902)	Local	19801	5400
.... *ramosa* Jackson, 1912	Local	19805	5420
Meioneta innotabilis (O.P. Cambridge, 1863)	Common	19903	5430
.... *rurestris* (C.L.Koch, 1836)	Common	19906	5440
.... *saxatilis* (Blackwall, 1844)	Common	19907	5460
.... *beata* (O.P. Cambridge, 1906)	Local	19901	5480
Microneta viaria (Blackwall, 1841)	Common	20001	5510
Centromerus sylvaticus (Blackwall, 1841)	Common	20313	5560
.... *prudens* (O.P. Cambridge, 1873)	Common	20311	5570
.... *dilutus* (O.P. Cambridge, 1875)	Common	20306	5600
Tallusia experta (O.P. Cambridge, 1871)	Common	20401	5700
Centromerita bicolor (Blackwall, 1833)	Common	20501	5710
.... *concinna* (Thorell, 1875)	Common	20502	5720
Saaristoa abnormis (Blackwall, 1841)	Common	20801	5750
.... *firma* (O.P. Cambridge, 1905)	Local	20802	5760
Macrargus rufus (Wider, 1834)	Common	20902	5770
Bathyphantes approximatus (O.P. Cambridge, 1871)	Common	21001	5790
.... *gracilis* (Blackwall, 1841)	Common	21002	5800
.... *parvulus* (Westring, 1851)	Common	21004	5810
.... *nigrinus* (Westring, 1851)	Common	21003	5820
Kaestneria dorsalis (Wider, 1834)	Common	21101	5840
.... *pullata* (O.P. Cambridge, 1863)	Common	21102	5850
Diplostyla concolor (Wider, 1834)	Common	21201	5860
Poeciloneta variegata (Blackwall, 1841)	Local	21301	5870
Drapetisca socialis (Sundevall, 1833)	Common	21401	5880
Tapinopa longidens (Wider, 1834)	Common	21501	5890
Floronia bucculenta (Clerck, 1757)	Local	21601	5900
Labulla thoracica (Wider, 1834)	Common	21801	5920
Stemonyphantes lineatus (Linnaeus, 1758)	Common	21901	5930
Bolyphantes luteolus (Blackwall, 1833)	Common	22002	5940
.... *alticeps* (Sundevall, 1833)	Local	22001	5950
Lepthyphantes nebulosus (Sundevall, 1830)	Local	22114	5960
.... *leprosus* (Ohlert, 1865)	Common	22110	5970
.... *minutus* (Blackwall, 1833)	Common	22113	5980
.... *alacris* (Blackwall, 1853)	Common	22101	5990
.... *obscurus* (Blackwall, 1841)	Local	22115	6010
.... *tenuis* (Blackwall, 1852)	Common	22119	6020
.... *zimmermanni* Bertkau, 1890	Common	22121	6030
.... *cristatus* (Menge, 1866)	Local	22105	6040
.... *mengei* Kulczynski, 1887	Common	22111	6050
.... *beckeri* Wunderlich, 1973 †	RDB K		6055

	Status	BRC	Index
Lepthyphantes flavipes (Blackwall, 1854)	Common	22108	6060
.... *tenebricola* (Wider, 1834)	Local	22118	6070
.... *ericaeus* (Blackwall, 1853)	Common	22106	6080
.... *pallidus* (O.P. Cambridge, 1871)	Local	22116	6090
.... *insignis* O.P. Cambridge, 1913	Notable: Nb	22109	6110
.... *midas* Simon, 1884	RDB 2	22112	6150
Helophora insignis (Blackwall, 1841)	Common	22201	6180
Pityohyphantes phrygianus (C.L. Koch, 1836)	Local	22301	6190
Linyphia triangularis (Clerck, 1757)	Common	22407	6200
.... *hortensis* Sundevall, 1830	Common	22403	6210
.... *montana* (Clerck, 1757)	Common	22405	6220
Linyphia clathrata Sundevall, 1830	Common	22401	6230
.... *peltata* Wider, 1834	Common	22406	6240
Microlinyphia pusilla (Sundevall, 1830)	Common	22502	6270
.... *impigra* (O.P. Cambridge, 1871)	Local	22501	6280
Allomengea vidua (L. Koch, 1879)	Local	22602	6300

† Notes:
Micrommata virescens recorded by Chalcraft (Rowley, 1897). See comments pages 36-37.
Tegenaria gigantea and **Tegenaria saeva** are synanthropic species, except in the south.
Tegenaria parietina recorded by Chalcraft and Mayes, Leicester.
Araneus marmoreus f. **pyramidatus** recorded by Chalcraft, Owston Wood.
Oedothorax gibbosus f. **tuberosus** (dimorphic males of *Oedothorax gibbosus*) have been recorded separately, these generally occurring with females. Since this information may be of some interest, *Oedothorax tuberosus* has been included in the checklist, but not counted as a separate species. These records are plotted on page 147.
Thyreosthenius biovatus recorded by Donisthorpe (1927) at Buddon Wood in nests of the ant *Formica rufa*, both sexes common in each nest examined. The ant is now extinct at what remains of Buddon Wood. *T.biovatus* is unknown elsewhere in the county.
Lepthyphantes beckeri new to Britain. See comments on pages 22, 50 and 74.

Euryopis flavomaculata f. (x8.5)

CHAPTER THREE

An Historical Review of Arachnology in Leicestershire and Rutland, Watsonian Vice-county 55

Introduction

Although the scientific study of spiders in Britain goes back to the seventeenth century and Martin Lister's catalogue of 34 species in 1678, no great interest was taken in this order in Leicestershire until the latter half of the twentieth century. Indeed, W.S. Bristowe (1939) states that the spider faunas of Leicestershire and Rutland were two of the least known in England and lists only 26 species for Rutland which he had recorded himself. There has been no serious attempt to improve this situation from within Rutland. No paper records or spirit collections are held by Oakham Museum, Uppingham School Museum or the Rutland Natural History Society, and spiders currently collected on an *ad hoc* basis are forwarded to Leicestershire Museums Service for identification or verification. All spider survey in Rutland has been initiated from Leicestershire as part of the vice-county strategy which is perpetuated in this publication, combining Leicestershire and Rutland in the single Watsonian Vice-county 55. All references to the 'county', therefore, apply to the vice-county, as strictly speaking do references to 'Leicestershire', but because of traditional usage references to 'Rutland' will occasionally be used and these are taken literally.

The history of arachnology in Leicestershire falls conveniently into two parts: firstly the period up to 1959 – before the influence of 'Locket & Millidge' – and secondly, post-1959, when the study of spiders was taken much more seriously and conducted on a systematic footing. Up to 1959 there was no co-ordination in arachnological studies, and prior to this date collecting was carried out on a casual and mostly superficial level. The stimulus provided by G.H. Locket and A.F. Millidge, with the publication of their two volumes of *British Spiders* in 1951 and 1953, did not take hold in Leicestershire until 1959, after which, field-work was organised and well documented. With few exceptions, records for the first period were only localised to county, whereas those from 1959 onwards were sufficiently detailed to allow plotting on a 2 km square (tetrad) grid. These latter records form the substance of the present atlas (chapter five). A summary of all first records for the county is given in the form of two tables (pages 56 to 61): the first, covering 88 records, mostly from published sources, where very little detail is available other than the species name and county; the second, covering the remaining 238 records, which are presented in date order with location and name of collector and determiner. In the first part of this chapter, current nomenclature is indicated by the use of bold italics, to assist in the interpretation of the rather confusing synonymy.

Whereas the revised county list had grown to 105 by 1955, some of the species mentioned for the first time in literature and other sources since 1889 must be viewed with caution, particularly the pre-war records, on account of inadequate taxonomic literature. Apart from H.St.J. Donisthorpe, the eminent entomologist, who was also a specialist on myrmecophilous spiders, and W.S. Bristowe, whose records are only cursory and mostly picked up from the *Victoria County History* (1907), we have had no specialist working on the Araneae until recently. Neither have we evidence that the early collectors referred their determinations to national specialists – which in the days before 'Locket & Millidge' was the only way to be certain of many difficult species. The early lists are therefore more of curiosity interest than practical value and all but five (*Micrommata virescens, Tegenaria parietina, Araneus marmoreus, Araneus alsine* and *Thyreosthenius biovatus*) have been re-discovered during recent years. What is revealed by this historical survey is the paucity of detail accompanying records and lack of voucher material, rendering them of limited usefulness.

The turning point came in 1953, with the second of the three Ray Society volumes on *British Spiders* by G.H. Locket & A.F. Millidge (and later P. Merrett). This definitive taxonomic tool enabled workers to identify their own material with more confidence and greater accuracy. Without it the revival of interest in spiders in Leicestershire would not have taken place.

Nationally, the formation of the Flatford Mill Spider Group in 1958, later to become the British Spider Study Group and finally the British Arachnological Society, owes its success to the growing number of competent arachnologists weaned on 'Locket & Millidge'. It is significant that two of Leicestershire's sons have made a major contribution to the succeeding ecological and taxonomic advancement of the science of arachnology, in the persons of Dr Eric Duffey and Dr Michael J. Roberts. Unfortunately, neither has collected extensively in Leicestershire since moving away from the county to pursue their individual careers, but the present authors owe a great debt to them both for encouragement and intellectual stimulation.

The first records which can be reliably mapped and which include location, sex and habitat details are museum specimens collected by T.A.Walden and others (1948-1953) and those from E. Duffey (1952-1955). These are shown on the accompanying maps as open circles; otherwise all modern records from 1955 onwards are shown as solid circles (black dots).

PART I 1795-1955

The usual starting point for an account of any aspect of the fauna of Leicestershire is George Crabbe's very useful *Natural history of the Vale of Belvoir* (Crabbe in: Nichols, 1795), but in the case of the 'Araneae', Crabbe is less than enlightening, dismissing them, along with several other groups of non-insect arthropods, as follows: "Nothing particular need be mentioned of these genera, the same species which are found in other counties will be met with here....". His not unreasonable excuse is "they are so imperfectly described, that it is very difficult, in many cases, to say if they be the individuals meant by the authors or not."

The Cradle of Natural Philosophy

In trying to establish where an interest in spiders began in Leicestershire, we are taken back to the Victorian institutions of science and philosophy founded by studious amateurs. Here it is, in the Leicester Literary and Philosophical Society, that we find the first stirrings of awareness of arachnids, amongst the enthusiastic entomologists of the late nineteenth century. Members of the Society and its Sections had access to an excellent Scientific Library which may well have acquired important volumes such as John Blackwall's *Spiders of Great Britain and Ireland* (1861-64) published by the Ray Society. So it is perhaps in order to go back to the establishment of this learned society and follow its wider influence through those early years. For this task it has been necessary to rely heavily on Lott, F.B. (1935) *The Centenary Book of the Leicestershire Literary and Philosophical Society*, from which the following notes have largely been extracted.

The Leicester Literary and Philosophical Society originated in the spring of 1835, with its inaugural meeting at the Medical Library in High Street in June of that year. The early preoccupation of the Society was to further the prospects of liberal education and the provision of a place where Conservatives and Liberals, Churchmen and Non-conformists could meet for friendly discourse – theological and political questions being excluded. It was emphasised that the term 'Philosophical' embraced not only Moral but also Natural Philosophy, and the Society determined in 1838 upon the desirability of a Natural History Museum. The Society's collection, housed in the New Hall, Wellington Street, had a strong bias towards the geological sciences, due to the great interest in the rocks of Charnwood Forest and was opened daily in 1841. Control was vested in twelve Trustees in 1845, eventually being transferred to the Town Council on 19th June 1849 when the collection comprised some 10 000 objects. These were housed in the converted Proprietary School, which had been erected in 1837 on the New Walk and later purchased as the Town Museum. The collections were enhanced by the presentation of scientific books and a large number of plants, shells and birds. By 1877 the number of objects catalogued in the Town Museum had risen to 22 000 and a Museum Library had been created.

Sectional Committees of the Society were formed in the same month in which the Museum was presented to the Town Council. These encompassed Geology, Zoology, Botany, Archaeology and Fine Arts. Many papers were read on the subject of Natural Science and these at a time when the pre-Darwinian theories of evolution featured in lectures to the Society. The Museum Committee drew up rules for the Management of the Museum which were adopted by the Town Council on 21st May 1872, and Honorary Curators were appointed for each of the departments.

We now follow the Zoology and Botany Section, which in 1874 was renamed the Natural History Section. In 1883 this was split into Section 'D' (Biology, Zoology and Botany), and Section 'E' (Zoology – the Leicestershire Fauna). Later, in 1894, Section 'F' (Entomology) was created and for about 14 years was very active, but ceased in 1919-20 for lack of support.

The first reported excursion reported by the Society was a picnic to Bradgate Park on 15th June 1861. This was very much a social event and appears to have set the standard for future excursions to Groby, Markfield, Bardon Hill, Buddon Wood, Outwoods, Beaumanor, Charley Hall, Quenby Hall, Tilton, Burrough Hill and Belvoir. However, these events opened up the private estates and properties to collectors, particularly the entomologists who found them rich hunting grounds.

Annual Transactions began in 1879-80, and in 1884 a volume of earlier transactions from 1835-1879 was produced; then in 1886 Transactions were published quarterly instead of with the Annual Report. Because of rising costs, publication became half-yearly after 1903 and continued till 1911 when production of the Transactions was again incorporated with the Annual Report of the Council.

At the first two meetings of Section 'E' (Zoology) it was resolved that much of the Section work would be done on the Fauna of Leicestershire and the Report for 1883-84 states that "the several orders of invertebrates are being fairly worked out ....", and this certainly applied to the Coleoptera. Section 'F' (Entomology) was appointed in January 1894. Before June it had twenty-five members and five meetings had been held, with an average attendance of nine. There had been special excursions to Bardon Hill, Narborough Bog and Buddon Wood, and the main interest of members seems to have been in establishing county records of Coleoptera and Lepidoptera. As we will see later, this had some bearing upon the spider work carried out by G.B. Chalcraft in 1897, with encouragement from Frank Bouskell, who was an active member of Section 'F' in its early days, and Honorary Minute Secretary for the Parent Body during 1896-99. This would have given Bouskell direct access to Chalcraft's lecture manuscript and notes on exhibits which were not detailed in the minute books.

On March 8th 1869 the first steps were taken towards the establishment of a Public Free Library, and in 1885 the provision of a really good Reference Library was advocated. The idea of a University College was first mooted from the Presidential Chair of the Society in 1884, and again in 1885 and 1886, but the proposal was not taken up publicly until 1912, when a considerable amount of interest was generated amongst the Press and influential men in the town. Despite all this backing, the new University College was not established until after the war, in 1920, with D.R.F. Rattrap as Acting Principal, later to be confirmed as Principal, and in 1924-25 elected President of the Society. Academic progress bypassed arachnological studies as, with the exception of the Acari, arachnids did not generally feature in the academic curriculum until after the Second World War, and little useful work has come out of Leicester University in this sphere of research until more recent times. No standard reference books were available for the identification of British spiders until 1951-53 and this undoubtedly influenced the absence of this group of invertebrates from the zoological syllabus.

The Great War curtailed the activities of the Sections and the main thrust of activities in the natural sciences during the succeeding inter-war years, and indeed until 1947, was in botany and commercial geology. In 1947, the Leicestershire Fauna Survey was launched as a joint venture by the Zoology Department at the University College, the Museum, and the Botany and Zoology Section of the Society, which temporarily regenerated interest in spiders.

Early Stirrings of Interest

At Leicester, the late nineteenth century was a period of tremendous activity amongst a fraternity of dedicated and competent entomologists, spurred on by the opening up of opportunities to study the world fauna, as collectors returned from Commonwealth collecting expeditions. The importance of this infectious enthusiasm cannot be overstated and Reports of Section meetings bear witness to the range of interests embraced by these specialists. In 1883, with the setting up of Section 'E', the first attempt was made to concentrate on the fauna of Leicestershire whilst Section 'D' continued to embrace wider zoological topics. Between this date and 1907, six names appear in the record, each to a greater or lesser degree contributing to the initial focus on spiders in the county. Firstly, we have the Rev William Agar, who had a strong leaning towards natural history and presented a general paper on Spiders to Section 'D' in 1889. Then, eight years later in 1897, George B. Chalcraft exhibited a collection of Leicestershire spiders at a meeting of Section 'F'.

Chalcraft was very much motivated by his entomological contemporaries, who made casual collections of spiders and frequently took part in discussions on the subject at meetings. These, in order of importance to Chalcraft, were Frank Bouskell, W.J. Kaye, C.B. Headly and H.St.J. Donisthorpe, all four Fellows of the Entomological Society.

William Agar

But where did it all start? We first come across the 45-year-old William Agar listed in the Leicester Literary and Philosophical Society roll of members for 1853-54. His name reappears frequently but not continuously up until 1865-66, when it is presumed he moved away from Leicester, possibly to qualify for and enter the non-conformist Ministry. In 1881 he was living with his 53-year-old wife Elizabeth at Blaby and, at 73 years of age, his occupation is recorded as 'Farmer Employer', with no occupation shown for the lady of the house. In 1888, William reappears in the Associate roll for Section 'D' Biology (Zoology & Botany) as Rev W.Agar, and the following year on December 12th 1889, at 81 years of age, he presented his paper on 'Spiders' to Section 'D'. The paper is of a general nature, dealing mainly with structure and life history. He quotes from J. Blackwall (1861-64) concerning silk spinning, and describes the habit of **Zygiella x-notata** "the common geometrical spider of our window-panes", mistakenly naming this *Epeira calophylla* (= **Zygiella atrica**?). He describes the distinctive egg cocoon of *Agelena brunnea* (= **Agroeca brunnea**) and relates his observations on *Epeira diadema* (= **Araneus diadematus**) around his house. Then he proceeds to talk about trap-door spiders of Jamaica, the fenland Raft Spider and the Diving Spider (**Argyroneta aquatica**) "said to be common in the vicinity of London", continuing with observations on topics such as gossamer. What is interesting here is that he describes in detail the gossamer on an early morning walk in November over the Victoria Park. This is a very creditable paper and reveals Agar's enthusiasm for spiders; it also shows him as a well-read exponent of his subject.

Agar continued to appear as Sectional Associate up to 1897 when he would have been 89. His last appearance in the record was on February 17th 1897 when he read a paper to the Section on 'Seaside Studies', an account of holiday experiences at Ramsgate the previous year. This was a very general talk – no spiders or other specifics were mentioned. The inference drawn from this sketchy background is that Agar, with a long-standing interest in natural history, was introduced to the subject of spiders some time during the period between 1866 and 1881 whilst serving as Minister of a non-conformist church, probably away from Leicester. He may even have been a member of the Ray Society and acquired a copy of John Blackwall's *Spiders of Great Britain and Ireland*. Agar was an observer rather than a collector and his paper reflects his general interest rather than any specialist inclinations. Nevertheless, the impact of this paper in 1889 should not be underestimated, particularly in the new climate of exploration and discovery.

As has been noted, the 1890s were a decade of fervent activity in Section 'F', with several members taking more than a cursory interest in spiders. For example, a Mr Burford remarked at one meeting that spiders were proof against strychnine and other alkaloids, but were speedily killed in ordinary entomological cyanide killing bottles. On another occasion Mr E.F. Cooper described a variety of experiments he had made in the preservation of spiders for the cabinet, no one of which proved satisfactory. At the meeting on January 15th 1890 "a fine example of the very large and uncommon British spider, *Tegenaria atrica*, with legs over two inches long, caught in a fowl-house in Wharf Street [Leicester] by Mr B. Headley [sic]" was exhibited. At least people were aware of spiders, but were quite obviously not very skilled in the study of this group of invertebrates. The description given above for Headly's *Tegenaria atrica* would more nearly apply to **Tegenaria parietina**, and this is discussed later since both **Tegenaria parietina** and *Tegenaria atrica* (= **Tegenaria gigantea**?) appear in the 1907 lists.

Charles Burnard Headly was a keen lepidopterist and an Associate of Section 'D' between 1888 and 1898, later transferring to Section 'F' (Entomology) where he was a Committee Member from 1896. Amongst other insects, he exhibited Coleoptera and Diptera. He was active until about the time of the First World War, after which his name does not appear in the Section minute books. Headly did not enjoy good health, and there is a suggestion that he was an epileptic. It is thought he died around 1920 at the age of 50.

W.J. Kaye was primarily active in Section 'F' as a Section Associate and Committee Member, but he also made contributions on geology to Section 'E'. His name appears in 1896 in connection with exhibits of Lepidoptera from Trinidad and New Zealand, together with notes on

Micro-lepidoptera taken at Kibworth and Owston Wood, and reports on excursions to the New Forest, Isle of Wight and Wicken Fen. In February of that year he exhibited a case of spiders "which with the species in tubes in spirit, mounted on porcelain, preserved the colour very well. Amongst those shown [not necessarily from Leicestershire] were *Epeira bicornis, E.umbratica, E.angulata, Drassus cupreus* and many other interesting species". By then, sufficient collecting had been carried out in Leicestershire to justify inclusion of an illustrated paper in the forthcoming Section 'F' programme the following June, to be presented by 'W.J. Kaye, F.E.S. and G.B. Chalcraft' entitled 'Notes on the Arachnida (Spiders) of Leicestershire'. This paper was eventually given on July 2nd 1897 by Chalcraft alone under the title 'The Aranidae (Spiders) of Leicestershire' (see abstract: Rowley, 1897). Chalcraft "drew attention to the comparatively small amount of study which these interesting animals seemed to have attracted, ... and also to the difficulty of preserving them in a satisfactory condition". The paper was of a general nature, but included a list of 33 spiders observed in Leicestershire and exhibited at the meeting with accompanying notes; the notes however were not included in the published abstract and we have to turn to Bouskell (1907) to add some of this information. Chalcraft concluded that this list "was very far from complete" and he "hoped to add considerably to it as his knowledge of the fauna of the county increased." It is recorded that a long discussion followed in which Messrs F.R.Rowley (Chairman), B. Headley [sic], F. Bouskell and H.St.J. Donisthorpe took part.

George B. Chalcraft

George B. Chalcraft picked up his interest in spiders some time around 1895-96. He was elected a Member of Section 'F' (Entomology) on February 18th 1896 and, as is often the case with someone discovering a new subject, threw himself into the study of this order with vigour. During 1896 he joined the Section excursions to the New Forest (June 3rd-9th) and Wicken Fen (August) and reported back at the autumn indoor meetings on the spiders he had collected. The species lists from both sites are disappointing, being of mostly large showy species, and his comments accompanying the lists give rise to the suspicion that everything was rather new to him. Throughout, Chalcraft uses Blackwall's (1861-64) nomenclature which supports the view that he had access to a copy of *The Spiders of Great Britain and Ireland*. The New Forest list comprises 13 species, all females, and includes an unidentified *Drassus* and several specimens of what Chalcraft took to be Blackwall's *Epeira ornata*. This is a mis-identification and is discussed below as it appears also on the Leicestershire list. The Wicken list of 9 species is similarly uninspiring for such a good site.

In the Leicestershire list *Tegenaria domestica* (= **Tegenaria parietina**) is included to the exclusion of *Tegenaria atrica* (= **Tegenaria gigantea**?). Did Headly realise he had made a mistake in his early identification, or is this a mistake by Chalcraft for the common **Tegenaria gigantea** in Leicester? *Epeira ornata* is an invalid species; it was unfortunately carried through into subsequent lists in error and was later dropped from the British list, so must be deleted here from Chalcraft's list. The inclusion of *Epeira lutea* (= **Araneus alsine**) poses a problem in the light of Chalcraft's inexperience, since this is a nationally rare spider, not subsequently rediscovered in Leicestershire. As we have seen, Agar, in describing **Zygiella x-notata**, referred to it by the wrong name (*Epeira calophylla*) which reappears in Chalcraft's list and is likely also to have crept through into both the 1907 lists. However, *Epeira calophylla* (= **Zygiella atrica**) has subsequently been shown to be widespread and common in the county, so it is probably safer to credit Chalcraft with this new species. *Epeira cucurbitina* is unacceptable due to possible mis-identification. Blackwall's drawing of the palp of *Epeira cucurbitina* (= **Araniella cucurbitina**) is quite distinctive, but there is no epigyne figured and it is unlikely, in the absence of a closely related species being described, that the early collectors would have looked further than at the general appearance of this spider, beautifully figured on plate 25 of Blackwall's *The Spiders of Great Britain and Ireland*. This confusion between two superficially identical species persisted, even after male genitalia of the closely related **Araniella opistographa** were figured by Locket & Millidge (1953) and female vulvae by Locket, Millidge & Merrett (1974), until the problem was highlighted by Roberts (1985). **Araniella cucurbitina** and **Araniella opistographa** are both common and widespread species, often occurring in the same habitat. It has been shown that in Leicestershire **A.opistographa** is more frequently met with than **A.cucurbitina**.

Chalcraft was elected a Member of Section 'E' (Zoology & Geology) on February 14th 1899 but appears not to have contributed anything else on spiders after his paper in 1897. His list is an important one and is reproduced here in full, but we have to turn to the *Victoria County History of Leicestershire* (Bouskell, 1907) for comments on the species. To avoid duplication, Bouskell's annotations are added here in table 1. Of the 33 species listed by Chalcraft, 31 represent new records for Leicestershire, added to the 4 established by Agar and Headly.

Table 1. List of Leicestershire spiders by G.B. Chalcraft, in the order of the original, with current nomenclature in bold italics, and annotations from Bouskell (1907) and current comments in square brackets.

Lycosa campestris: [***Trochosa ruricola***] — Very common. Found running on the ground.

L. saccata: [***Pardosa amentata***] — Very common. Always found in large numbers together; found in profusion in such places as dung-heaps and places where straw is laid.

Dolomedes mirabilis: [***Pisaura mirabilis***] — Found running on the ground among long grass, and carries its egg sac under the sternum. Owston Wood (G.B.C.).

Salticus scenicus — Probably all over the county. Taken off wall covered with ivy, Leicestershire: also Swithland Wood (G.B.C.).

Thomisus cristatus: [***Xysticus cristatus***] — Very common in the county; one of the spiders which has the power of launching itself into the air, and often travels considerable distances in this manner.

T. citreus: [***Misumena vatia***] — Found on flowers of valerian, from whence it captured Lepidoptera sitting there. Owston Wood (F.B. & W.J.K.).

Sparassus smaragdulus: [***Micrommata virescens***] — [no commentary; unconfirmed record]

Drassus cupreus: [***Drassodes cupreus***] — Taken at Buddon Wood.

Clubiona accentuata: [***Anyphaena accentuata***] — Very rapid in its movements; found in woods. Taken at Buddon Wood.

Ciniflo atrox: [***Amaurobius fenestralis***] — Very common; found under stones and loose bark; hunts at night. Varies considerably in the colour of the markings of the abdomen. Leicester (G.B.C.) October.

C. ferox: [***Amaurobius ferox***] — [no commentary].

Tegenaria domestica: [***Tegenaria gigantea***?] — Found in houses; very common. Taken in Leicestershire (G.B.C.). [Blackwall's *T.domestica* = ***Tegenaria parietina***, but this description does not fit: this is almost certainly ***Tegenaria gigantea***].

T. civilis: [***Tegenaria domestica***] — Found in houses; common. Taken in Leicester (G.B.C.).

Coelotes saxatilis: [***Coelotes atropos***] — Frequents dark hiding places. Common at Bardon Hill and probably elsewhere (G.B.C.).

Textrix lycosina: [***Textrix denticulata***] — [no commentary].

Theridion quadripunctatum: [***Steatoda bipunctata***] — Found in cracks in walls and disused houses *etc.*, fairly common. Leicester (G.B.C.).

T. novosum [sic; = *T.nervosum*: ***Theridion sisyphium***] — Taken at Swithland.

T. guttatum [sic; = *T.guttata*: ***Crustulina guttata***] — Fairly distributed. Leicester (G.B.C.).

Neriene bicolor: [***Centromerita bicolor***] — Common; chiefly by beating [doubtful].

Walckenaera punctata: [***Lophomma punctatum***] — Taken at Aylestone (G.B.C.).

W. pratensis: [***Baryphyma pratense***] — Taken at Buddon Wood [doubtful].

Epeira cucurbitina: [***Araniella cucurbitina/opistographa***] — Very common [species uncertain].

E. patagiata: [***Larinioides patagiatus***] — [no commentary; occurs at Buddon Wood.]

E. scalaris: [***Araneus marmoreus*** (form *pyramidatus*)] — Taken at Owston Wood.

E. umbratica: [***Nuctenea umbratica***] — Generally considered uncommon, but probably that is on account of its retiring habits. Taken at Blaby under willow bark.

E. lutea: [***Araneus alsine***] — Common. Leicester (G.B.C.) [mistaken identity].

E. diadema: [***Araneus diadematus***] — Common (G.B.C.). [recorded by Rev W. Agar]

E. calophylla: [***Zygiella atrica***] — Fairly common. Leicester (G.B.C.).

E. ornata: — One taken. [invalid species: Bristowe (1941) remarks 'identity uncertain' – removed from British list].

E. antriada: [***Metellina merianae***] — At Saddington Reservoir (G.B.C.).

E. inclinata: [***Metellina segmentata***] — Taken in Leicester and Buddon Wood.

E. conica: [***Cyclosa conica***] — Taken in Buddon Wood.

Tetragnatha extensa — Taken in Owston Wood and Swithland.

The Victoria County History

The publication in 1907 of two more lists of Leicestershire spiders, by Bouskell and Horwood, consolidated the nineteenth century efforts of the small group of entomologists which culminated in Chalcraft's 1897 list (table 1). Neither of the new lists adds any new species, but each is interesting as they are produced independently from the same source. Bouskell makes an historical statement with his inclusion of the 'lost' species notes, and Horwood provides a projection into the twentieth century with the adoption of revised nomenclature based on the new British checklist (O.P.-Cambridge, 1900).

Frank Bouskell, F.E.S., F.R.H.S. was a keen horticulturalist from his early years and a leading authority and judge at Britain's chief flower shows including Edinburgh, Shrewsbury and Chelsea. He qualified as a solicitor in 1894 and later moved from Leicester to Market Bosworth, where he lived for more than 55 years until his death in 1952 at the age of 82. He was an Associate of Section 'D' in 1888 when he was eighteen years old. Active in local entomological circles between 1890-1920, Bouskell became Secretary of Section 'F' (Entomology) on its foundation in 1894, and later prominent as Chairman and Vice-chairman up to 1921. He was a frequent contributor of papers, reports, discussions and exhibits at Section 'F' meetings, including papers on 'The extinction of certain species of Lepidoptera and Coleoptera: its cause and suggestions for the protection of the rarer species' and 'Parthenogenesis in Insects'. On one occasion he exhibited four cases of eggs of British Birds. From his reports on excursions and field meetings, and notes on a wide range of allied topics, it is clear that Bouskell was a dynamic character and responsible, in part at least, for the success of the Entomology Section, which in 1897-98 reported that "throughout the year, many specimens of all orders have been exhibited .... several new species of Coleoptera and Lepidoptera have been added to the county list .... a system of permanent county records has been established .... the Section is in an exceedingly flourishing condition....". He was interested in spiders and collected anything which appeared new to him. He was obviously a great motivator and clearly encouraged Chalcraft in his work on spiders, as is borne out by his participation in discussions and by the inclusion in his invertebrate lists of Chalcraft's list of spiders to which he himself had contributed.

Bouskell was responsible for the chapter on Insects in the Zoology section of *The Victoria County History of Leicestershire* (Page, 1907) and appended two pages on Spiders (Bouskell, 1907). In the introduction to the chapter he makes the following comment on Leicestershire: "The greater part of the county is arable and pasture land in a high state of cultivation, but on the whole it is well wooded. Charnwood Forest, which includes well-known localities like Buddon Wood, Bardon Hill and Bradgate Park, is perhaps the richest district in the matter of records, probably because it has been more worked than other districts. Owston Wood, on the Rutland border, with its varied flora, produces a number of species [of insects] not found in other parts of the county. Seal Wood, Grange Wood, and the Ambion Wood, in the neighbourhood of Sutton Cheney, are all good collecting ground". In the introduction to the chapter, acknowledgement is made to "Mr W.J. Kaye, F.E.S., Mr C.B. Headly, F.E.S., Mr G.B. Chalcraft, ... and others" for notes on various orders of insects.

The Arachnida-Spiders section comprised an annotated list of 30 species, lifted straight from Chalcraft's list, using the same outdated nomenclature. However, the value of Bouskell's list lies in the remarks on the species "taken by Messrs.W.J. Kaye and C.B. Chalcraft [sic]" which were omitted from the original list, but were tantalisingly acknowledged in the report of the meeting at which the spiders were exhibited. No trace can be found of Chalcraft's original manuscript or the 'full notes' of the species listed, but Frank Bouskell knew Chalcraft quite well and was at the presentation of his paper and almost certainly obtained a copy of the notes which have been used here. For some reason, three of Chalcraft's species are omitted: *Epeira patagiata* (= **Larinioides patagiatus**), *Sparassus smaragdulus* (= **Micrommata virescens**) and *Ciniflo ferox* (= **Amaurobius ferox**). Of these, only **Micrommata virescens** is in doubt, since the others have been established as legitimate Leicestershire species. However, other evidence would suggest that **Micrommata virescens** was probably present in ancient county woodland in the past. It is such a distinctive spider that it would not have escaped the notice of keen-eyed entomologists and could not be mistaken for anything else. **Larinioides patagiatus** is still established at Buddon Wood, to which many Section excursions were made; perhaps **Micrommata** was also resident there before the clear-felling, burning and quarrying of the post-1940 era, or even still exists at Owston Wood!

The list adds nothing to that published in 1897 and raises doubts about *Tegenaria domestica* (= **Tegenaria parietina**) which is said to be "very common in houses in Leicestershire". Whereas **Tegenaria parietina** may well have been present in Leicester (see Mott, 1890), it is considered that in this list it is a mistaken identity for the very common **Tegenaria gigantea** (*Tegenaria atrica* in Horwood's 1907 lists, but subsequently shown to be a separate species not represented in Leicestershire). Confusion has arisen in the past over the presence of *Tegenaria civilis* (= **Tegenaria domestica**) in both 1897 and 1907 lists, which is correctly stated to be "found in houses; common, taken in Leicester". This is a sound record. In summary:—

> *Tegenaria domestica* and *Tegenaria civilis* stand as separate species; *Theridion novosum* is corrected to *T.nervosum*; and doubts remain about the following species:
>
> *Tegenaria domestica* is assumed to refer to our contemporary **Tegenaria gigantea**.
>
> *Sparassus smaragdulus* (= **Micrommata virescens**) is unconfirmed, but probably sound.
>
> *Neriene bicolor* Blackwall (= **Centromerita bicolor**) "common; chiefly obtained by beating". This is a grass-roots spider and one would not expect to take it by beating.
>
> *Walckenaera pratensis* Blackwall (= **Baryphyma pratense**) "taken at Buddon Wood". A rare spider, but possible from damp meadows or streamside marshes at Buddon.
>
> *Epeira cucurbitina* Clerck (= **Araniella cucurbitina**) "very common". Could be mistaken for **Araniella opistographa**, which is equally common! Species uncertain.
>
> *Epeira lutea* Koch (= **Araneus alsine**) "common, Leicester". A very rare spider and not at any time common. Almost certainly a mis-identification by Chalcraft.
>
> *Epeira ornata* is an invalid species and as such has no place on this list.

Also published in 1907 (July) was a list of 32 spiders compiled by **A.R. Horwood** for the *British Association Guide* (Horwood, 1907). Although he does not give any indication of the source of his information, the list is basically Chalcraft's in content and order, but with the nomenclature revised in compliance with the new Checklist of British Spiders (O.P.-Cambridge, 1900). It would appear that Horwood scrutinised the Leicester Literary and Philosophical Society Reports and Transactions as his source. Working independently of Bouskell, Horwood retains **Amaurobius ferox**, **Larinioides patagiatus** and **Micrommata virescens**, and repeats Chalcraft's record of the mysterious *Epeira ornata*, but omits **Textrix denticulata** and **Metellina segmentata**. However he introduces *Tegenaria atrica* (= **Tegenaria gigantea**?), presumably from his scrutiny of L.L.P.S. literature. In view of the longstanding confusion over the nomenclature of the *atrica* group, the species recorded here as *Tegenaria atrica* is taken to be **Tegenaria gigantea** which is widespread and common in Leicestershire. Horwood states, reasonably enough, that since "over 500 species of spiders are known to inhabit different stations in the British Isles ... the list must be taken as purely preliminary".

At the time of his contribution to the B.A. Guide, Horwood was serving as Sub-Curator at Leicester Museum and it is perhaps worth mentioning that their published accession lists contain just two references to spiders between 1891 and 1938. These relate to 'banana spiders' acquired in 1907, and to a specimen of *Epeira umbratica* (= **Nuctenea umbratica**) collected at Aylestone on 30th April 1908 by Mr L.S. Biggs. As Sub-Curator (1902-1922), Horwood would have had to interest himself in matters zoological (or at least after the departure of Montagu Browne in 1907), but his notes on 'Zoology' in the 1907 B.A. volume appear to be his only publication in this subject area. His other publications are all botanical and geological. For further details see biographical notes on A.R. Horwood F.L.S. (1879-1937), on pp.76-78 in Primavesi, A.L. & Evans, P.A. (1988) *Flora of Leicestershire*.

For completeness, it should also be mentioned that the brief account of invertebrates, other than insects, in the V.C.H. for Rutland (Douglas, R.N. 1908), totally ignores spiders.

The Early Twentieth Century

The next contribution to the history of Leicestershire arachnology is a hitherto unpublished manuscript in the files of the Leicestershire Museums Service, in the distinctive hand of **Dr E.E. Lowe** (Director of Leicester Museum 1907-1941 and also City Librarian at this time – a unique double appointment). A competent entomologist, with a special interest in Diptera and aculeate Hymenoptera, Lowe's interests were curtailed by his professional responsibilities. The manuscript lists the contents of 47 tubes of spiders collected by him in Leicestershire, identified by him and given in 1912 to W.H. Barrow (pers.comm. W.H. Barrow 29.9.1961). Although the specimens no longer exist and the records are unlocalised, the list is of some interest since, of the

34 specimens identified to species, no less than 20 (CR in table 2) are first records for the county. The order is that in which they appear in the original.

Table 2. List of spiders in the collection made by E.E. Lowe.

X. cristatus:	[**Xysticus cristatus**]	
Araneus diadema:	[**Araneus diadematus**]	
Drassus lapidicolens:	[**Drassodes lapidosus**]	CR
Trochosa picta:	[**Arctosa perita**]	CR
Zilla x-notata:	[**Zygiella x-notata**]	
Theridion lineatum:	[**Enoplognatha ovata** (form *lineata*)]	CR
Helophanes flavipes:	[**Heliophanus flavipes**]	CR
Tetragnatha extensa		
Clubiona grisea:	[**Clubiona stagnatilis**]	CR
Araneus cucurbitina:	[**Araniella cucurbitina/opistographa**]	
Meta merianae:	[**Metellina merianae**]	
Linyphia bimaculatum:	[**Theridion bimaculatum**]	
Linyphia clathrata		CR
Theridion nervosum:	[**Theridion sisyphium**]	
Coelotes atropos		
P. aureolus:	[**Philodromus aureolus**]	CR
Meta segmentata:	[**Metellina segmentata**]	
D. mirabilis:	[**Pisaura mirabilis**]	
Lycosa amentata:	[**Pardosa amentata**]	
Hasarius falcatus:	[**Evarcha falcata**]	CR
Philodromus dispar		CR
A. quadratus:	[**Araneus quadratus**]	CR
Harpactes hombergii:	[**Harpactea hombergi**]	CR
Neriene livida:	[**Robertus lividus**]	CR
Linyphia pusilla:	[**Microlinyphia pusilla**]	CR
Clubiona stagnalis:	[**Clubiona stagnatilis**]	
E. truncatus:	[**Episinus angulatus**]	CR
T. bimaculatum:	[**Theridion bimaculatum**]	CR
Pachygnatha Degeeri:	[**Pachygnatha degeeri**]	CR
Linyphia triangularis		CR
Agelena labyrinthica		CR

Also listed, with queries against them, are:

Epeira angulata = *Araneus angulatus* (see commentary below)		
Tegenaria campestris:	[**Tegenaria silvestris**]	CR
Pachygnatha listeri		CR

There is no evidence to suggest that Lowe submitted specimens to any specialist for a second opinion. Neither would he have known much about the distribution of species which have subsequently been shown to be very rare and out of their range in Leicestershire. It seems right to omit records of such species from the County Check List, but to include a commentary on them in case further evidence should come to hand. Where reasonable doubt still exists, such records have been included in the checklist with qualifying comments on their validity.

Lowe's list is a respectable one, mostly of larger species likely to be encountered and noticed by an entomologist, and appears to have been based on Pickard-Cambridge's nomenclature (*Spiders of Dorset* 1879-1881, and his definitive British checklist 1900). Three species on the list are unacceptable and two others remain doubtful for reasons stated below. Thus we have a list of 31 species, of which 20 are new records for Leicestershire and have subsequently been taken at various sites in the county.

Linyphia bimaculata C.L.K. is a synonym of **Theridion bimaculatum** and since both species are listed, the former is suppressed. *E(pisinus) truncatus*, as listed, appears to be another case of mistaken identity (= *Theridion angulatum*) and is in fact **Episinus angulatus**, subsequently discovered by M.G. Crocker in heather at The Brand on 15th May 1966, and later at Buddon Wood – very likely Lowe's original site. Bristowe (1941 p.510) remarks that "until 1906 two species of *Episinus* were confused under one name. When separated by O.P.-Cambridge in 1906, *angulatus* was incorrectly called *truncatus* and the true *truncatus* called *lugubris*".

*Zilla x-notata = Epeira similis = **Zygiella x-notata***. Although this is the first correctly named reference to this species for Leicestershire, Agar (1889) described this "common geometrical spider of our window panes" but attributed to it the name *Epeira calophylla*, rightly belonging to the less common rural ***Zygiella atrica***, often associated with gorse bushes. As mentioned earlier this error seems to have been perpetuated, possibly right through into Bristowe's 1939 checklist for Leicestershire and Rutland, where both species are recorded.

Lowe's list again raises the sensitive question of accuracy and the competence of the authors of such lists. At this time there were many errors in the literature which was in any case very limited and inadequate. It was necessary in those days, as indeed it is today, to submit specimens to the reigning authorities for confirmation; in Britain such names as Blackwall, Pickard-Cambridge, Carpenter, Pocock, Jackson, Hull, Falconer – and later Bristowe, La Touche, Locket, Millidge and Merrett. There appears no evidence that any such confirmations were sought and the confusion in the early lists seems to substantiate this point of view.

Epeira angulata (= ***Araneus angulatus*** Clk, 1757) is listed by Lowe, but this is a very rare spider, found only in woods along the South Coast. Even O. Pickard-Cambridge (1881, p.272) commented upon its rarity. Blackwall's figures are not particularly helpful either, and as he correctly observes "the descriptions of *Epeira angulata* given by arachnologists are, in general, so brief and imperfect as to render any attempt to reconcile the perplexed synonyma of this species almost hopeless" (Blackwall, 1861-1864 p.361). Confusion with synonyms prevailed even after O. Pickard-Cambridge attempted to resolve these problems with the publication of a definitive checklist at the turn of the century (Pickard-Cambridge, 1900). This record is therefore unacceptable.

Another case of mistaken identity or mis-spelling occurs in the appearance on Lowe's list of two 'species' *Clubiona grisea* and *Clubiona stagnalis*. The first, *Clubiona grisea* Thorell 1873, is a synonym of ***Clubiona stagnatilis*** Kulczynski 1897, and is acceptable as such, but the second, *Clubiona stagnalis* has never appeared on the British list. The description of this Finnish 'species' entered the literature in 1901 (Odenvall & Jarvi). There are various other references to *Clubiona stagnalis* up to 1937, but no others before Lowe's list (1912). Bonnet (1956) gives priority to Kulczynski's **stagnatilis** and suggests that *stagnalis* is suppressed as a synonym of the former. *Clubiona stagnalis* is therefore an invalid species. Whether Lowe had intended that these names should have applied to two different species we will never know: it may be that *stagnalis* was a mis-spelling of **stagnatilis**, and that *grisea* referred to ***Clubiona reclusa***; in either case the conjecture is academic.

It is interesting that despite this list of spiders attributed to Lowe, none of these was mentioned in the *Scientific Survey of Leicestershire and District* (Bryan, 1933) produced for the British Association for Science Annual Meeting at Leicester in 1933, to which both E.E. Lowe and W.E. Mayes contributed.

As mentioned, Lowe's collection of spiders was given to W. Hubert Barrow in 1912 but the location of this material is unknown if indeed it has survived. It is therefore apposite to relate something of what is known of Barrow, for the record. Born around 1890, he was a regular visitor to Leicester Museum from his childhood right up to his eighties. Barrow inherited the family brickworks but retired early to devote himself to his ornithological interests. Between 1959-72 he had a bench in the Taxidermy Room at the Museum where he prepared study skins for the Wildfowl Trust, arriving every morning at 7.30 a.m. He was a collector all his life, particularly of birds and their eggs, which he exchanged with other collectors and museums, notably that at Liverpool. Barrow published a few notes in the 1920s on birds and corresponded extensively with other collectors, but his interest in spiders seems to have been confined to the collection he received from Lowe.

H. St.J. K. Donisthorpe

The next records of spiders for Leicestershire are those of Donisthorpe (1927) published in his book *The Guests of British Ants*. Donisthorpe, an eminent entomologist and distinguished coleopterist from a well known Leicestershire family, writes authoritatively on spiders and was a correspondent of Rev O. Pickard-Cambridge and Dr Randell Jackson. He was an active and long-standing member of the Leicester Literary and Philosophical Society, and collected spiders in Bradgate Park and at Buddon Wood during his studies of ants, recording the following two species new to the county:—

Mastigusa macrophthalma (Kulczynski, 1897) recorded as *Cryphoeca recisa* O.P-C. Two females in nests of the ant *Formica fusca*, Bradgate Park, 3rd May 1909.

Thyreosthenius biovatus (O.P.-Cambridge, 1875). Widespread in ants' nests, especially *Formica rufa*. He states "I have personally found both sexes of *T.biovatus* in every *rufa* nest I have examined, from the Highlands to the South Coast of England, and in every month of the year". Both sexes in *Formica rufa* nests, Buddon Wood, 4th May 1909.

W.P. Mayes. So far as is known, E.E. Lowe took no further interest in spiders after he disposed of his spider collection, such time as he could spare from his responsibilities as Director of Museums and Libraries being devoted to the study of Diptera and Hymenoptera. However, in the early thirties, he may have encouraged an interest in spiders in W.Philip Mayes, a Diploma student employed by the Museum (W.P. Mayes was the son of Walter E. Mayes who succeeded A.R. Horwood as Curator). Mayes (in litt. 23.2.1960) "collected spiders and woodlice – because neither group had been worked to death – for [his] Diploma work and collected all over Leicestershire, with an emphasis on the Thurnby area [where he lived] and Rutland and the Lake District". The specimens collected by Mayes are the earliest still extant in the collections of the Leicestershire Museums Service. They were contained originally in 33 tubes, each labelled with locality, month and year of collection (February–July 1934) and an attempt at an identification. There is in the Museum files a list, in Mayes' hand, of the tube numbers and localities, many of which are indeed in Leicestershire, but others in Berkshire, Hertfordshire, Surrey and Northants. A few specimens are recorded as having been collected by Alec Bonner and C.J. Lane. There is also what appears to be a later list gathering all the available information together. In 1962 I.M. Evans examined and re-identified the specimens and attempted a reconciliation of all the available information on labels and lists. In a number of cases the number and identity of the specimens bore so little relationship to the data on the labels that the information relating to these tubes was set aside. In other cases it was unambiguous and five of these are of interest:

Scotophaeus blackwalli:	female, Thurnby, indoors, April 1934 (tube 9)	CR
Theridion pallens:	female, Swithland Woods, June 1934 (tube 22)	CR
Arctosa perita:	female, Waltham Quarry, 23rd June 1934 (tube 24)	
Pardosa pullata:	female with ova, Waltham Quarry, 23rd June 1934 (tube 25)	CR
Dysdera crocata:	female, west Leicester (City), August 1934, coll. Alec Bonner (tube 27)	CR

There is in addition, associated with the collection, a male specimen of *Tegenaria parietina* with the data 'Leicester Nov. 1935 W.E. M[ayes]'. W.P. Mayes left Leicester to join the staff of Exeter Museum towards the end of 1934. Later, to use his own words, he "moved over to the art side of museum work, as being likely to offer the best chances of advancement"; eventually he became Keeper of the Art Department at the Imperial War Museum.

The Comity of Spiders (W.S. Bristowe, 1939-41)

This is the first comprehensive account of the distribution of spiders, county by county, throughout the British Isles, and together with the comprehensive coverage of the biology and behaviour of spiders, established the foundation for a new era of arachnological activity in Britain. Volume I (1939) presents a list of species for the British Isles of some 556 spiders excluding sub-species and species from the Channel Islands. Bristowe comments on the relative paucity of records from Leicestershire and Rutland "of the English counties, Rutland (24) [sic; 26 listed] and Monmouth (29) are least known, followed by Leicester, Bedford, Hunts and Northants, all with fewer than 100 species." Whilst having picked up the obvious published records, Bristowe was not aware of the additional unpublished records included in this historical review. *The Comity* adds 17 species to the list for VC55 and includes all 26 Rutland records. As to the source of these new records, Bristowe stated (in litt. April 1960): "When I published my lists for Leicestershire and Rutland in Vol.1 of *The Comity of Spiders*, there was only the list for Leicestershire in the V.C.H. and no records from Rutland at all. I had collected near Market Harborough (Leic) and near Uppingham so the additional species for Leicestershire were mine, (published for the first time) and those for Rutland also."

In fact, after the addition of his own 17 new records, the total list for vice-county 55 comprising Leicestershire and Rutland still amounted to only 80 species, and no more new records for either Leicestershire or Rutland occur among the 100 or so additional records in volume II of *The Comity* (1941). Several of Chalcraft's species are missing from Bristowe's list of 56 species, since Bouskell's (1907) list was used as his baseline.

Table 3. Species listed for Leicestershire and Rutland in Bristowe's *Comity* (1939): Leicestershire 55 species, Rutland 26 species, those which constitute first records for the separate counties are starred *; they include all the records for Rutland. The 17 new county records for VC55 are shown as CR. Current nomenclature, (see checklist – chapter two) is shown in bold italics.

Dictyna uncinata		L*R *	CR
Ciniflo similis Blk., 1845	[***Amaurobius similis***]	L*R*	CR
C.fenestralis Stroem, 1768	[***A.fenestralis***]	LR	
Drassodes lapidosus		L	
Scotophaeus blackwalli		L	
Clubiona phragmitis		L*R*	CR
C.compta C.L.Koch, 1839	[***Clubiona comta***]	L*R*	CR
Anyphaena accentuata		L	
Misumena calycina (Linn., 1758)	[***Misumena vatia***]	L	
Xysticus viaticus (Linn., 1758)	[***Xysticus cristatus***]	L	
Oxyptila flexa O.P.-Cambr., 1895	[***Ozyptila brevipes***]	L*	CR
Philodromus aureolus cespiticolis Walck., 1825			
	[***Philodromus cespitum***]	L*R*	CR
Salticus scenicus		L	
Lycosa saccata (Linn., 1758)	[***Pardosa amentata***]	LR*	
Trochosa ruricola		L	
Pisaura listeri (Scopoli, 1763)	[***Pisaura mirabilis***]	L	
Tetrilus arietinus ssp. *macrophthalmus* (Kulcz., 1891-7)			
	[***Mastigusa macrophthalma***] L		
Amaurobius atropos (Walck., 1825)	[***Coelotes atropos***]	L	
Tegenaria atrica (C.L.Koch, 1843)	[***Tegenaria gigantea***?]	L [see note 1]	
T.domestica		LR*	
Textrix denticulata		L	
Tetragnatha extensa		L	
T.obtusa		LR*	CR
Pachygnatha degeeri		LR*	
P.clerki		LR*	CR
Meta reticulata (Linn., 1758)	[***Metellina segmentata***]	LR*	
M.merianae (Scopoli, 1763)	[***Metellina merianae***]	L	
Cyclosa conica		L	
Aranea raji betulae Sulz., 1776	[***Araneus marmoreus***]	L [see note 2]	
A.alsine Walck., 1802	[***Araneus alsine***]	L	
A.diadema Linn., 1758)	[***Araneus diadematus***]	LR*	
A.cucurbitina Linn., 1758	[***Araniella cucurbitina/opistographa***] L		
A.sexpunctata Linn., 1758	[***Nuctenea umbratica***]	L	
Zygiella litterata (Oliver, 1789)	[***Zygiella x-notata***]	LR*	
Z.atrica		L*	
Theridion notatum (Linn., 1758)	[***Theridion sisyphium***]	L [see note 3]	
T.varians		L*	CR
T.tinctum		L*R*	CR
T.pallens		L	
T.redimitum (Linn., 1758)	[***Enoplognatha ovata***]	LR*[see note 4]	
Stearodea bipunctata (Linn., 1758)	[***Steatoda bipunctata***]	LR*[see note 5]	
Crustulina guttata		L	
Robertus lividus		R*	
Baryphyma pratensis (Blk., 1861)	[***Baryphyma pratense***]	L	
Thyreosthenius biovatus		L	
Typhocrestus digitatus		L*	CR
Hypomma cornuta (Blk., 1833)	[***Hypomma cornutum***]	LR*	CR
Oedothorax retusus		LR*	CR
Lophomma punctata (Blk., 1841)	[***Lophomma punctatum***]	L	
Bathyphantes concolor (Wid., 1834)	[***Diplostyla concolor***]	L*R*	CR
B.gracilis		LR*	CR
Linyphia montana (Linn., 1758)	[***Linyphia triangularis***]	LR*[see note 6]	
L.hortensis		L*R*	CR
L.pusilla Sund., 1829	[***Microlinyphia pusilla***]	LR*	
Lepthyphantes tenuis		LR*	CR
Centromerus bicolor (Blk., 1833)	[***Centromerita bicolor***]	L	

Note 1. See nomenclatural note on page 49
Note 2. = *Aranea raji betulae* H.Wiehle, 1931. This is the form *pyramidatus* Cl., 1757 which is now considered the type of *marmoreus*.
Note 3. = *Theridion notatum* H.Wiehle, 1937
Note 4. = *Theridion redimitum* H.Wiehle, 1937. Form with red folium.
Note 5. = *Stearodea* F.O.P.-C., 1902, synonym of **Steatoda** Sund., 1833.
Note 6. In 1757 Clerck named two Linyphia species (*Araneus triangularis* and *Araneus montanus*). However, Linnaeus also named the former species *Aranea montana* in 1758. Clerck's names have priority and *A.montana* Linn., 1758 is suppressed as a synonym of **Linyphia triangularis** (Clerck, 1757). Clerck's *A.montanus* also stands as **Linyphia montana** (Clerck, 1757).

A.M. Wild was born in Sheffield in 1916, took a degree in chemistry at Imperial College and spent 36 years in the Scientific Civil Service, retiring in 1976. He began his serious studies on spiders in 1943 after purchasing copies of Blackwall (1861; 1864) and Bristowe (1939; 1941). Wild, (1952) presents detailed results of his own spider collecting during holidays in Wales, together with a list of spiders taken by W.S. Bristowe in other counties. Dates for Wild's own records (not in Leicestershire) range from 1945 to 1950, but no dates or localities are given for the 13 Leicestershire and Rutland records attributed to Bristowe, of which 8 are new county records. In the absence of precise details, it is assumed that these spiders were collected by Bristowe between 1941 (*The Comity*, Vol.II) and 1950 and therefore take precedence over later records associated with the Leicestershire Fauna Survey.

It is noted that whereas Bristowe included *Linyphia montana* Linn., 1758 in his 1939 list, this name is a synonym for **Linyphia triangularis** (Clerck, 1757), previously recorded for Leicestershire by Lowe (1912). In Wild's list (table 4.) the species referred to is **Linyphia montana** (Clerck, 1757), recorded here for the first time.

Table 4. Species listed for Leicestershire and Rutland in Wild (1952): Leicestershire 9 species, Rutland 11 species: those which constitute first records for the separate counties are starred *. The 8 new county records for VC55 are shown as CR. Current nomenclature (see checklist – chapter two) is shown in bold italics.

Segestria senoculata		L*R*	CR
Tibellus oblongus		L*	CR
Pisaura mirabilis		R*	
Amaurobius atropos	[***Coelotes atropos***]	R*	
Tegenaria atrica	[***Tegenaria gigantea***?]	R*	
Tetragnatha extensa		R*	
Araneus cornutus	[***Larinioides cornutus***]	L*	CR
Theridion denticulatum	[***Theridion melanurum/mystaceum***]	LR	
Ero furcata		L*R*	CR
Labulla thoracica		L*R*	CR
Linyphia montana		L*R*	CR
Stemonyphantes lineata	[***Stemonyphantes lineatus***]	L*R*	CR
Floronia bucculenta		L*R*	CR

To this list Bristowe has added **Oonops domesticus** L* CR (Bristowe, 1963), presumably taken sometime in the 1950s, although again there is no information on the precise locality or date.

Leicestershire Fauna Survey (1947-1953)

In 1947 the Leicestershire Fauna Survey was launched. This was a joint venture by the Zoology Department at the University College of Leicester (under Prof. H.P. Moon), the Museum (under T.A. Walden) and the Botany and Zoology Section of the Leicester Literary and Philosophical Society. There are, in the Museums Service collections, a small number of specimens of spiders collected in the early years of this enterprise, between 1947 and 1953. Their identity was checked by I.M. Evans in 1962. Amongst them are first records for Leicestershire of the following four species:

Zelotes latreillei (female, Waltham-on-the-Wolds 3.7.48, coll. T.A.Walden)
Alopecosa pulverulenta (female, Waltham-on-the-Wolds 3.7.48, coll. T.A.Walden)
Pirata piraticus (female with ova, Scraptoft, 16.6.48, coll. T.A.Walden)
Erigonidium graminicola (= **Hylyphantes graminicola**) (female, Wistow, 2.7.53, coll. D.S. Fieldhouse)

T.A. Walden joined Leicester Museum as a trainee in the 1930s, leaving for active service in the Navy during the Second World War. He returned to the Museum as Keeper of Biology in 1947, becoming Deputy Director around 1950 and Director in 1952. He left to become Director of Glasgow City Art Galleries and Museums in about 1973, where he died 'in harness'. Trevor Walden was very active in the Museums Association and a greatly respected member of the profession, receiving the O.B.E., and later C.B.E., for his work. One of the originators of the Leicestershire Fauna Survey in 1947 and a founder member of the Leicestershire Trust for Nature Conservation in 1956, he had a broad interest in natural history but did not specialise.

In 1953 **J.A.L. Cooke** and **G.P. Lampel** (Cooke & Lampel, 1953) published a list of eight records thought to be new to Rutland, "taken on a roadside near Stamford in early September" which were "the results of a few minutes [collecting] while going through by car" (in litt. Lampel, 16.4.1960). Of these, **Clubiona corticalis** and **Clubiona terrestris** are new county records. It should perhaps be pointed out that, although Stamford is in Lincolnshire, the county boundary with Rutland lies very close to the western and northern boundary of the town.

The first substantial paper on Leicestershire spiders, listing 11 species taken at two sites on Charnwood Forest is that by **Eric Duffey** (Duffey, 1955). The significance of this paper is in the accurate reporting of habitat details and location, sex and number of specimens collected, as this makes such records of much greater value. However, the substance of this paper was not co-ordinated with any arachnological survey at that time and must therefore fall into the category of random observations, concluding the first half of this historical appraisal, which covers the overall period 1795-1995. Because ecological and distributional details are given by Dr Duffey, these records are entered in the Leicestershire database and are included in the accompanying atlas, where they are shown as open circles (unless these coincide with subsequent filled circles). **Lepthyphantes whymperi**, the main subject of this paper, is essentially a mountain spider, with its main strongholds at high altitudes in the north of Scotland and west Wales. Because of the national significance of this species on Charnwood Forest it has been a matter of concern to establish its true status at High Sharpley. Over some fifteen years, assiduous hand collection and pitfall trapping has failed to rediscover **L.whymperi** at this site. Unlike most of the previous records covered by this review where no voucher material has been retained, it has in this case been possible to re-examine these specimens, still held by Dr Duffey (pers.comm. 11.5.95), and it is now conclusively established that the determination of **Lepthyphantes whymperi** was incorrect. The female specimen collected at High Sharpley on 28.12.53 is in fact **Poeciloneta variegata** (Blk., 1841), and therefore **Lepthyphantes whymperi** is struck off the Leicestershire list. The following 7 species from Duffey's revised list are new County Records

Oonops pulcher	(Beacon Hill)
Zora spinimana	(Beacon Hill)
Walckenaeria acuminata	(Beacon Hill)
Poeciloneta variegata	(High Sharpley)
Cnephalocotes obscurus	(Beacon Hill)
Lepthyphantes zimmermanni	(Beacon Hill and High Sharpley)
Lepthyphantes ericaeus	(High Sharpley)

After the publication of the above paper, Duffey supplied (in litt. 14.4.1960) records of other collections made by him in 1952 and 1955. They include the first records for Leicestershire of the following 5 species:

Oedothorax fuscus	(The Brand, SK 5313, 14.4.52)
Zelotes apricorum	(The Brand, SK 5313, 14.4.52)
Trochosa terricola	(Loddington Reddish, SK 7702, 5.5.55)
Diplocephalus picinus	(Loddington Reddish, SK 7702, 5.5.55)
Diplocephalus latifrons	(Loddington Reddish, SK 7702, 5.5.55)

PART II 1959–1995

In Part I, current nomenclature has been indicated by the use of bold italics to assist in progress through the maze of synonyms. Up to this point, new county records from literature and museum sources, localised to county, can be traced through the chronological listing in table 5 at the end of this chapter. Hereafter, there will be no attempt to list new county records within the text, and normal italics will be adopted for all latin names, unless bold typeface is necessary for clarification, as previously. After 1948, all new county records which have plottable data are listed in table 6, in date order. This list, based on field records, overlaps table 5 chronologically between 1948 and 1963, but this is unavoidable because of the lack of detail in the published records. The entire database of records, where four-figure map references are available, is plotted on the maps in chapter five.

The Post-1959 Period

Three people stand out as the main activists in arachnological survey in Leicestershire and Rutland during the 1959-1995 era – I.M. Evans, J. Crocker and more recently J.T. Daws.

Collecting spiders in the early post-war years by staff of Leicester Museum and dealing with a steady stream of local enquiries from the public, often with specimens in a poor state of preservation, was a sporadic business with no clear direction. No real advancement was made in an understanding of the Leicestershire spider fauna until the arrival at the Museum of Ian Evans in 1959 and the revitalisation of the renamed Natural History Section of the Leicester Literary and Philosophical Society. The next event which had a major influence on spider survey in Leicestershire was the formation of the Loughborough Naturalists' Club in 1960. One of the first objectives of the Loughborough Club was to get to know the flora and fauna of Charnwood Forest better, and to this end individual members were encouraged to undertake specialist studies of their chosen subjects. The Club was fortunate in having good coverage of some of the invertebrate orders and John Crocker, as Organising Secretary, chose to specialise in the Arachnida, building on a nine-year general but uninspired interest in spiders. With mutual support, both Crocker and Evans built up an arachnological expertise and encouraged other field workers in their respective organisations to collect spiders, harvestmen and pseudo-scorpions, which they undertook to get identified. So it was that the 1960s and 1970s saw a tremendous improvement in our understanding of the spider fauna of the county, and regular articles, reports and records appeared in the newsletters and transactions of not only the Club and Section, but also the Flatford Mill Spider Group, British Spider Study Group and British Arachnological Society, of which both Crocker and Evans were active members. By 1980, other responsibilities prevented continuation of work on spiders at the same pace and it was not until 1991 that a new focus was provided by Derek A. Lott through his fieldwork on beetles. Lott was an active member of the Loughborough Naturalists' team and in his capacity as Keeper of Biology for the Leicestershire Museums Service employed Jonathan T.Daws to assist in sorting pitfall trap catches, primarily separating the beetles and spiders. This resurgence of activity, and Daws' dedicated commitment to spiders, eventually stimulated the production of this present publication.

Ian M. Evans took up his appointment as Keeper of Biology at Leicester Museum in July 1959. He first collected spiders on the 14th of that month and interest was further aroused by a collection of spiders made by Michael Watts, son of a doctor living at Ibstock, and brought into the museum as an enquiry in September. From late 1959 Evans began to encourage staff of the Museum, local naturalists and members of the Museum Biology Club to bring in spiders. Amongst staff who contributed were R.D. Abbott, G.A. Chinnery, J.W. Greaves, R. Lee, Miss J.B. McKellar and T.H. Riley. During the early 1960s, Derek Foxwell, Trainee Taxidermist in the Biology Section, took a particular interest in the spiders and other fauna found in the banana warming rooms beneath the fruit and vegetable market in the centre of Leicester. The larger species amongst those imported also occasionally appeared as enquiries from greengrocers and their customers. Specimens were identified by Evans, with assistance from the arachnologists at the British Museum (N.H.), and included *Tegenaria pagana*, *Tegenaria parietina*, *Achaearanea tepidariorum*, the sub-tropical heteropodids *Heteropoda venatoria* and *Torania variata* and, very rarely, the tropical ctenid *Phoneutria fera*.

Amongst local naturalists collecting spiders were W.H. Barrow, H.A.B. Clements, M.J. Leech and R.D. Osborne. S.B. Scargill, in particular, encouraged the pupils of Manor House Special School in Haddenham Road, Leicester, to collect specimens, and Tony Squires collected much material, most of it outside Leicestershire. Biology Club members were regular contributors, including Graham A.C. Bell, David Bland, John C. Clarke, Keith Edwards, Donald Goddard, Rodney Higgins, Derek Roff, David Scruton, Alan Smith and Geoffrey Smith.

Evans attended spider courses at Flatford Mill in September 1960 and April 1961, which visited a number of prime localities in Essex and Suffolk. There he met David Mackie on the first course and joined the Flatford Mill Spider Group, of which Mackie was Secretary. Locally, he took part in the L.N.C. survey of Bradgate Park in June 1962 and continued to collect spiders 1962-1964, and less assiduously 1965-1967. For ten years or so after that the only spiders he identified were those brought in to the Museum as enquiries and occasional specimens collected in the course of more general fieldwork. Evans served on the Council of the British Arachnological Society and as its Reprint Librarian until 1978, during which year his interest revived when he identified the spiders collected in the course of a survey of the proposed North-east Leicestershire Coalfield (Evans, 1979). He also identified spiders collected by D.G. Goddard in a series of contract invertebrate surveys in Leicestershire 1979-1982, notably River Soar (1980), Ashby Canal (1980) and Rutland Rivers (1982). After this he was only involved in casual collecting of spiders in Leicestershire until retirement and relocation to Sutherland in 1991.

John Crocker had begun to take an interest in the larger spiders in 1953 after he moved to Loughborough, but with the new responsibilities of marriage, setting up house, National Service and completion of academic studies he had little time for this pursuit. His early interest was centred mainly on producing water colour paintings of the spiders he found around the house and garden, ignoring the small 'money spiders'. It was not until 1959, when copies of the two volumes of *British Spiders* were acquired, that he realised how many species there were and how difficult it was to identify them with certainty without a microscope. He set about trying to identify those he had painted using a 10x hand lens, but within a short time it became obvious this was unsatisfactory and that a microscope was needed. A monocular microscope was made available on loan from Leicester Museum until he purchased his own stereoscopic instrument.

In 1960 Crocker made a serious commitment to the study of arachnids and since then this has been his main interest in the natural sciences, with a bias towards spider ecology. With the founding of the Loughborough Naturalists' Club in 1960, and its emphasis on field-work, a focus was provided for spider study on Charnwood Forest. Thereafter, a group of specialists concentrated on building up detailed knowledge of the flora and fauna of this area north-west of Leicester and south of Loughborough. At about this time, through contacts with the North-west Naturalists' Union and David Mackie, he joined the Flatford Mill Spider Group and from the beginning took an active part in all aspects of its administrative and scientific affairs, resulting in the greater part of his subsequent ecological work being carried out for the Nature Conservancy Council outside Leicestershire.

A strong bond of co-operation was built up between Loughborough and Leicester, such that many of Crocker's early specimens were sent to Leicester Museum for their collections, whilst a steady stream of specimens arrived at Loughborough from the Museum for confirmation. From 1962 Crocker began building up a reference collection which is discussed in chapter four.

There is no doubt that the appearance of the first two volumes of *British Spiders* (Locket & Millidge, 1951; 1953) stimulated a great deal of interest in spiders, with the numbers of serious workers growing rapidly during the 1950s and 1960s. G.H. Locket and A.F. Millidge were at that time the leading British authorities on spider taxonomy and gradually Peter Merrett assumed this mantle which is shared today with Michael J. Roberts.

A recognised and long-standing procedure for verification of records had been established long before the days of Locket & Millidge, where specimens were sent to the authorities of the day – such eminent names as J. Blackwall, O. Pickard-Cambridge, A.R. Jackson and A.A.D. La Touche. In more recent times, processing the rising volume of referrals became a very time-consuming task as the number of workers increased, so a filtering system at county level was introduced by the British Spider Study Group to ease the workload at the top, and encourage new specialists to expand their contacts and influence. John Crocker has been County Recorder for spiders since 1964, working as part of a national team under the umbrella of the British Spider Study Group, later to become the British Arachnological Society.

Literature – Local Publications

As with the pre-1959 literature, no cohesive interpretation of the overall status of spiders within the county has subsequently been attempted. Significant advances have been made nationally in the gathering of distributional data and in the provision of identification tools. Thus the status of many species is becoming clearer and by the end of the century this intelligence will be published. In the meantime, individual and collective effort is being directed at local level towards this objective. This collective approach to the study of spiders has been stimulated by the formation of national and local study groups, each of which has published its own newsletters, but much of the content of these is ephemeral or incomplete, comprising mostly short notes. However, they do provide a valuable resource for the historian in tracing the expansion of interest in spiders at this time. With the setting up of recording centres at national and local level, casual references in newsletters were covered in more detail in the permanent records – in Leicestershire held by Crocker and Leicestershire Museums Service. References to individual species and species lists in newsletters and survey reports – of Loughborough Naturalists' Club (*Heritage, etc.*), Leicester Literary & Philosophical Society (Natural History Section), Flatford Mill Spider Group, British Spider Study Group and the Spider Recording Scheme – will therefore be ignored where they are embraced elsewhere. Attention is drawn to the following short papers and notes which have a relevance to Leicestershire spiders and are discussed elsewhere. Full titles will be found in the List of References at the end of this book.

Heritage (Quarterly Bulletin of the Loughborough Naturalists' Club)
From the first issue of the bulletin in the spring of 1961, spider records – primarily from the Charnwood Forest area, collected by Club members – were included in the Arthropod section. The unnamed bulletin was first graced with a standard heading and title 'Heritage' on its twenty-first issue (Spring 1966), published 18th April, but the title has been applied in retrospect to all issues back to number one. Often the same records appeared in the newsletters of the Flatford Mill Spider Group and, later the British Spider Study Group, whilst those originating from Leicester Lit.& Phil. Natural History Section excursions in East Leicestershire found their way into the Section newsletters. None of these records are strictly considered as 'published' records and appear in greater detail in the written field records of Evans and Crocker, subsequently incorporated into the Leicestershire spider database.

Crocker, J. (1962a) 'Two rare spiders from Leicestershire'
This short note records the re-discovery of *Mastigusa macrophthalma* at Bradgate Park, Donisthorpe's (1927) original location, and establishes *Urozelotes rusticus* as new to Leicestershire.

Crocker, J. (1962b) [1] '*An initial survey of the spiders of Bradgate Park and Cropston Reservoir margins*'
The significance of this report is that it was an attempt to draw together as much as was known about the natural history of a specific site, and contains the first annotated list of spiders for a site in Leicestershire. The survey also embraced the principle of team-work and involved specialists from the British Museum (N.H.), Leicester Museum and members of the L.N.C. Seventeen of the 54 species listed were new to the county. These appear in table 6 at the end of this chapter, against the dates 24th April, 5th & 16th May, and 1st August 1962.

Crocker, J. (1963) 'Spiders of Charnwood Heath'.
Early work on this SSSI, later superseded by more detailed studies, yielded some interesting material on the wide colour variation in a colony of *Drapetisca socialis*, which is well camouflaged on lichen-covered tree trunks.

Crocker, J. (1966) 'Some observations on the habitat requirements of *Dysdera* in Leicestershire'
Dysdera crocata and *Dysdera erythrina* are discussed, with some background notes. The conclusions are endorsed by subsequent records of these two species in Leicestershire; see the distribution map and comments, p.81.

---

[1] This duplicated report, produced for general sale in limited edition, was the first in a series, later to include Grace Dieu, Swithland Wood, Ulverscroft N.R., Groby Pool and Benscliffe Wood, with other titles in preparation. Although some of these give lists of spiders, such records are incomplete and are here treated as *Heritage* Records.

Crocker, J. (1973) 'The habitat of *Tetrilus macrophthalmus* (Kulczynski) in Leicestershire and Nottinghamshire'.

The paper discusses the contrasting habitats of the rare spider *Tetrilus macropthalmus* (= **Mastigusa macrophthalma**) in Sherwood Forest and Charnwood Forest. Distribution in Leicestershire was known from five sites (now eight) on Charnwood and one at Donington Park on the Derbyshire border. Observations on breeding populations at Bradgate Park, Ives Head, High Sharpley, Beacon Hill and Bardon Hill are presented.

Arnold, G.A. & Crocker, J. (1967) '*Arctosa perita* (Latr.) from colliery spoil heaps in Warwickshire and Leicestershire'.

This short paper reports the discovery of very dark specimens of *Arctosa perita* on widely separated, weathered, dirt banks – often with little or no vegetation. These melanic forms had adapted to the dark blackish-grey substrate on which normal light-coloured specimens would have been at a considerable disadvantage.

Evans, I.M. (1963) 'Native or Immigrant ? Some spiders from banana warehouses, Leicester'.

The interest in this short note is in the occurrence of *Tegenaria parietina* inside a heated warehouse in the centre of Leicester. Many foreign creatures imported with fruit occur in such places, and it has long been a preoccupation of certain herpetologists to trace the origins of such aliens. Thus we happen to know that the regular importations of bananas at this time came from Trinidad. Whereas *T.parietina* has been recorded from South Africa, where it could have been introduced from the Mediterranean or Portugal, there is no evidence that the species is established in Trinidad. *T.parietina* was probably more common in Leicester in the days when there were many old and derelict buildings in the city, a habitat favoured by this large spider. This particular record is certainly not to be dismissed as a foreign introduction – as undoubtedly is *Tegenaria pagana*, found in a similar situation – and may be a remnant of a more local population.

Evans, I.M. (1979) 'Spiders' in: Evans, I.M. (ed) *North-east Leicestershire Coalfield. Report of a biological survey*.

The report includes a systematic list of spiders collected in the north-east of Leicestershire as part of a biological survey of three proposed deep-mine development sites in 1978, and is presented with location, date, gender and habitat details. Sixty-one species were represented in the material that could be identified, including two previously unrecorded for Leicestershire. Comparisons between the faunas of the three sites are shown, development of only one of which took place, that of the new Asfordby Mine.

1. the Asfordby site yielded 30 species, of which 10 were restricted to it.
2. the Hose site yielded 35 species of which 17 were restricted to it.
3. the Saltby site yielded 28 species of which 12 were restricted to it.

As with all post-1959 lists and individual records, details have been entered in the database and records plotted on the distribution maps. The two new county records, *Ozyptila sanctuaria* and *Robertus neglectus*, are included in the table at the end of this chapter.

Literature – the National Scene

In 1964 the Biological Records Centre – originally set up under the Nature Conservancy, then part of the Institute of Terrestrial Ecology – produced a Spider List Card which was the forerunner of the Spider Record Cards in use today. However, there was not a great deal of enthusiasm among the arachnological field workers of the day, who each seemed to be operating their own system of recording. Dr Peter Merrett, at I.T.E., had started collecting spider records for the whole of Britain and Ireland in 1961, and continued until he was able to publish these results plotted on pre-1974 county maps (Merrett, 1964-65; 1967; 1969; 1974). The distribution of 612 species throughout the British Isles was clearly demonstrated, and supplementary lists of county records have since been published (Merrett, 1975; 1982; 1989; 1995) to enable the county maps to be updated. In 1986, the British Arachnological Society agreed to liaise with B.R.C. with the aim of reviving the Spider Recording Scheme which was based on the widely used 10 km national grid format. The scheme was set up and Clifford Smith appointed National Organiser, a position he held until forced to resign in 1993 through ill-health (Smith, 1987). Current details of the Spider Recording Scheme are given at the end of this book.

Merrett, P. (1974) 'Distribution Maps of British Spiders'.

The publication of the long awaited third volume of *British Spiders* included distribution maps for all the species on the British list. It showed 247 species for VC55 (Leicestershire = 245; Rutland = 56), mostly from records submitted since 1960 through the national recording schemes operated by the British Spider Study Group and later the British Arachnological Society. Also included were records extracted from earlier literature, six of which require further comment.

Micrommata virescens	Unconfirmed, but acceptable
Tegenaria parietina	Acceptable
Araneus marmoreus	Unconfirmed, but acceptable
Araneus alsine	Unconfirmed, mis-identified
Thyreosthenius biovatus	Acceptable
Lepthyphantes whymperi	False record, mis-identified

Since 1974, *Oedothorax gibbosus* and *Oedothorax tuberosus* have been shown to be conspecific, and with the deletion of *Araneus alsine* and *Lepthyphantes whymperi* from the Leicestershire list, the species total for the old county of Leicestershire should be 242, and 244 for VC55. *Thyreosthenius biovatus* and *Lepthyphantes whymperi* have been discussed elsewhere; the other four species are from C.B. Chalcraft's original list of 33 species for Leicestershire in 1897. Although the difficulties surrounding these species have already been mentioned at the beginning of the chapter, the debate on these historical records requires to be concluded with a consolidation of the evidence available to us.

Araneus marmoreus (= *Epeira scalaris* = *Araneus marmoreus pyramidatus*) is a widespread species, fairly common in suitable habitats around the Midlands. It is a well-marked distinctive spider with adequate illustrations in contemporary literature. The species is likely to be present in the east of the county which still requires adequate survey coverage. *Araneus marmoreus pyramidatus* is considered to be a colour form of *A.marmoreus* with a wide range of intermediates. The distribution of *pyramidatus* is much more widespread than the type form, therefore subspecific status is not justified and has been suppressed (Merrett *et al.*, 1985).

The dilemma surrounding **Micrommata virescens** (= *Sparassus smaragdulus*), which is equally distinctive, but much rarer, arises out of the fact that it is included in Chalcraft's 1897 list, but omitted from Bouskell's 1907 list. Bouskell knew Chalcraft quite well and was present at his lecture, taking part in the long discussion which followed. Did Bouskell omit this species (together with *Epeira patagiata* and *Ciniflo ferox*, both of which are legitimate species on the original list) because he had doubts about their veracity? *Micrommata* used to be more widespread than it is today and was almost certainly established in suitable Leicestershire habitats. Many of the prime sites regularly visited by the Victorian entomologists would have had richer faunas than today. An example is Buddon Wood, where **Larinioides patagiatus** (= *Epeira patagiata*) is still found in remnants of the old woodland. *Micrommata* is therefore retained on the list, but recorded as unconfirmed.

Mention has already been made (page 34), of the early confusion surrounding the true identity of species belonging to the genus *Tegenaria*, and this opportunity is taken to review the situation. Regarding **Tegenaria parietina** (= *Tegenaria domestica*), Chalcraft is almost certainly referring to what we know today as *Tegenaria gigantea* (though *Tegenaria parietina* would seem to be a legitimate Leicestershire species from other evidence); the main evidence here is provided by Bouskell in his 1907 list where this species is stated to be "Found in houses; very common. Taken in Leicestershire (G.B.C.)". The comment against **Tegenaria domestica** (= *Tegenaria civilis*) "Found in houses, common. Taken in Leicester (G.B.C.)" is apposite, but the other statement cannot apply to *Tegenaria parietina*. What authority was Chalcraft using to identify his specimens? In Victorian times entomologists were more likely to determine a species on its general appearance, and one would ask what literature was generally available to the late nineteenth century collectors. In the 1960s one of our national museums was still using Blackwall (1861-64) *A history of the Spiders of Great Britain and Ireland* to identify spider material submitted by the public; so early collectors can be forgiven if they determined their captures on the basis of coloured pictures! Blackwall's figures of *Tegenaria domestica* (= **Tegenaria parietina**) and *Tegenaria atrica* (= **Tegenaria saeva**) are excellent but may not have been available to Chalcraft. The descriptive text could lead the reader to believe that "*T.atrica*" would be unlikely to occur in Leicester, unlike the inference that "*T.domestica*" was the common house spider. However, Blackwall's drawings are so good that there should have been no confusion. Staveley's (1866) popular *British Spiders* illustrates only *T.domestica* (= **T.parietina**), whilst *Tegenaria civilis* (= **T.domestica**) and *T.atrica* (= **T.gigantea/saeva**) are poorly described without illustrations.

It is of further interest that *Tegenaria saeva*, which is at present displacing and interbreeding with *Tegenaria gigantea* in Yorkshire, seems not yet to have become established in Leicestershire. Its occurrence in the county is based on one confirmed and one unconfirmed record.

It appears that *Tegenaria parietina* was established in Leicester in the past, probably synanthropically. Headly's 1890 *atrica* description fits *parietina* which could have been introduced via river transport to nearby buildings. The species is associated with old buildings in its present range, and considering the dramatic changes in this type of habitat during the latter half of this century, there is a possibility that one of the factors in its demise is the loss of most of our old buildings. *Tegenaria parietina* could therefore have been established in Leicestershire at some time in the past, but if this was the case, it seems to have been displaced by the now widespread and common opportunist species *Tegenaria gigantea* which is indeed "found in houses" and is "very common". Evans (1963) refers to *Tegenaria parietina* from banana warehouses in Leicester, but this is unacceptable as evidence of a free living independent community. However, Headly's historic record and the specimen from Leicester in Mayes collection (see page 40), together with Evans' contribution would indicate that *Tegenaria parietina* is a legitimate Leicestershire species.

Two problem species remain, *Epeira lutea* and *Epeira ornata*. Both are described by Blackwall (1864, pp.345-46), but only *lutea* (= **Araneus alsine**) is figured (Pl.XXV fig.249), whilst neither is illustrated in Staveley and the descriptions are totally inadequate to make an accurate diagnosis. Blackwall (1850) described a new species of *Epeira* from a 'showy' female taken in the south of England in April 1848. This he named *Epeira ornata*, and it would appear to be the only specimen ever seen. The specimen was further described by Blackwall (1864, p.346) but not figured. Staveley (1866, p.244) gives an unsatisfactory description without any illustrations and has obviously 'lifted' this information from Blackwall. Pickard-Cambridge (1881, p.531) also gives a short description based on that of Blackwall, and states "I have not seen this spider". He includes *E.ornata* in his 1900 checklist of 532 British species, but Bristowe (1939, p.54) states "the identity of *Epeira ornata* Bl. is doubtful, so it has been dropped from the 1900 list".

Epeira lutea Koch (= **Araneus alsine**) recorded by Bouskell (1907) from Leicester as "common", and attributed to G.B. Chalcraft, is clearly a mistaken identification, for nowhere in Britain can this species be considered common. It is a rare spider, often associated with damp, shady places, woods *etc.*, and sometimes lives in the same type of habitat as *Araneus quadratus*. In this instance Chalcraft possibly confused his *lutea* with very red specimens of *quadratus* which are indeed common around Leicester.

The following four revisions to the 1974 maps have been published by the British Arachnological Society. These give lists of new counties in which each species has been recorded since the last update. Since no other data are presented in these papers, the relevance to Leicestershire is given below.

Merrett, P. (1975) 'New County Records of British Spiders'.

No new records are shown for Leicestershire or Rutland as all current records were included on the 1974 published maps, after which date the two counties were merged.

Merrett, P. (1982) 'New County Records of British Spiders'.

Twenty-seven species are added to the 'county' list for pre-1974 'Leicestershire', and none for 'Rutland'. These records were submitted by Crocker including *Cercidia prominens* which was based on an immature specimen from Buddon Wood. Since neither details of this record, nor the specimen can be located, this record is disallowed. The species has subsequently been taken at two sites in Rutland; also at Barnack Hills and Holes (Lincolnshire) and Sherwood Forest (Nottinghamshire).

Merrett, P. (1989) 'Twelve hundred new County Records'.

Three new records are listed for 'Leicestershire' and four for 'Rutland'. All four of the Rutland species have previously been recorded for the vice-county of Leicestershire. Unfortunately the origin of two of the Leicestershire records (*Xysticus erraticus* and *Euryopis flavomaculata*) cannot be traced, but these species have subsequently been recorded from several sites in the county since 1992.

Merrett, P. (1995) 'Eighteen hundred new County Records'.

No new records are listed for Leicestershire (or Rutland).

Rare species account for about a third of the British spider fauna. As stated elsewhere, it is not the purpose of this book to concern itself with the national status or distribution of spiders, but some statement is necessary concerning the nationally rare species which occur in Leicestershire. For this we have to rely on two published accounts – Merrett (1990) deals with *Nationally Notable* spiders, and Bratton (1991) with *Red Data Book* species. Merrett (pers.comm. 3.2.96) suggests that two of the nationally notable species in the 1990 list – *Philodromus praedatus* (Nb) and *Pityohyphantes phrygianus* (Na) – should be down-graded to 'local' status, since they appear now to be well established and increasing their range in Britain. The present position in Leicestershire is that, after the above revision, we have three Red Data Book species, one notable (Na) and thirteen notable (Nb) species; these are indicated in the annotated check list (chapter two).

Merrett, P. (1990) *'A Review of the Nationally Notable Spiders of Great Britain'*.
 This was the first baseline document on the status of rare and notable spiders in Britain, discussing 232 species from the British list of 625 (93 Red Data Book, 42 Notable **a**, 97 Notable **b**). The criteria for defining nationally rare and notable species are discussed in relation to the British fauna, and parameters proposed for borderline species.

Bratton, J.H. (1991) *'British Red Data Books No.3: Invertebrates other than Insects'*.
 Procedures for the selection of Red Data Book species are given; in the case of spiders, levels of biological recording in Britain have been sufficient for the status of the species to be assessed by examination of recent and historical records. It is stressed that a species with few records is not necessarily classified as rare, and a rare species is not necessarily threatened.
 The three Red Data Book categories represented in Leicestershire are **vulnerable** (*Lepthyphantes midas*), **rare** (*Mastigusa macrophthalma*) and **insufficiently known** (*Lepthyphantes beckeri*). These categories are defined as follows: Vulnerable (**RDB2**): Taxa believed likely to move into the Endangered category (**RDB1**) in the near future because of decreasing populations caused by habitat destruction or other environmental factors. Rare (**RDB3**): Taxa with small populations in localised sites within a restricted geographical range. Insufficiently Known (**RDBK**): Taxa suspected of being endangered, vulnerable or rare, but about which there is insufficient information to be certain.
 Lepthyphantes midas is known from a colony inhabiting a single decaying ancient oak tree in old open parkland. Although there are a number of such trees in the park, intensive searching has failed to reveal another colony. There is an interval of several generations between the over-mature oaks and the younger trees, so that eventually the continuity of suitable habitat will be broken. Unlike *L.midas*, which is threatened in Leicestershire, *Mastigusa macrophthalma* is very local on Charnwood Forest, but is widespread in a number of different sites, mostly stable habitat on SSSIs. The greatest threat to this species is roadstone quarrying should SSSI protection be removed from its main stronghold on Bardon Hill (Crocker, 1973). Despite these comments about *Mastigusa*, no new sites have been discovered for this species since 1974, other than on Charnwood Forest, and it must be considered a declining species due to loss of habitat. *Lepthyphantes beckeri* Wunderlich, 1973 is a species new to Britain, recorded from a single female collected in March 1970 at Bradgate Park, a SSSI in public ownership. Only one other specimen is known, another female from Germany taken in similar circumstances to that of the Bradgate specimen. This spider is obviously extremely rare and hardly anything is known about its biology or ecology; it must therefore be placed in the category RDBK. See also chapter four, pages 70-75.

Rutland

The people of Rutland have fought hard to resist the abrogation of their county and there is much sympathy with their stand. Nevertheless, setting aside these administrative and emotional considerations, there is little or no advantage in retaining this small area as a national biological recording unit. No one from Rutland has taken any serious interest in its spiders and there is no precedent for maintaining the singularity of this area. The uniformity and versatility of the national grid plotting system has assured its widespread adoption, but it seems likely – as in the case of this publication – that Watsonian Vice-county units will continue to be used in conjunction with the 10 km square grid. For this reason, the Rutland boundary is included on the main vice-county distribution maps but the maintenance of separate county records for Rutland is not recognised after 1974, when Rutland ceased to be an independent county. All records received for Rutland after that date, with satisfactory grid references, are included on the maps, and where these are first occurrences for the vice-county will be known as Leicestershire county records. This statement is made to clarify the position should Rutland regain county status.

Bristowe (1939) lists 26 species of spider for Rutland. Wild (1952) adds another 10 species and Cooke & Lampel (1953) contribute a further 2, giving a total of 38 species in 1953.

Merrett (1974) shows 56 species for Rutland including *Oedothorax gibbosus* and *O.tuberosus* as separate species. It has since been shown that these are but a single species, reducing the list to 55, of which *Erigonidium graminicola* (= **Hylyphantes graminicola**) and *Maso gallica* (= **Maso gallicus**) were not recorded for Leicestershire.

Further casual records have been published for Rutland (Dalingwater, 1984; Merrett, 1989), and are identified from: site A, collected by P.A. Selden; and sites B & C collected by P.A. Selden and J.E. Dalingwater. All are included on the current distribution maps.

Site A 02.10.83	Whitwell SK925088	aeronaut	*Oedothorax fuscus* m.	
Site B 22.10.83	Barnsdale SK901094	grubbing, roadside	*Diplocephalus cristatus* m. *Lepthyphantes tenuis* f.	
Site C 22.10.83	Cottesmore Wood SK913128	grubbing, woodland edge	*Clubiona lutescens* *Ero cambridgei* *Oedothorax apicatus* m. *Savignia frontata* m. *Erigone dentipalpis* f.	*Meioneta rurestris* f. *Bathyphantes gracilis* f. *Diplostyla concolor* f. *Lepthyphantes zimmermanni* m. *Linyphia clathrata* f.

Fieldwork in Rutland has been initiated from Leicester by the University of Leicester, Leicestershire Museums Service and the present authors. Oakham Museum has no natural history specimens and no paper records of spiders. Uppingham School Museum has no spirit collections (and flora specimens were transferred to Leicestershire Museums herbarium). Rutland Natural History Society (founded 1965) has no collections or paper records relating to spiders, but recently small scale collecting has been carried out by R. Harvey, one of their members, and material submitted to the authors for verification.

Important Identification Aids

Roberts, M.J. (1985; 1987; 1993) *'The Spiders of Great Britain and Ireland'*.
  The publication of this three volume work marked another milestone in the history of British arachnology and has provided not only the stimulus for a new generation of field workers but also a valuable tool for all who are involved with the identification of British spiders.
  Volume I – published in 1985, covering the families Atypidae to Theridiosomatidae, was accompanied by Volume III – a magnificent collection of colour plates depicting 322 of Britain's most impressive spiders. Volume II – Linyphiidae, appeared in 1987 and contained additional colour plates, including one of a female *Lepthyphantes midas* from Donington Park, Leicestershire. 1993 saw the publication of a Compact Edition of the complete work in two volumes, Vols I & II being combined into Part 1 and the plates appearing as Part 2. A new appendix was included in Part 1, covering corrections, alterations and additions to the original volumes.

Roberts, M.J. (1995) *'Spiders of Britain & Northern Europe'*.
  Following the tremendous success of his *magnum opus*, the author has produced a worthy companion in this field guide. Beautifully produced and of supremely high technical standard, this 'popular' spider book is destined to capture the imagination of a new generation of naturalists as did Bristowe's New Naturalist title *The World of Spiders* in 1958.

Michael J. Roberts, born in Leicestershire, has been interested in natural history for as long as he can remember. His maternal grandfather, who lived at Ellistown, had a passing interest in spiders and must have sowed the seeds of curiosity for this neglected order of invertebrates. Michael left home in 1963 for Sheffield University where, as a medical student, he began illustrating flowers, birds and butterflies. These pursuits led him naturally to develop an interest in spiders, largely influenced by W.S. Bristowe's popular *The World of Spiders*. Qualifying in 1968, he took up an appointment at Scunthorpe and District War Memorial Hospital where he continued to perfect his technique for illustrating spiders. Dr Roberts has maintained his contacts with Leicestershire, where his family still live, and has further contributed to local arachnological knowledge through field work and advice on taxonomic problems. The genitalia of *Mastigusa macrophthalma* figured on page 167 (Roberts, 1985 Vol.I), and also the plate of the male (Vol.III pl.95), are from Bardon Hill specimens, Leicestershire.

Garden and House Spiders

As with any habitat, a good list of spiders will reveal a lot of information about a garden but, as with any piece of detection work, additional clues are helpful in building up a complete picture. The longer the period over which collecting has taken place, the greater will be the number of species established for that particular study area, since new spiders are continually 'dropping in' or being brought in unknowingly. If there is no suitable micro-habitat for these new arrivals their sojourn will be short-lived, and obviously a single male or unmated female cannot establish a new colony. The sampling frequency will be another factor and sampling methods will also have an influence on the range of species collected. Over a period of time, hand collecting alone will produce a good list but beating trees and shrubbery, sweeping rank vegetation, litter sifting and pitfall trapping are certain to give a better representation of the garden fauna. In one garden, the use of a Malaise trap over a long period of time – effective in catching aeronaut spiders – has produced a very good list of species, many of which are rare encounters and need not necessarily be resident. The examination of obscure niches such as culverts and the underside of manhole covers may produce species otherwise missed, as will collecting inside the house and around outbuildings. Also, a garden pond or permanent marshy ground will boost the species list with their own specialist spiders. The age of the garden is important and likewise the range of habitats represented. A small, open, intensively cultivated garden, will contain less variety in its resident populations than a large mature garden that has had relatively little disturbance. In another garden the presence of an ancient parish boundary hedge has considerably enhanced the species list, which emphasises the significance of siting. Rural gardens will be influenced by adjacent semi-natural habitats, as those in urban areas will be by the proximity of industrial sites or centres of human activity.

In appendix IIg, lists of spiders are given for five gardens and brief descriptions for these appear in the gazetteer, appendix Ib. Gardens 1 & 2 are urban sites in Leicester, hand sampled by grubbing and beating by Daws over only a few years (1992-1995). Determined collecting forays have been undertaken and the list from garden 1 reflects this activity. Gardens 4 & 5, on the southern edge of Loughborough, have each been hand-sampled over a period of fifteen years by Crocker. Garden 4 (1960-1975) was partially cultivated but with a short section of an ancient hedge and had mature shrubberies dating back to the early 1930s. Although garden 5 is small – only half the size of garden 4 – it has two ponds, several rockeries, a patio and a small lawn, constructed on previously open pasture land in the mid-1970s. It cannot be said that either of these two gardens has been worked systematically for spiders, as most of the records are from casual encounters, where the trained eye has been helpful in knowing which specimens to collect and which to ignore. Garden 3, on the other hand, at the north-eastern edge of Leicester, has been the subject of a long-term general ecological study (Owen, 1991) in which spiders have been collected, primarily by a Malaise trap set up in the back garden each year from April to September, but also by occasional hand collecting and pitfall trapping. This post-war garden is in a suburban setting and has a border of evergreen shrubs and trees at the front of the house, including box, holly and conifers, within which are herbaceous borders and a car parking area. There is a pond close to the house. The back garden is more open and is centred on a medium-sized lawn with a fringe of flower beds and paved areas around the house. Around the perimeter is a range of deciduous shrubs and trees that generally screen the garden from adjacent property. It is interesting to see *Lathys humilis* in each of the Leicester gardens. *Clubiona phragmitis* is obviously associated with the pond in garden 3, whilst *Nesticus cellulanus* and *Lessertia dentichelis*, from garden 5, are typically from the underside of foul-drain manhole covers – not a well-worked habitat! *Misumena vatia* and *Ozyptila atomaria* are certainly introductions, and species such as *Oonops domesticus*, *Pholcus phalangioides*, *Tegenaria domestica* and *Theridion melanurum* are indoor house spiders. Amongst the remainder are many spiders known to be common aeronauts. This dispersal technique will account for most of the rarer species and where conditions are suitable, some of these may become established.

Owen, J. (1991) Spiders, in: *'The Ecology of a Garden, the first Fifteen Years'*.

This is garden 3 discussed above and included in appendix IIg with an updated species list. The spider section in Dr Owen's book is a general introduction to the order Araneae and is accompanied by an annotated list of 64 species, taken mostly in pitfall and Malaise traps over the fifteen years the author has been monitoring the flora and fauna of her garden. No quantities, sex or dates are given, and voucher specimens have not been retained. The problem with such a wide-ranging study is in the identification of critical specialist subjects, such as spiders. These difficulties have already been demonstrated with early published lists and are aggravated by the absence of suitable voucher material. Unfortunately this list presents its own problems which must be addressed if Dr Owen's study is to have any value. The immediate significance of this particular study is that it is comparable with similar work done in other gardens in Loughborough and Leicester, but using different collecting techniques; so it is important to reconstruct this list as accurately as possible.

The author states: "A total of 64 species of garden spiders were identified by R.F. Owen using Locket & Millidge (1951): I.M. Evans confirmed the identification of a few of the more difficult species". In personal correspondence with I.M.E. (20.2.95), it is established that the only specimen checked by him was a female *Micrargus herbigradus* from a pitfall trap, collected during 1982, brought in to Leicestershire Museum by Richard Owen and identified on 28.10.82. He goes on to say "I have no recollection of personally checking any other material .... There is no list of spiders [from this site] in the Museum files .... [Dr Owen] has not kept any of the material. Neither does she have any records on paper except the entries in her species index of invertebrates from the garden". Of the 64 published species, 34 have been confirmed by Clifford J. Smith. A further 20 are ubiquitous species which are acceptable without verification, and a further one, *Misumena vatia,* is acceptable with qualification. Nine of the published records are unacceptable for reasons stated below and should be deleted from the list.

As County Recorder for spiders, Crocker received routine B.R.C. Record Cards (RA65) for Leicestershire from C.J. Smith, who was National Organiser for the B.A.S./B.R.C. Spider Recording Scheme. Among these were a number of cards for specimens he had identified for Owen from her garden. Arising out of correspondence with Smith (6.1.95), it was established that he had determined material submitted to him between 1986 and 1994, adding to the 34 pre-1991 species a further 17 identified between 1991 and 1994. Since then Daws has continued this service and has identified the 1995 material, adding a further two species to the site checklist. This raises the total of *confirmed* species to 53, to which can be added the 21 acceptable but unconfirmed species. The final checklist of spiders for this garden, up to January 1996, therefore comprises 74 species, including 9 local species, as shown in appendix IIg, garden 3, at the end of this book. The 19 species confirmed since 1991, and so not on the published list, are tabulated below:

Clubiona phragmitis	*Oedothorax fuscus*
Ozyptila praticola	*Silometopus reussi*
Philodromus dispar	*Porrhomma microphthalmum*
Philodromus cespitum	*Meioneta saxatilis*
Heliophanus flavipes	*Microneta viaria*
Tegenaria gigantea	*Bathyphantes gracilis*
Theridion mystaceum	*Lepthyphantes mengei*
Metellina mengei	*Lepthyphantes pallidus*
Entelecara acuminata	*Microlinyphia pusilla*
Dismodicus bifrons	

Qualification:

Misumena vatia. A single occurrence of this distinctive spider on *Chrysanthemum frutescens* flower in 1980. Almost certainly introduced involuntarily into the garden; previously recorded from Owston Wood on valerian, also from Pickworth Great Wood.

Deletions:

Textrix denticulata "common under stones". More usual in stone walls and dry stone piles; common on Charnwood Forest, easily confused with *Pardosa* species or immature *Tegenaria*. Unconfirmed record. A determined search has recently been made for evidence of this species but nothing found.

Tegenaria saeva "Pitfall trap". This is almost certainly a misidentification as *T.saeva* does not appear to be established in Leicester at present. If correct, this record would be of national significance and a voucher specimen is required for confirmation. Species in the *Tegenaria atrica* Group require very careful comparison and can be difficult to separate. The literature has been somewhat confusing on this point, but taxonomic problems now seem to be resolved. Smith (pers.comm. 6.1.95) has confirmed females of *T.gigantea* in 1989 and 1990, and this record is probably of the same species.

Theridion melanurum "Common on vegetation". Probably misidentified. Very close to *T.mystaceum* confirmed by Smith. *T.melanurum* is a common spider and almost certainly occurs on the buildings at Scraptoft; but there is a conflict here since *T.mystaceum* is more likely to be "common on vegetation".

Nesticus cellulanus. This spider lives in dark wet places such as culverts, sewers or under piles of damp stones; it is widespread and quite distinctive. Recorded here as "Common on buildings". A determined search has recently been made in the garden, in the spider's natural habitat, by Jon Daws, but no evidence of its presence was found. Since the stated habitat is completely wrong the published record is unacceptable without voucher specimens.

Araneus cucurbitinus (= ***Araniella cucurbitina***) "fairly common on vegetation". Easily confused with *Araniella opistographa* which is equally common. No voucher specimen.

Oedothorax agrestis and *O.retusus*, unconfirmed. Difficult species to separate. Smith has confirmed *O.fuscus* both sexes 1986, 1987 and 1988.

Erigone promiscua "on vegetation". Tiny money spider not described in Locket & Millidge (1951). Unconfirmed. This would be a first record for Leicestershire; voucher specimen required before acceptable.

Porrhomma convexum "on vegetation, pitfall trap and Malaise trap". A cave spider, and rare in the county; unconfirmed, voucher specimen required. This is almost certainly *Porrhomma pygmaeum*, an ubiquitous species.

Revival

Following the active years of the 1960s and 1970s in Leicestershire, a period of relative inactivity pervaded the arachnological scene during the 1980s.

In 1983 Derek A. Lott, an Oxford chemistry graduate, joined Leicestershire Museums Service following six years as a science teacher in Leicestershire, to work in the Biology Department. Lott's main interest was in beetles – primarily water beetles at that time – which expanded with his work at the Museum where he was later appointed Keeper of Biology. In 1985 he was elected to Fellowship of the Royal Entomological Society and in 1991 started part-time PhD research at Newcastle University, concerned with riparian Coleoptera.

At that time, following a period working with environmental consultants Humphries Rowell Associates on amphibian re-location, **Jon Daws** undertook a short commission for Lott at the Museum, hand collecting beetles from Lockington Marshes as part of an invertebrate survey (Gilbertson, 1991). Lott had been working this site during the 1980s, assisted by Helen Ikin who had retained all spider material, which was identified by Crocker.

The same year, Lott extended his fieldwork to grassland sites – initially restored opencast workings in the north-west of the county at Lount and Coalfield West – and retained the spiders from his pitfall traps. This led to an extensive study of established SSSI grassland, urban derelict sites and old rural industrial sites, during which time (1992 and 1993) Daws – still working with Lott – attended various spider identification courses run by members of the British Arachnological Society. Though the main thrust of this grassland work was directed towards studying beetles (Lott & Daws, 1993), spiders, harvestmen and woodlice were also collected and identified by Daws, who then began working on the backlog of araneological material produced since 1991. He systematically referred identifications for verification, which attention to accuracy has considerably enhanced the value of these important records.

Associated with Lott's PhD research, new wetland survey work in the county was initiated in 1993. This included sites away from the main rivers such as Newton Burgoland and Tinwell Marshes, Great Bowden Pit and Rutland Water, where the previous policy of collecting spiders along with the beetles was continued. By now Daws was working virtually full-time with

Leicestershire Museums on invertebrate studies, which were extended, in 1994, to include pitfall trapping in riparian habitats along the River Soar within the city of Leicester. On his own initiative he carried out further pitfall trapping at Seaton Meadows in the Welland valley, Ketton Quarry and other sites.

Also in 1994, Emma L. Caradine, a University of Exeter biology graduate, completed her post-graduate MSc dissertation at Leicester University on spider and harvestman communities of urban and urban fringe riverside wetlands (Caradine, 1994; Caradine et al., 1995). This research analyses wetland sites along the Soar valley north and south of the city of Leicester, adding further spider records to those of Lott and Daws. Physical aspects of the vegetation, frequency of habitat disturbance and successional age are suggested as being important determinants of spider and harvestman faunal composition.

Finally, mention must be made of current survey work initiated by Lott on Leicestershire and Rutland Trust for Nature Conservation reserves, where pilot pitfall trapping schemes, for collecting beetles (and spiders), were carried out in 1994 at Egleton, Lea Meadows, Bloody Oaks and Stonesby Quarries. The best effort was made by Christine Kirk, convenor for Stonesby, and Jane Wilson who have continued this work – before and after scrub clearance and other management activities – throughout 1995. New trapping lines have been put down at other Trust Reserves during 1995, including Charnwood Lodge Nature Reserve, but much of the current material collected has yet to be identified.

First Spider Records for VC55

Table 5: This table comprises the first 88 county records, taken from published and unpublished sources up to 1963, where in each case, insufficient detail is given to localise at other than county level. Since, in most of these, no specific dates are given, they are numbered for ease of cross reference. Further details of the source of these records will be found in the first part of this chapter and in the References. The fourth record in this table, published by Mott, is credited to C.B. Headly, whilst records 5-34 are G.B. Chalcraft's, as reviewed by Rowley. The Lowe and Mayes sources are from manuscripts and collections held by Leicestershire Museums Service. It will be noted that there is an overlap of dates, in tables 5 and 6, between 1948 and 1963, resulting from a lack of detail in the published records at the end of table 5. This table also includes some contemporary published records which are not sufficiently detailed to contribute to the Leicestershire distribution maps, but are nonetheless valid county records. All but four species (*Micrommata virescens, Tegenaria parietina, Araneus marmoreus* and *Thyreosthenius biovatus*) on this list have been rediscovered in recent years.

No.	Source	Species	Recorded as	Remarks
1	Agar (1890)	Agroeca brunnea	Agelena brunnea	
2	Agar (1890)	Araneus diadematus	Epeira diadema	
3	Agar (1890)	Zygiella x-notata	Epeira calophylla	wrong identity, see commentary p.34
4	Mott (1890)	Tegenaria parietina	Tegenaria atrica	wrong identity, see commentary p.48
5	Rowley (1897)	Trochosa ruricola	Lycosa campestris	
6	Rowley (1897)	Pardosa amentata	L. saccata	
7	Rowley (1897)	Pisaura mirabilis	Dolomedes mirabilis	
8	Rowley (1897)	Salticus scenicus		
9	Rowley (1897)	Xysticus cristatus	Thomisus cristatus	
10	Rowley (1897)	Misumena vatia	T. citreus	
11	Rowley (1897)	Micrommata virescens	Sparassus smaragdulus	unconfirmed, but acceptable
12	Rowley (1897)	Drassodes cupreus	Drassus cupreus	
13	Rowley (1897)	Anyphaena accentuata	Clubiona accentuata	
14	Rowley (1897)	Amaurobius fenestralis	Ciniflo atrox	
15	Rowley (1897)	Amaurobius ferox	C. ferox	
16	Rowley (1897)	Tegenaria gigantea	Tegenaria domestica	wrong identity, see commentary p.48
17	Rowley (1897)	Tegenaria domestica	T. civilis	
18	Rowley (1897)	Coelotes atropos	Coelotes saxatilis	
19	Rowley (1897)	Textrix denticulata	Textrix lycosina	
20	Rowley (1897)	Steatoda bipunctata	Theridion quadripunctatum	
21	Rowley (1897)	Theridion sisyphium	T. novosum	
22	Rowley (1897)	Crustulina guttata	T. guttatum	doubtful identification but found later
23	Rowley (1897)	Centromerita bicolor	Neriene bicolor	doubtful identification but found later
24	Rowley (1897)	Lophomma punctatum	Walckenaera punctata	

No.	Source	Species	Recorded as	Remarks
25	Rowley (1897)	*Baryphyma pratense*	*W. pratensis*	doubtful identification but found later
26	Rowley (1897)	*Araniella cucurbitina*	*Epeira cucurbitina*	doubtful, could be *opistographa*
27	Rowley (1897)	*Larinioides patagiatus*	*E. patagiata*	
28	Rowley (1897)	*Araneus marmoreus*	*E. scalaris*	unconfirmed, see commentary p.48
29	Rowley (1897)	*Nuctenea umbratica*	*E. umbratica*	
30	Rowley (1897)	*Zygiella atrica*	*E. calophylla*	
31	Rowley (1897)	*Metellina merianae*	*E. antriada*	
32	Rowley (1897)	*Metellina segmentata*	*E. inclinata*	
33	Rowley (1897)	*Cyclosa conica*	*E. conica*	
34	Rowley (1897)	*Tetragnatha extensa*		
35	Lowe (1912)	*Drassodes lapidosus*	*Drassus lapidicolens*	
36	Lowe (1912)	*Harpactea hombergi*	*Harpactes hombergii*	
37	Lowe (1912)	*Clubiona stagnatilis*	*Clubiona grisea*	
38	Lowe (1912)	*Philodromus dispar*		
39	Lowe (1912)	*Philodromus aureolus*		
40	Lowe (1912)	*Heliophanus flavipes*	*Helophanes flavipes*	
41	Lowe (1912)	*Evarcha falcata*	*Hasarius falcatus*	
42	Lowe (1912)	*Arctosa perita*	*Trochosa picta*	
43	Lowe (1912)	*Agelena labyrinthica*		
44	Lowe (1912)	*Tegenaria silvestris*	*Tegenaria campestris*	
45	Lowe (1912)	*Episinus angulatus*	*E. truncatus*	
46	Lowe (1912)	*Theridion bimaculatum*		
47	Lowe (1912)	*Enoplognatha ovata*	*Theridion lineatum*	
48	Lowe (1912)	*Robertus lividus*	*Neriene livida*	
49	Lowe (1912)	*Pachygnatha listeri*		
50	Lowe (1912)	*Pachygnatha degeeri*		
51	Lowe (1912)	*Araneus quadratus*		
52	Lowe (1912)	*Linyphia triangularis*		
53	Lowe (1912)	*Linyphia clathrata*		
54	Lowe (1912)	*Microlinyphia pusilla*	*Linyphia pusilla*	
55	Donisthorpe (1927)	*Mastigusa macrophthalma*	*Cryphoeca recisa* O.P.-C.	
56	Donisthorpe (1927)	*Thyreosthenius biovatus*		
57	Mayes (1934)	*Dysdera crocata*		
58	Mayes (1934)	*Scotophaeus blackwalli*		
59	Mayes (1934)	*Pardosa pullata*		
60	Mayes (1934)	*Theridion pallens*		
61	Bristowe (1939)	*Amaurobius similis*	*Ciniflo similis*	
62	Bristowe (1939)	*Dictyna uncinata*		
63	Bristowe (1939)	*Clubiona phragmitis*		
64	Bristowe (1939)	*Clubiona comta*	*Clubiona compta*	
65	Bristowe (1939)	*Ozyptila brevipes*	*Oxyptila flexa*	
66	Bristowe (1939)	*Philodromus cespitum*	*P. aureolus cespiticolis*	
67	Bristowe (1939)	*Theridion varians*		
68	Bristowe (1939)	*Theridion tinctum*		
69	Bristowe (1939)	*Tetragnatha obtusa*		
70	Bristowe (1939)	*Pachygnatha clercki*		
71	Bristowe (1939)	*Hypomma cornutum*	*Hypomma cornuta*	
72	Bristowe (1939)	*Oedothorax retusus*		
73	Bristowe (1939)	*Typhocrestus digitatus*		
74	Bristowe (1939)	*Bathyphantes gracilis*		
75	Bristowe (1939)	*Diplostyla concolor*	*Bathyphantes concolor*	
76	Bristowe (1939)	*Lepthyphantes tenuis*		
77	Bristowe (1939)	*Linyphia hortensis*		
78	Wild (1952)	*Segestria senoculata*		
79	Wild (1952)	*Tibellus oblongus*		
80	Wild (1952)	*Ero furcata*		
81	Wild (1952)	*Larinioides cornutus*	*Araneus cornutus*	
82	Wild (1952)	*Floronia bucculenta*		
83	Wild (1952)	*Labulla thoracica*		
84	Wild (1952)	*Stemonyphantes lineatus*	*Stemonyphantes lineata*	
85	Wild (1952)	*Linyphia montana*		
86	Cooke & Lampel (1953)	*Cubiona corticalis*		
87	Cooke & Lampel (1953)	*Clubiona terrestris*		
88	Bristowe (1963)	*Oonops domesticus*		

Table 6: Together with the foregoing 88 early published and manuscript records in table 5, the following list of 238 detailed **field records**, presented in date order, make up a total of **326** county records for Leicestershire. The overlap of dates between 1948 and 1963 has been mentioned at the head of the previous table, where the reason is given. All records in table 6 have sufficient detail to enable a plot to be made on the distribution maps (chapter five). Those between 1948 and 1959 are shown with open circles as, for the purpose of this book, they are treated as historical records. Many of these have been rediscovered in the same tetrad and therefore the original open circle will be filled. Post-1959 records are shown as solid circles. The 16 flagged species (*) in table 6 are those known in the county from only a single record – that shown in this table.

Date	Species	Location	Collected	Determined
03.07.48	Zelotes latreillei	Waltham-on-the-Wolds	T.A.Walden	T.A.Walden
03.07.48	Alopecosa pulverulenta	Waltham-on-the-Wolds	T.A.Walden	T.A.Walden
16.06.48	Pirata piraticus	Scraptoft	T.A.Walden	T.A.Walden
14.04.52	Oedothorax fuscus	The Brand, Woodhouse	E.Duffey	E.Duffey
14.04.52	Zelotes apricorum	The Brand, Woodhouse	E.Duffey	E.Duffey
02.07.53	Hylyphantes graminicola	Wistow	D.S.Fieldhouse	I.M.Evans
28.12.53	Oonops pulcher	Beacon Hill, Woodhouse	E.Duffey	E.Duffey
28.12.53	Zora spinimana	Beacon Hill, Woodhouse	E.Duffey	E.Duffey
28.12.53	Walckenaeria acuminata	Beacon Hill, Woodhouse	E.Duffey	E.Duffey
28.12.53	Lepthyphantes zimmermanni	High Sharpley, Whitwick	E.Duffey	E.Duffey
28.12.53	Lepthyphantes ericaeus	High Sharpley, Whitwick	E.Duffey	E.Duffey
28.12.53	Poeciloneta variegata	High Sharpley, Whitwick	E.Duffey	E.Duffey
28.12.53	Cnephalocotes obscurus	Beacon Hill, Woodhouse	E.Duffey	E.Duffey
05.05.55	Trochosa terricola	Loddington Reddish	E.Duffey	E.Duffey
05.05.55	Diplocephalus picinus	Loddington Reddish	E.Duffey	E.Duffey
05.05.55	Diplocephalus latifrons	Loddington Reddish	E.Duffey	E.Duffey
18.07.59	Phrurolithus festivus	Breedon Cloud Quarry	I.M.Evans	D.J.Clark
18.07.59	Erigone atra	Puddledyke, Newtown Linford	I.M.Evans	D.J.Clark
18.07.59	Prinerigone vagans	Puddledyke, Newtown Linford	I.M.Evans	D.J.Clark
26.08.59	Sitticus pubescens	Ibstock, The Limes	M.A.H.Watts	I.M.Evans
21.01.60	Argyroneta aquatica	Grand Union Canal, Kibworth	I.M.Evans	I.M.Evans
06.02.60	Diplocephalus cristatus	Birstall, 46 Bramley Road	G.& A.Smith	I.M.Evans
06.02.60	Lepthyphantes nebulosus	Birstall, 46 Bramley Road	G.& A.Smith	I.M.Evans
06.02.60	Lepthyphantes leprosus	Birstall, 46 Bramley Road	G.& A.Smith	I.M.Evans
13.02.60	Theridion melanurum	Scraptoft, Hall Road	G.Smith	I.M.Evans
03.03.60	Erigone dentipalpis	Scraptoft, Hall Road	R.Lee	I.M.Evans
16.03.60	Tegenaria agrestis	Leicester, West Humberstone	J.Clarke	I.M.Evans
05.04.60	Maso sundevalli	Leicester, Haddenham Road	S.B.Scargill	I.M.Evans
12.06.60	Tetragnatha montana	Eyebrook Reservoir	I.M.Evans	I.M.Evans
12.06.60	Tetragnatha nigrita	Eyebrook Reservoir	I.M.Evans	I.M.Evans
13.05.61	Clubiona reclusa	Grand Union Canal, Gumley	I.M.Evans	I.M.Evans
13.05.61	Clubiona neglecta	Grand Union Canal, Gumley	I.M.Evans	I.M.Evans
27.05.61	Pocadicnemis pumila	Countesthorpe	J.Crocker	J.Crocker
30.05.61	Linyphia peltata	Swithland Wood	J.Crocker	J.Crocker
31.05.61	Cryphoeca silvicola	Coleorton Hall Woods	J.Crocker	J.Crocker
31.05.61	Theridion mystaceum	Coleorton Hall Woods	J.Crocker	J.Crocker
06.06.61	Hypomma bituberculatum	Groby Pool	I.M.Evans	I.M.Evans
16.06.61	Neon reticulatus	Swithland Wood	P.H.Gamble	J.Crocker
16.06.61	Pardosa prativaga	Swithland Wood	P.H.Gamble	J.Crocker
22.06.61	Euophrys frontalis	Coleorton Churchyard	J.Crocker	J.Crocker
24.06.61	Pirata latitans	Bradgate Park	I.M.Evans	I.M.Evans
29.06.61	Meioneta rurestris	Coleorton Hall	J.Crocker	J.Crocker
15.07.61	Bolyphantes luteolus	Charnwood Lodge N.R., Charley	J.Crocker	J.Crocker
15.07.61	Lepthyphantes cristatus	Charnwood Lodge N.R., Charley	J.Crocker	J.Crocker
07.08.61	Gonatium rubens	Hose Grange	J.Crocker	J.Crocker
26.08.61	Tapinopa longidens	Heyday Hayes Wood, Newtown Linford	A.E.Squires	J.Crocker
01.09.61	Araeoncus humilis	Hugglescote, Coalville	D.B.Forgham	J.Crocker
02.09.61	Pirata hygrophilus	Wardley Wood	J.Crocker	J.Crocker
04.09.61	Urozelotes rusticus	Shepshed, Oakley Road Tip	J.Crocker	J.Crocker
10.09.61	Oedothorax apicatus	Pignut Spinney marsh, Loughborough	J.Crocker	J.Crocker
10.09.61	Lepthyphantes minutus	Pignut Spinney marsh, Loughborough	J.Crocker	J.Crocker
14.10.61	Metellina mengei	Leicester, Castle Gardens	D.G.Goddard	I.M.Evans
08.11.61	Centromerus sylvaticus	Bosworth Park	H.I.James	I.M.Evans
25.02.62	Hahnia helveola	Outwoods, Loughborough	J.Crocker	J.Crocker
08.04.62	Centromerus prudens	Charnwood Lodge N.R., Charley	J.Crocker	J.Crocker
08.04.62	Macrargus rufus	Charnwood Lodge N.R., Charley	J.Crocker	J.Crocker
24.04.62	Pardosa nigriceps	Bradgate Park	J.Crocker	J.Crocker
24.04.62	Ceratinella brevis	Bradgate Park	J.Crocker	J.Crocker

Date	Species	Location	Collected	Determined
24.04.62	*Savignia frontata*	Bradgate Park	J.Crocker	J.Crocker
24.04.62	*Porrhomma microphthalmum*	Bradgate Park	J.Crocker	J.Crocker
24.04.62	*Bathyphantes nigrinus*	Owston Wood	H.A.B.Clements	J.Crocker
24.04.62	*Lepthyphantes alacris*	Owston Wood	H.A.B.Clements	J.Crocker
24.04.62	*Lepthyphantes mengei*	Bradgate Park	J.Crocker	J.Crocker
05.05.62	*Xysticus ulmi*	Cropston Reservoir Waterworks	I.M.Evans	I.M.Evans
05.05.62	*Pardosa lugubris*	Bradgate Park	P.C.Jerrard	P.C.Jerrard
05.05.62	*Walckenaeria nudipalpis*	Cropston Reservoir	P.C.Jerrard	P.C.Jerrard
05.05.62	*Gnathonarium dentatum*	Cropston Reservoir	I.M.Evans	I.M.Evans
05.05.62	*Erigonella ignobilis*	Cropston Reservoir	I.M.Evans	I.M.Evans
05.05.62	*Diplocephalus permixtus*	Cropston Reservoir	P.C.Jerrard	P.C.Jerrard
05.05.62	*Porrhomma pygmaeum*	Bradgate Park	P.C.Jerrard	D.G.Clark
05.05.62	*Bathyphantes approximatus*	Cropston Reservoir	I.M.Evans	I.M.Evans
13.05.62	*Walckenaeria dysderoides*	Charnwood Lodge N.R., Charley	J.Crocker	J.Crocker
13.05.62	*Gongylidiellum vivum*	Charnwood Lodge N.R., Charley	J.Crocker	J.Crocker
13.05.62	*Lepthyphantes obscurus*	Charnwood Lodge N.R., Charley	J.Crocker	J.Crocker
16.05.62	*Dismodicus bifrons*	Cropston Reservoir Waterworks	J.Crocker	J.Crocker
16.05.62	*Micrargus herbigradus*	Cropston Reservoir Waterworks	J.Crocker	J.Crocker
16.05.62	*Kaestneria pullata*	Cropston Reservoir	J.Crocker	J.Crocker
20.05.62	*Gongylidium rufipes*	Loddington Reddish	J.Crocker	J.Crocker
20.05.62	*Oedothorax gibbosus*	Tugby Wood	J.Crocker	J.Crocker
26.05.62	*Clubiona lutescens*	Shepshed, Oakley Road Tip	J.Crocker	J.Crocker
26.05.62	*Monocephalus fuscipes*	Breedon Hill, Bulwarks	J.Crocker	J.Crocker
26.05.62	*Meioneta saxatilis*	Breedon Hill, Bulwarks	J.Crocker	J.Crocker
26.05.62	*Bathyphantes parvulus*	Breedon Hill, Bulwarks	J.Crocker	J.Crocker
02.06.62	*Anelosimus vittatus*	Buddon Wood, Quorn	P.H.Gamble	J.Crocker
02.06.62	*Atea sturmi*	Buddon Wood, Quorn	P.H.Gamble	J.Crocker
03.06.62	*Agyneta conigera*	Skeffington Wood	J.Crocker	J.Crocker
03.06.62	*Kaestneria dorsalis*	Skeffington Wood	J.Crocker	J.Crocker
04.06.62	*Walckenaeria unicornis*	Shepshed, Oakley Road Tip	J.Crocker	J.Crocker
11.06.62	*Dictyna arundinacea*	Shepshed, Brick Hole	H.A.B.Clements	J.Crocker
11.06.62	*Araniella opistographa*	Shepshed, Brick Hole	H.A.B.Clements	J.Crocker
23.06.62	*Clubiona brevipes*	Buddon Wood, Quorn	P.H.Gamble	J.Crocker
29.06.62	*Entelecara acuminata*	Loughborough, 66 Outwoods Drive	J.Crocker	J.Crocker
30.06.62	*Micaria pulicaria*	Charnwood Lodge N.R., Charley	H.A.B.Clements	J.Crocker
01.07.62	*Salticus cingulatus*	King Lud's Entrenchments, Sproxton	I.M.Evans	I.M.Evans
01.07.62	*Antistea elegans*	Holwell Mouth, Ab Kettleby	I.M.Evans	I.M.Evans
01.07.62	*Entelecara erythropus*	Loughborough, 66 Outwoods Drive	J.Crocker	J.Crocker
01.07.62	*Pocadicnemis juncea*	King Lud's Entrenchments, Sproxton	I.M.Evans	I.M.Evans
01.07.62	*Micrargus subaequalis*	Loughborough, 66 Outwoods Drive	J.Crocker	J.Crocker
07.07.62	*Maso gallicus*	Shacklewell Hollow, Empingham	J.Crocker	J.Crocker
01.08.62	*Euophrys erratica*	Bradgate Park	A.E.Squires	J.Crocker
13.08.62	*Saaristoa abnormis*	Coleorton Hall Woods	J.Crocker	J.Crocker
01.09.62	*Clubiona trivialis*	Charnwood Lodge N.R., Charley	J.Crocker	J.Crocker
01.09.62	*Clubiona diversa*	Charnwood Lodge N.R., Charley	J.Crocker	J.Crocker
01.09.62	*Drapetisca socialis*	Charnwood Lodge N.R., Charley	J.Crocker	J.Crocker
01.09.62	*Helophora insignis*	Charnwood Lodge N.R., Charley	J.Crocker	J.Crocker
02.09.62	*Bolyphantes alticeps*	Charnwood Lodge N.R., Charley	I.M.Evans	I.M.Evans
08.09.62	*Lepthyphantes flavipes*	Farnham's Wood, Quorn	J.Crocker	J.Crocker
13.09.62	*Dysdera erythrina*	Bradgate Park	J.Crocker	J.Crocker
07.10.62	*Silometopus reussi*	Loughborough, 66 Outwoods Drive	J.Crocker	J.Crocker
07.10.62	*Ostearius melanopygius*	Loughborough, 66 Outwoods Drive	J.Crocker	J.Crocker
13.10.62	*Pholcomma gibbum*	Charnwood Lodge N.R., Charley	J.Crocker	J.Crocker
13.10.62	*Gonatium rubellum*	Charnwood Lodge N.R., Charley	J.Crocker	J.Crocker
14.10.62	*Larinioides sclopetarius*	Eyebrook Reservoir	I.M.Evans	I.M.Evans
11.11.62	*Hahnia montana*	Charnwood Lodge N.R., Charley	H.A.B.Clements	J.Crocker
11.11.62	*Microneta viaria*	Charnwood Lodge N.R., Charley	H.A.B.Clements	J.Crocker
06.03.63	*Walckenaeria cuspidata*	Shepshed, Oakley Road Tip	H.A.B.Clements	J.Crocker
14.04.63	*Tallusia experta*	Woodwell Head, Wymondham	I.M.Evans	I.M.Evans
25.05.63 *	*Moebelia penicillata*	Outwoods, Loughborough	J.Crocker	J.Crocker
05.06.63	*Ballus chalybeius*	Buddon Wood, Quorn	P.H.Gamble	J.Crocker
08.06.63	*Clubiona pallidula*	Buddon Wood, Quorn	P.H.Gamble	J.Crocker
29.06.63	*Walckenaeria antica*	Bardon Hill	J.Crocker	J.Crocker
29.06.63	*Lepthyphantes pallidus*	Bardon Hill	J.Crocker	J.Crocker
14.07.63	*Walckenaeria furcillata*	Charnwood Lodge N.R., Charley	J.Crocker	J.Crocker
14.07.63	*Dicymbium tibiale*	Charnwood Lodge N.R., Charley	J.Crocker	J.Crocker
14.07.63	*Centromerus dilutus*	Charnwood Lodge N.R., Charley	J.Crocker	J.Crocker
30.07.63	*Pirata uliginosus*	Holwell Mouth, Ab Kettleby	H.A.B.Clements	J.Crocker
30.07.63	*Peponocranium ludicrum*	Bardon Hill	J.Crocker	J.Crocker
14.09.63	*Ozyptila praticola*	Shepshed, Tyler Brigg Marsh	H.A.B.Clements	J.Crocker

Date		Species	Location	Collected	Determined
15.09.63		*Monocephalus castaneipes*	Charnwood Lodge N.R., Charley	J.Crocker	J.Crocker
20.10.63		*Theonoe minutissima*	Charnwood Lodge N.R., Charley	J.Crocker	J.Crocker
20.10.63		*Tiso vagans*	Charnwood Lodge N.R., Charley	J.Crocker	J.Crocker
20.10.63		*Porrhomma pallidum*	Charnwood Lodge N.R., Charley	J.Crocker	J.Crocker
20.10.63		*Centromerita concinna*	Charnwood Lodge N.R., Charley	J.Crocker	J.Crocker
08.03.64		*Leptorhoptrum robustum*	Swithland Wood	I.M.Evans	I.M.Evans
22.03.64		*Walckenaeria cucullata*	Charnwood Lodge N.R., Charley	J.Crocker	J.Crocker
22.03.64		*Dicymbium nigrum*	Charnwood Lodge N.R., Charley	J.Crocker	J.Crocker
12.04.64		*Tapinocyba praecox*	Charnwood Lodge N.R., Charley	D.W.Mackie	D.W.Mackie
12.04.64		*Micrargus apertus*	Charnwood Lodge N.R., Charley	J.Crocker	J.Crocker
24.04.64		*Saloca diceros*	Owston Wood	J.Crocker	J.Crocker
24.04.64		*Erigonella hiemalis*	Owston Wood	J.Crocker	J.Crocker
14.05.64		*Tmeticus affinis*	River Soar, Normanton	H.A.B.Clements	J.Crocker
14.05.64		*Oedothorax agrestis*	River Soar, Normanton	H.A.B.Clements	J.Crocker
30.05.64		*Hypsosinga pygmaea*	King Lud's Entrenchments, Sproxton	J.Crocker	J.Crocker
30.05.64		*Troxochrus scabriculus*	King Lud's Entrenchments, Sproxton	J.Crocker	J.Crocker
24.06.64		*Nesticus cellulanus*	Beacon Hill, Woodhouse	J.Crocker	J.Crocker
16.08.64		*Pardosa palustris*	Charnwood Lodge N.R., Charley	J.Crocker	J.Crocker
29.08.64		*Allomengea vidua*	Saddington Reservoir	J.Crocker	J.Crocker
15.11.64	*	*Walckenaeria capito*	Charnwood Lodge N.R., Charley	J.Crocker	J.Crocker
03.02.65		*Psilochorus simoni*	Coleorton Hall	J.Crocker	J.Crocker
15.05.65		*Meta menardi*	Blackbrook Reservoir	J.Crocker	J.Crocker
20.06.65		*Silometopus elegans*	Charnwood Lodge N.R., Charley	M.G.Crocker	J.Crocker
17.07.65		*Euophrys aequipes*	Cropston Reservoir	J.Crocker	J.Crocker
17.07.65		*Agyneta decora*	Bradgate Park	J.Crocker	J.Crocker
16.10.65		*Microctenonyx subitaneus*	Charnwood Lodge N.R., Charley	J.Crocker	J.Crocker
17.10.65		*Thyreosthenius parasiticus*	Charnwood Lodge N.R., Charley	J.Crocker	J.Crocker
23.03.66		*Lessertia dentichelis*	Coleorton Hall	J.Crocker	J.Crocker
15.05.66		*Heliophanus cupreus*	The Brand, Woodhouse	M.G.Crocker	J.Crocker
30.07.66		*Lepthyphantes tenebricola*	Belvoir, Briery Wood	J.Crocker	J.Crocker
11.09.66		*Ozyptila trux*	Swithland Wood	J.Crocker	J.Crocker
01.01.67		*Tapinocyba insecta*	Swithland Wood	J.Crocker	J.Crocker
25.02.67		*Asthenargus paganus*	Nailstone Wiggs	J.Crocker	J.Crocker
30.05.67		*Lathys humilis*	Swithland Wood	J.Crocker	J.Crocker
30.05.67		*Pelecopsis nemoralis*	Swithland Wood	J.Crocker	J.Crocker
09.06.67		*Porrhomma convexum*	Swithland Wood	J.Crocker	P.Merrett
22.07.67		*Agroeca proxima*	Ulverscroft N.R.	J.Crocker	J.Crocker
22.07.67		*Walckenaeria atrotibialis*	Ulverscroft N.R.	J.Crocker	J.Crocker
22.07.67	*	*Baryphyma trifrons*	Ulverscroft N.R.	J.Crocker	J.Crocker
02.09.67		*Drepanotylus uncatus*	Grace Dieu, Cademan Moor	J.Crocker	J.Crocker
22.01.68		*Ero cambridgei*	Coleorton, Blue Door Meadow	J.Crocker	J.Crocker
11.05.68		*Evansia merens*	Bardon Hill	J.Crocker	J.Crocker
08.06.69		*Haplodrassus signifer*	Newbold Verdon	M.J.Roberts	M.J.Roberts
08.06.69		*Theridion impressum*	Newbold Verdon	M.J.Roberts	M.J.Roberts
15.02.70	*	*Porrhomma egeria*	Bardon Hill	J.Crocker	J.Crocker
22.03.70	*	*Lepthyphantes beckeri*	Bradgate Park	J.Crocker	J.Crocker
12.04.70		*Saaristoa firma*	Bardon Hill	J.Crocker	J.Crocker
30.07.71		*Tetragnatha striata*	Groby Pool	J.Crocker	J.Crocker
08.08.71		*Lepthyphantes midas*	Donington Park, Castle Donington	J.Crocker	J.Crocker
12.07.72		*Tetragnatha pinicola*	Anstey	O.H.Black	J.Crocker
23.07.72		*Milleriana inerrans*	Donington Park, Castle Donington	J.Crocker	J.Crocker
01.07.73		*Theridion pictum*	Buddon Wood, Quorn	D.B.Forgham	J.Crocker
02.09.73		*Agyneta subtilis*	Donington Park, Castle Donington	J.Crocker	J.Crocker
07.10.73		*Halorates distinctus*	Barrow Gravel Pits, River Soar	A.E.Squires	J.Crocker
07.10.73		*Microlinyphia impigra*	Barrow Gravel Pits, River Soar	M.G.Crocker	J.Crocker
16.06.74		*Walckenaeria vigilax*	Grace Dieu, Cademan Moor	M.G.Crocker	J.Crocker
25.05.75		*Haplodrassus silvestris*	Buddon Wood, Quorn	M.G.Crocker	J.Crocker
01.06.75		*Theridion simile*	Buddon Wood, Quorn	J.Crocker	J.Crocker
01.06.75		*Enoplognatha thoracica*	Buddon Wood, Quorn	M.G.Crocker	J.Crocker
07.06.75		*Tapinocyba pallens*	Buddon Wood, Quorn	M.G.Crocker	J.Crocker
07.06.75		*Agyneta ramosa*	Buddon Wood, Quorn	M.G.Crocker	J.Crocker
22.06.75		*Ceratinella brevipes*	Buddon Wood, Quorn	J.Crocker	J.Crocker
20.06.76	*	*Pardosa monticola*	Iveshead, Shepshed	J.Crocker	J.Crocker
19.06.77		*Porrhomma campbelli*	Narborough Bog	J.Mathias	J.Crocker
02.07.77		*Achaearanea lunata*	Buddon Wood, Quorn	J.Crocker	J.Crocker
10.06.78	*	*Xysticus lanio*	Sheet Hedges Wood, Newtown Linford	J.Crocker	J.Crocker
10.06.78		*Gibbaranea gibbosa*	Sheet Hedges Wood, Newtown Linford	J.Crocker	J.Crocker
14.09.78		*Robertus neglectus*	Welby Osier Beds	LMARS	I.M.Evans
22.09.78		*Ozyptila sanctuaria*	Cooper's Plantation, Croxton Kerrial	LMARS	I.M.Evans
28.03.79		*Euophrys lanigera*	Ashby-de-la-Zouch	J.Crocker	J.Crocker

CHAPTER THREE

Date		Species	Location	Collected	Determined
01.01.83		*Porrhomma errans*	Scraptoft	I.M.Evans	I.M.Evans
31.05.83		*Erigone arctica*	Lockington Meadows	H.Ikin	J.Crocker
05.02.84		*Diaea dorsata*	Wymondham Rough N.R.	I.M.Evans	I.M.Evans
15.04.89	*	*Tegenaria saeva*	Leicester, 450 Braunstone Lane	B.E.Wills	I.M.Evans
16.05.90		*Pholcus phalangioides*	Leicester, 33 Perkyn Road	J.Woodhead	I.M.Evans
26.06.91		*Lepthyphantes insignis*	Coalfield West, Ravenstone	D.A.Lott	J.Daws
16.08.91		*Meioneta beata*	Lount Grassland, Worthington	D.A.Lott	J.Daws
23.11.91		*Cicurina cicur*	Lount Grassland, Worthington	D.A.Lott	J.Daws
20.05.92		*Panamamops sulcifrons*	River Soar, Loughborough Meadows	J.Daws	J.Daws
01.06.92		*Hahnia nava*	St Mary's Allotments, Leicester	J.Daws	J.Daws
01.06.92		*Pelecopsis parallela*	CEGB Rawdykes, Leicester	J.Daws	J.Daws
03.06.92		*Xysticus erraticus*	High Sharpley, Whitwick	J.Daws	J.Daws
03.06.92		*Euryopis flavomaculata*	Moira Junction, Ashby Woulds	J.Daws	J.Daws
03.06.92		*Ceratinopsis stativa*	Lount Meadow No 2, Coleorton	J.Daws	J.Daws
08.06.92		*Drassyllus pusillus*	The Drift, Croxton Kerrial	J.Daws	J.Daws
08.06.92		*Ceratinella scabrosa*	King Lud's Entrenchments, Sproxton	J.Daws	J.Daws
01.07.92		*Metopobactrus prominulus*	Bardon Hill	J.Daws	S.Dobson
01.07.92		*Agyneta cauta*	Bardon Hill	J.Daws	S.Dobson
07.07.92		*Walckenaeria incisa*	King Lud's Entrenchments, Sproxton	J.Daws	J.Daws
04.09.92		*Achaearanea simulans*	Moira Junction, Ashby Woulds	J.Daws	P.Merrett
19.11.92		*Cheiracanthium virescens*	Moira Junction, Ashby Woulds	J.Daws	J.Daws
21.03.93	*	*Trochosa robusta*	Ketton Quarry	J.Daws	M.J.Roberts
30.05.93		*Ozyptila atomaria*	Ketton Quarry	J.Daws	J.Crocker
18.06.93	*	*Minyriolus pusillus*	Norris Hill, Ashby Woulds	J.Daws	J.Crocker
22.04.94		*Xysticus audax*	Luffenham Heath Golf Course	J.Daws	J.Daws
28.05.94	*	*Philodromus albidus*	Essendine Church	J.Daws	J.Daws
28.05.94		*Theridion blackwalli*	Great Casterton Church	J.Daws	J.Daws
31.05.94	*	*Clubiona subtilis*	Seaton Meadow	J.Daws	J.Daws
01.06.94		*Gongylidiellum latebricola*	Ketton Quarry	J.Daws	J.Daws
10.06.94		*Dictyna latens*	Luffenham Heath Golf Course	J.Daws	J.Daws
10.06.94		*Philodromus collinus*	Luffenham Heath Golf Course	J.Daws	J.Daws
22.06.94	*	*Meioneta innotabilis*	Burbage Wood	J.Daws	J.Daws
26.06.94	*	*Pityohyphantes phrygianus*	Pickworth Great Wood	J.Daws	J.Daws
02.07.94		*Philodromus praedatus*	Croxton Park, Croxton Kerrial	J.Daws	J.Daws
04.07.94		*Pardosa agrestis*	Seaton Meadow	J.Daws	J.Crocker
20.07.94		*Erigone longipalpis*	River Soar, Cotes Bridge	D.A.Lott	J.Daws
14.11.94		*Alopecosa barbipes*	Geeston Quarry, Ketton	E.Caradine	J.Daws
17.04.95	*	*Ozyptila scabricula*	Geeston Quarry, Ketton	I.Phillips	D.R.Nellist
11.08.95		*Cercidia prominens*	Luffenham Heath Golf Course	J.Daws	J.Daws
15.10.95	*	*Agalenatea redii*	Ketton Quarry	H.N.Ball	J.Crocker

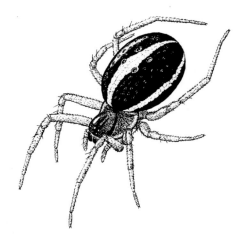

Hypsosinga pygmaea f. (x8)

CHAPTER FOUR

Material and Methods, Voucher Specimens, and Red Data Book Spiders

Material

Sources of spider records used in the preparation of this book are from literature, museum archives and specimens, and directly from fieldwork or observation. Many of the early Leicestershire records of spiders are unsupported by voucher material, and where pre-1960 specimens are available inaccuracies have been experienced in determinations. However, this has not been a real problem since, with a few documented exceptions, these early records have been re-established by later work. Because precise locations were not given for most of these early records it has not been possible to include them on the distribution maps. Hence, the substance of this publication is based on dedicated post-1960 field work, with relatively little contribution from literature sources or museum collections which are examined in chapter three. Voucher material supports all recent records of interest.

Field work has fallen into two principal categories: that undertaken by individuals, often on a random and unco-ordinated basis but occasionally as longer-term personal projects; and organised studies directed with particular objectives in mind, usually with institutional backing. In the first category Drs Eric Duffey, John Mathias, Jenny Owen and Michael J. Roberts have all made contributions to our knowledge of the Leicestershire spider fauna. The second category falls into five sub-divisions.

The *Leicestershire Fauna Survey* (1947 onwards) – a joint project established by the University of Leicester Zoology Department, the Leicester Museum Biology Section and the Botany and Zoology Section of the Leicester Literary and Philosophical Society – yielded only a few spider records, mostly by Trevor A. Walden. These were, however, well documented and supported by specimens held at the Museum.

The *Charnwood Forest Survey*, a major initiative by the Loughborough Naturalists' Club, launched in 1962, was supported by Leicester Museum and the Natural History Section of the L.L.P.S. It embraced comparative field work by John and Marcene Crocker away from Charnwood at Sherwood, Epping, Windsor and the New Forest to establish 'yardstick' evaluations of spider populations. Similar work in north-west and east Leicestershire formed the basis for county recording and mapping of spiders. The main participants in the arachnological element of this undertaking were J. and M.G. Crocker, Harry A.B. Clements, Peter H. Gamble and Helen Ikin, with Ian M. Evans, Colin Green, P.C. Jerrard, D.W. Mackie, D.R. Nellist and F.R. Wanless making significant contributions.

Museum activities. Enquiries from the general public handled by the Biology Section of the former Leicester Museum (since 1974 the Leicestershire Museums Service), have yielded many useful spider records, particularly from the Leicester City area. In the period 1959-1991 all such material was identified by I.M. Evans. Also, a number of the biological surveys carried out by the Museums Service under the direction of I.M. Evans yielded good spider lists from hand collecting and/or pitfall trapping. Notable amongst these were the 1978 survey for the proposed North-east Leicestershire Coalfield (Evans, 1979), and a series of surveys of the invertebrates of aquatic habitats in the county undertaken by Dr D.G. Goddard. Material retained from these surveys is lodged in the collections of the Biology Section. As part of a recent research programme into the beetle fauna of wetland and other habitats in the county during the last five years being carried out by Derek A. Lott, Keeper of Biology at the Museum, assisted by Jonathan Daws, extensive pitfall trapping and hand collecting have been undertaken. Jon Daws has also pursued his own independent collecting over this period aimed at under-worked areas in the county. Altogether, this recent work has yielded over eight thousand spider records.

Invertebrate survey work involving spiders was also undertaken by environmental consultants Humphries Rowell Associates Ltd on Loughborough Big Meadow, alongside the River Soar, during the summer season of 1993. Pitfall traps yielded some interesting results which were followed up with supplementary field work by Jon Daws on behalf of the Leicestershire and Rutland Trust for Nature Conservation.

Undergraduate studies were carried out by Loughborough University students Lorna Slade and Robert Wild on several hedgerow sites near Loughborough during 1980 when representative collections of spiders were made. Leicester University post-graduate student Emma L.Caradine spent a year studying the spiders and harvestmen of wetland sites along the River Soar valley for her MSc degree in 1994, and is currently researching life history strategies of spiders for her doctorate. Also, a useful collection of spiders was made by Leicester University post-graduate student Ian Phillips whilst pitfall trapping for beetles at Geeston Quarry, during 1994-5.

Participation

The following lists give a breakdown of the 144 individual collectors who have contributed over eighteen thousand records to the Leicestershire Spider Survey since 1948, excluding the literature/museum sourced records (1890-1962) which are discussed in chapter three and comprise an additional 14 collectors: William Agar, L.S. Biggs, Alec Bonner, Frank Bouskell, W.S. Bristowe, G.B. Chalcraft, J.A.L. Cooke, H.St.J. Donisthorpe, C.B. Headly, W.J. Kaye, G.P. Lampel, E.E. Lowe, W.E. Mayes & W.P. Mayes.

Major collections (over 100 records)
Emma Caradine (583), H.A.B. Clements (243), John Crocker (4694), Marcene G. Crocker (654), Jon Daws (7586), J.E. Dawson (131), I.M. Evans (270), P.H. Gamble (178), D.G. Goddard (233), Helen Ikin (411), D.A. Lott (381), I. Phillips (1058), A. Sanderson for Humphries Rowell Associates (237) and Jane Wilson (187).

Significant collections (10 – 100 records)
G.A. Chinnery (10), Eric Duffey (37), D.B. Forgham (10), M. Frankum (27), Colin Green (42), R. Harvey (44), R.C. Higgins (16), P.C. Jerrard (60), D.W. Mackie (59), J. Mathias (92), F.M. & J. Murphy (26), S.D. Musgrove (17), D.R. Nellist (34), R.D. Osborne (12), Jenny Owen (72), Bob Parker (11), Eleanor Parker (18), J.R. Parker (15), R.R. Parker (15), D. Payne (12), M.J. Roberts (46), D. Scruton (13), L. Slade & R. Wild (89), G. & A. Smith (34), A.E. Squires (33), M. Walpole (12), F.R. Wanless (74), Katherin Ward (74), M.A.H. Watts (12), M.B. Webster (12).

Casual collections (1 – 9 records)
R.D. Abbott, H.N. Ball, R.J. Barnett, W.H. Barrow, A.E. Bath, Mrs A. Beer, G.A.C. Bell, W.J. Binns, O.H. Black, D. Bland, H. Bradshaw, Ian Branscomb, S. Brister, G. Brooks, E. Broughton, D. Brown, Mrs J. Buchanan, B.J. Bullwer, P. Busby, J.C. Clarke, J.D. Cooper, H. & E. Day, K. Dearden, Stan Dobson, K. Edwards, A.H. Evans, P.A. Evans (including 6 as P.A. Candlish), D.S & J. Fieldhouse, A. Finch, J.D. Foxwell, O. Freer, I.B. Gamble, Miss Gibson, C. Goddard, G. Grass, J. Greaves, F.R. Green, D.T. & R. Grewcock, S. Grover, D. Halliday-Greene, D.H. Hall-Smith, K. Harper, Andrew Heaton, M. Hilder, R. Hoyland, P.W. Jackson, H.I. James, Mrs I. Jones, L. Jones, J.C. Jordon, S. Judd, Mrs Kealing-Rogers, R. Lee, M.J. Leech, Walter Lemon, J. Mann, R. Marsdon, N.A. Martin, J.B. McKellar, Glen McPhail, J. Mickleburgh, H.J. Mousley, A. Myers, P. Nicholson, D.A. Ottey, J.C. Oswin, R. Owen, S.M. Pearce, N.W. Porter, T.H. Riley, D. Roff, A.J. Rundle, A. Russell, B. Russell, S.B. Scargill, P.A. Selden, M. & J. Singer, J.R. Small, J. Smith, John Stacey, M. Stew, G.N. Syer, J. Talbot, L.A. Taylor, Master Tite, T.A. Walden, J. Ward, S. Ward, D. Whiteley, M.M. Whittle, B.E. Wills, J. Woodhead, J.A. Wrifir.

Identifications have been carried out in most cases by the major collectors and where necessary determinations have been verified by specialist referees, notably D.J. Clark, G.H. Locket and P. Merrett,

Main determinations by:-
Emma Caradine (582), John Crocker (6637), Jon Daws (9339), Stan Dobson (39), Eric Duffey (37), I.M. Evans (885), P.C. Jerrard (53), D.W. Mackie (58), P. Merrett (11), F.M. Murphy (25), D.R. Nellist, (41), M.J. Roberts (57), A. Sanderson for Humphries Rowell Associates (233), C.J. Smith (51), F.R. Wanless (78).

Additional determinations by:-
D.J. Clark, H.A.B. Clements, C. Felton, D.G. Goddard, R. Harvey, Andrew Heaton, P. Kirby, G.H. Locket, D.A. Lott, J. Mann, J. Murphy, R.F. Owen, P.A. Selden, A. Smith and T.A. Walden.

Methods

For a general introduction to the structure and biology of spiders, and an excellent companion to this present work, the reader is directed to the recently published (1995) Collins Field Guide – *Spiders of Britain and Northern Europe* by Michael J. Roberts, in which over 450 species are illustrated. Dr Roberts also gives some useful advice on collecting, identifying and preserving spiders in this modestly priced classic publication. For the beginner, the little booklet *How to begin the study of Spiders* (Mackie, 1989) is recommended. Also Dr Duffey's paper *Ecological survey and the arachnologist* (Duffey, 1972) will be of interest to field workers and describes the methods, techniques and apparatus for ecological survey which are available to the amateur not normally having access to laboratory facilities. Further guidance is given in this chapter on good practice and forms the basis upon which the authors have approached their collecting and recording activities. Likewise, comments on the spirit collections are personal and reflect methods adopted by the authors.

Collecting

Apart from the work of Donisthorpe (1927), pre-1960 collecting of spiders in Leicestershire was carried out to establish faunal lists for the county. Thereafter, most collecting had the object of determining the ecological status of individual sites. Long-term stability of land-use and species richness have been important early indicators of valuable wildlife areas, often confirmed by further detailed work on rare or local species and their micro-habitat requirements. There are three basic questions which elementary ecological survey attempts to answer:

(1) which species, and how many, are found in the habitat or site being studied?
(2) how numerous are the individuals of each species and what is the seasonal variation?
(3) what are the habitat preferences and ecological tolerances in relation to environmental factors?

These were the questions fundamental to the Charnwood Forest Survey, but further emphasis has been directed, during post-1960 collecting, to establishing the wider distributional coverage of species in the entire vice-county. The four main collecting categories used in the preparation of this publication are:

(a) County distribution: to discover the geographical distribution of species within the county.
(b) Unit area studies: qualitative surveys to establish species lists for well defined sites which may comprise a range of habitat types, and to reveal habitat preferences of the various species.
(c) Ecological survey: detailed investigation of the ecological relationships of groups of spiders in a particular habitat over a wide area, involving many sites.
(d) Species profiles: research into the biology and ecology of individual species.

Collecting techniques for all of these categories are related to three basic vegetation zones:

(i) Ground zone: vegetation and ground structures up to 15 cm in height.
(ii) Field layer: 15 cm to 1 metre in height.
(iii) Canopy: above 1 metre, including trunk and branches of scrub and trees.

Each of these three zones requires different collecting methods. The ground zone is the most diverse in structure and is by far the richest in species. It produces the maximum species list for a given amount of time and effort. Hand collecting provides a rapid response towards assessing the natural history value of a site. This method is particularly valuable where a group of people is able to concentrate on one site, on a given day, when a wide cross section of habitats and micro-habitats can be sampled. Grubbing amongst stony ground and short vegetation or sieving litter over a white plastic sheet are the usual methods adopted when collecting in the ground zone. One of the most useful techniques established by Duffey (1972) is the timed collection whereby a set time, ideally two hours but more practically one hour, is spent hand sorting litter or grubbing in a specific micro-habitat, confining the area worked to that particular ecological niche. This system can be used effectively to compare yields from different sites, even in different parts of the country. For longer-term sampling, pitfall trapping is a useful collecting method and has been used widely in the present Leicestershire Spider Survey; but, whilst being convenient, it has limitations and should be used in conjunction with hand collecting in the field layer and canopy. It is also useful to spend a little time hand sorting litter from the ground zone when setting out or servicing traps, since

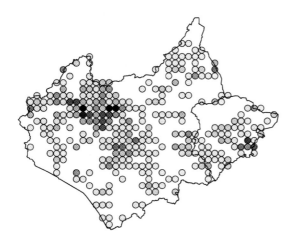

Figure 1 This map shows species coincidence but emphasises also spider collecting activity within the Vice-county. Charnwood Forest has been the most thoroughly worked area and the large number of unmarked tetrads indicates where no spider collections have been made.

some species are less mobile than others and may not be caught in the pitfalls. Pitfall trap-ping is more valuable if continued at the same site for long periods, over at least 13 months to sample seasonal populations. A method adopted less frequently has been extraction of inverteb-rates from ground litter, especially leaf litter and grass turfs, by Tullgren Funnel, a heat extraction process utilising heat from an electric bulb in a closed sheet steel funnel. This is a slow but thorough process used in conjunction with hand collecting and was found to yield very small species such as *Theonoe minutissima*, often missed by other techniques.

Collecting in the field layer has been carried out largely by sweeping but catching efficiency varies considerably with time of day and weather conditions. Hand collecting by observation, on the other hand, is usually very effective in the field layer and produces much more information about individual species than sweeping. Sweeping techniques involve sampling stands of the same vegetation type over as large an area as possible, then reworking productive patches of vegetation by hand.

Sampling in the canopy has been undertaken almost exclusively with the use of a beating tray but brushing tree trunks with a soft hand brush has also been successful. Examination of foliage by hand was the best method of finding the jumping spider *Ballus chalybeius* which spends most of its time in a silken tent-like retreat on the upper surface of oak leaves. Often, beating failed to dislodge *Ballus*, unless individuals were out on hunting forays or involved with courtship. A less efficient way of beating foliage is by using a plastic sheet placed on the ground, but is better than nothing if a beating tray is unavailable.

Another useful method of collecting is by direct observation at night when many species of spider are active and come out of their day-time retreats. A powerful torch is necessary and enables spiders on walls, paths, stonework, fences and tree trunks to be taken in their webs or in the open searching for prey. Stone turning is a useful procedure in quarries and on rocky ground but, if a site is disturbed too often, can be detrimental to species such as *Mastigusa macrophthalma* that prefer embedded rocks. There are other techniques not used in any of the Leicestershire sampling exercises due to lack of availability, as in the case of the Dietrick Vacuum Sampler (D-Vac), or difficulty in maintaining regular visits, such as aerial (sticky) traps and tree-trunk traps.

Pitfall trapping was used on a limited number of occasions during 1963-1973 in the early Charnwood Forest Survey field work, but became the main method of collecting for the 1978 North-east Leicestershire Survey, Loughborough Meadows 1993 Survey, D.A.Lott's and J.Daws' fieldwork 1991-1995, E.Caradine's 1994 wetland studies (Caradine *et al.*, 1995) and the Geeston Quarry study 1994-1995. Details vary, but the general principle has been to sink one or more lines of up to eight traps – 0.5 litre plastic beakers with a mouth diameter of around 8.5 cm, or small glass jam-jars – at various well defined sites. The traps are charged with approximately 3 cm of ethyl glycol to which is added a few drops of household detergent to reduce the surface tension of the liquid and enable small spiders to sink into the preservative. A suitable cover is provided, either made of small-mesh wire netting or a flat lid supported 12 mm above the rim of the jar to protect against rain and prevent small mammals falling into the trap. Contents of the trap are generally unaffected by rainfall and waterlogging since the preservative is heavier than water and remains submerged at the bottom of the trap with the catch; however, if flooding occurs early after setting up, specimens falling into the water disintegrate if left for any length of time. For this reason also, water traps – shallow white plastic dishes 22 cm diameter – require frequent attention. The frequency of emptying pitfall traps in Leicestershire has ranged from weekly to monthly servicing, according to the location, number of traps set and time available to service the trap lines.

Whether collecting is undertaken on a casual or an organised basis, it is necessary to work in a methodical way such that prolonged collecting expeditions can be easily accommodated and are able to incorporate the collections of several other individuals. If opportunities are not to be missed advance preparation is desirable. The basic requirements are an adequate supply of strong glass tubes (50 mm x 16.5 mm dia) with plastic closures, each containing a pre-numbered label (prepared with spirit-proof ink) and half filled with 70% alcohol, a small pocket book with a soft pencil for field notes and a good site map. Full details of the collecting site are entered in the notebook, with date, locality, grid reference, altitude and habitat entered for each site before collecting commences. Then as each micro-habitat is worked spiders are put into the numbered tubes and the number entered in the notebook with details of the micro-habitat and collecting method. Other information such as weather conditions, observations on behaviour, biological details and the collector's name can be added. The importance of full and accurate field notes, made on the spot, cannot be over-emphasised; as also the need to separate collections from different sites, micro-habitats, vegetation types and collecting methods. It is essential to know exactly from where a specimen has been obtained in the event of a follow-up being required.

Overfilling of tubes with spirit is to be avoided; half to a third full is quite adequate since, on a hot day, stoppers can pop off and the contents be lost if there is too much alcohol. Indian ink and good quality paper should be used for the tube number labels and these can be used many times. Bold ink-stencilled labels are ideal so that they can be seen amongst the contents. Pencil labels and poor quality paper should be avoided – if the tubes cannot be identified the contents are useless. It is advisable to record the details for each tube on a separate page of the notebook and to establish a six figure map reference for each tube. These procedures will avoid the frustrations involved in venturing into the field without adequate preparation. The author uses a squared-up shoulder haversack with partitions for tubes and compartments for the other standard collecting equipment (Crocker, 1967).

As soon as possible on return from a collecting excursion the material should be cleaned, relabelled and written up in the permanent notebook. It will not always be convenient to do this straight away and the longer field collections are unattended the more valuable will be the details recorded in the field. Alcohol in the tubes used to kill the spiders will not only lose its strength due to dilution and evaporation but specimens can begin to deteriorate. If a lot of collecting is undertaken, replenishment of preservative is expensive if the alcohol is not recovered; this is achieved by filtering the discoloured spirit into a glass receptacle containing activated carbon. After several days the cleared alcohol can be syphoned off for re-use in the field tubes. Reclaimed spirit should not be used for permanent storage of specimens as the concentration of alcohol will be below the required 70%.

A permanent Collection notebook is central to the organisation of the collection, and can also be used as the work book during identification of specimens to record taxonomic features and evidence for decisions taken. It should be written up in spirit-proof ink as a permanent record of the collection; black 0.3 mm fibre tip pens are suitable for this purpose. At an early point in any serious spider work it is advisable to standardise on a particular size and construction of permanent notebook in the interests of uniformity. This becomes more important as the number of books increases over the years. These need to be small enough to sit comfortably on the work bench and strong enough to stand a lot of rough handling. Hardback notebooks with stitched bindings and narrow ruled pages are ideal. Some people prefer loose-leaf binders but these do not stand up to heavy use and inevitably pages get lost – usually the most important ones!

Many workers keep supplementary notebooks in addition to the Collection notebook, such as a Genus book, a Site Register or a Cross Reference notebook, but nowadays these are being superseded by computer data storage. Even with a computer the Collection book is an essential part of keeping control of the system and should not be taken out into the field. The Collection book will carry a unique serial number for each Site Collection and will contain a fair copy of all details from the Field notebook, including lists of identified species against each temporary tube number, and any cross reference to the work book if this is separate. As material is identified it is transferred to permanent storage, either to the Main Spirit Collection or to the Reference Collection (Crocker, 1969).

Spirit Collections

The Main Collection comprises one or more tubes from each Site Collection in serial number order, each tube containing a label with serial number and location. Large site collections can be broken down as required and the serial number given a suffixed identification. It is useful to include the month and year of the collection in the serial number, for example 782b.9.84. All material from one site, other than that selected for the Reference Collection, is put into individual Site Collection tubes (50 mm x 16.5 mm dia) with plastic closures and containing 70% alcohol. The tubes are placed upright in screw-topped museum jars labelled with the range of serial numbered tubes inside, topped up with alcohol and stored in a standard filing cabinet. Access to the specimens is simple and maintenance easy.

The Reference Collection is intended for individual specimens, and access must be quick and easy. Specimens are stored in species-number order and, except for the larger spiders, are kept in double tubes, one inside the other to minimise preservative loss due to evaporation. Each tube has its own plastic closure and is topped up with 70% alcohol (or other suitable preservative). The inner tube, containing the specimen also holds a detailed label written on good quality acid-free paper in spirit-proof black ink. The label needs to carry all the important information about the specimen and has therefore to be very compact; for this purpose a 0.1 mm fibre tip pen is often used, but the ink must be spirit-proof. If the paper is not of a high quality the ink on the label can disintegrate and will be attracted to the specimen, coating it with a layer of black particles. Labels written in pencil are similarly unsuitable as the writing soon becomes illegible. Each tube must clearly identify the specimen inside and contain sufficient information to establish where and when it was collected, by whom and who identified it. Details should be entered in the Collection book to show that a specimen has been placed in the Reference Collection, and reference back from the specimen label to notes in the Collection book can be followed through the serial number.

Recording

Until 1964, card indexing was universally used for local, county and national recording of spiders. In October 1964, the Institute of Terrestrial Ecology Biological Records Centre issued the first spider recording card which was based on vice-counties and the Ordnance Survey national grid (six figure map references). As described in chapter three, the national distribution of spiders was plotted on a county basis by Dr Peter Merrett of the I.T.E. and published in *British Spiders* Vol.III (Locket, Millidge & Merrett, 1974). In 1964 the Spider Recording Scheme was one of the first of its kind in Britain, but by 1985 participation in this project had fallen off considerably, and the following year the British Arachnological Society in conjunction with the B.R.C. launched a major new initiative, which would culminate in a draft atlas of British spiders in 1997 and publication two years later. This programme is well on target. Nomenclatural changes and additions to the British list necessitated new recording cards which were issued in March 1987. Whilst still being based on the vice-county, the national recording unit was now the 10 km square, as established by Professor A.R. Clapham for the *Atlas of the British Flora* (Perring & Walters, 1962) and subsequently adopted by most British invertebrate atlases. The success of this scheme has enhanced considerably our understanding of the British spider fauna and has been the main stimulus for the present publication.

At county level, a distribution map for Leicestershire (excluding Rutland) was prepared in 1964 using the 1 km square as the recording unit. The card index was replaced by the more convenient Kalamazoo Visible Binder system and special 10" x 6" species record sheets were printed with columns for record number, date, grid reference, sex, frequency, locality, habitat, remarks and collector. The record sheets, each accommodating 42 records, were filed in book form, overlapping vertically so that only the species name along the bottom of each sheet was visible. Plotting at 1 km on the county map proved very tedious and this approach was abandoned in favour of the 10 km plot; but this was also unsatisfactory and eventually, with the advent of computer mapping, the tetrad (2 km squares) was adopted for county recording in

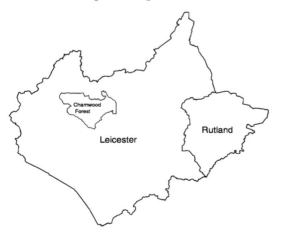

Figure 2 The boundary of Charnwood Forest as defined by the Loughborough Naturalists' Club (Crocker, 1981).

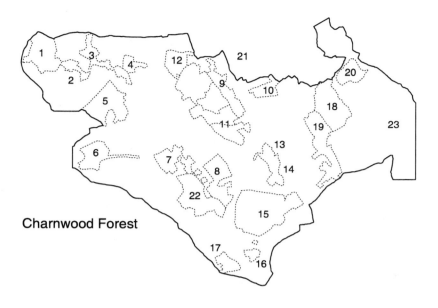

Figure 3 Areas of wildlife interest

line with botanical precedent (Primavesi & Evans, 1988).

By 1964 most of the effort in collecting spiders in Leicestershire was directed at Charnwood Forest, with occasional fieldwork excursions to east Leicestershire and further afield organised by the Loughborough Naturalists' Club and the Natural History Section of the Leicester Literary and Philosophical Society. However, the main thrust of spider survey work during the 1960s and 1970s undertaken by Crocker was at national level on research sponsored by the British Spider Study Group, British Arachnological Society and Nature Conservancy Council. This undoubtedly had a detrimental effect on both the quantity and quality of local spider collecting. Nevertheless, as Director of the Charnwood Forest Flora and Fauna Survey for L.N.C., Crocker was actively involved in all aspects of fieldwork on the Forest, working closely with Leicester Museum and the Leicestershire & Rutland Trust for Nature Conservation. Aerial surveillance and mapping formed an important part of early studies of the area resulting in the preparation of detailed physical/vegetation maps of each unit area of natural history interest. These were drawn with a plotting grid of 0.1 km squares, whilst the overall plotting scale for the Charnwood Forest area was established on a 1 km square grid. The large-scale unit survey maps were provided to all field workers and enabled grid references to be determined in relation to physical boundaries and individual habitat types. This approach helped to unify results from all participants and encouraged the accurate use of grid references.

The 1980s were relatively inactive years as far as spider collecting in Leicestershire was concerned, but as we have seen in chapter three, with the commencement in 1991 of a programme of intensive pitfall trapping for beetles, by D.A. Lott, a new dimension was introduced into local spider survey. Up to this time most of the records for Leicestershire were based on hand collecting and little attention had been given to Rutland. Now Rutland began to receive attention and pitfall trapping expanded rapidly. During 1991, Lott had set out his own traps and employed J. Daws as research assistant to sort out the pitfalls and separate the beetles and spiders. Daws then undertook to identify the mass of spider material so accumulated, and was thereafter also

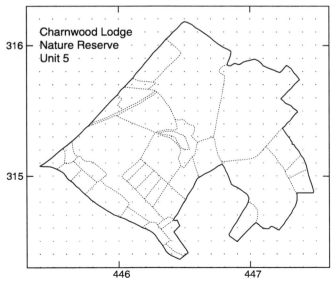

Figure 4 Unit Survey map, 0.1 Km square plot

deployed setting and servicing traplines throughout the county, with particular emphasis on sites on the Lincolnshire limestone and areas of disturbed habitat such as derelict land.

At this time the need for suitable computer recording software for spiders was discussed by the British Arachnological Society Council and, following a survey of the then available programs, it was decided that a simple database front-end utility should be developed to ensure uniformity in the presentation of records. The main drawbacks with the B.A.S./B.R.C. RA65 recording card were the lack of space to indicate sexes, limited habitat categories and the inability to make detailed comments on habitat, biology and collecting method for individual species. These problems were overcome by the development of *SPIREC*, a data input program with a built-in checklist of British spiders, to be used in conjunction with a standard commercial database.[1] This was adopted by the British Arachnological Society and the Biological Records Centre in 1993 as an alternative to the RA65 cards, enabling electronic media to be used for submission of records to the national spider recording scheme; all Leicestershire records were then computerised during 1993/1994. The computer entry screen offers the same fields as the Kalamazoo record sheet, with the addition of fields for type of record (literature, museum and field), a user reference, vice-county name and number, altitude, notes, and determiner.

Voucher Material

Any serious work on spiders requires access to reference material, particularly specimens. Whether such work is of a critical taxonomic nature or involves identification of species for biological research or distribution mapping, the need for accuracy is of paramount importance. Since araneology is a developing science taxonomic changes are inevitable, and it is often necessary to go back to earlier material for re-assessment. It is therefore essential that suitably organised spirit collections are maintained and made available to other workers in support of any published research.

The practice of storing spiders in alcohol is relatively recent but, unless these have been carefully maintained on a regular basis, they dry out and the specimens become damaged or destroyed. Many valuable collections are held in private hands, and have accumulated to present a formidable maintenance task, which can often be neglected due to pressure of other work or oversight. As far as Leicestershire is concerned, these problems are addressed and the background given to action taken to ensure the material on which this publication is based is preserved and available for examination.

Most of the Leicestershire spider material is in private collections, the main body belonging to the authors. During the compilation of this work a large amount of material has been re-examined and where necessary critical species verified by referees. A record of this process of verification is embodied in the computer database where authorities are given. The database will be made available to interested parties and will be integrated into the national databank when this can be arranged. Also, the spirit collections of the authors, as described earlier, containing the majority of post-1961 Leicestershire voucher specimens are being transferred to Manchester University Museum. The remaining material is held in the collections of the Leicestershire Museums' Biology Section, and the private collections of E. Caradine, E. Duffey, I.M. Evans, D.R. Nellist, M.J. Roberts and F.R. Wanless. Access to the collections held at Leicester can be arranged through the Keeper of Biology, Leicestershire Museums, Arts and Records Service, The Rowans, College Street, Leicester LE2 0JJ.

The spirit collection of spiders held by the Biology Section of the Leicestershire Museums Service was loaned to I.M. Evans in January 1994 for the purpose of maintenance and documentation. This material excluded the A.M. Wild spirit collection, the North-east Leicestershire Coalfield Survey material, all post-1990 accessions, and a small number of freeze dried spiders which are retained mostly for display purposes.

The main spirit collection consists of some 2800 tubes, housed in 126 jars and covering a date range 1930-1989, 95.5% of which (2673 tubes) are from the period 1960-1969. These tubes contain individual specimens or series of either the same or a number of species with the same site data. The overall total of specimens, likely to be in excess of 5000, is preserved in 70% ethyl alcohol with glycerine, but about 110 tubes in five jars (4%) show signs of damage due to desiccation at some

---

[1] Further details of SPIREC can be obtained from: Stanley Dobson, Moor Edge, Birch Vale, via Stockport. SK12 5BX

time in the past. In all, an estimated 2150 tubes are from localities in Leicestershire and Rutland and include accidentally imported tropical material from the banana ripening rooms in the centre of Leicester in the early 1960s (Evans, 1963). The remaining 650 tubes are from localities elsewhere in Britain.

Some 912 records have been extracted from this institutional collection and the personal collection of I.M. Evans (mainly 1980s material), and entered into the Leicestershire database from which the distribution maps and species notes published here are derived. All of these records have been verified and it is anticipated that the remaining unrecorded data from this collection will inevitably be of the commoner species in or around the city of Leicester.

In 1966, A.M. Wild transferred his collection of British spiders to the Biology Section of Leicester Museum on indefinite loan. The collection had been made by him during the period 1944-1966, and comprised some 600 'gatherings' (of one or more specimens) representing over 400 species. It contained the voucher material for records in some seven papers (1950-1962), including new county records in his 1952 paper. Wild also published two further papers in 1969 on aspects of spider behaviour . Late in 1982 he made an outright gift of the collection to the Leicestershire Museums Service (accessioned as Z11.1983). He also transferred to the Museums Service his notebooks containing details of all his records.

Several major spider collections are deposited at Manchester University Museum, notably those of G.H. Locket, A.A. La Touche and D.W. Mackie; and more are anticipated including that of J. Crocker which contains the Leicestershire Reference Collection, together with other material from England, Scotland, Wales and the Republic of Ireland. At Manchester there is a comprehensive reference collection of British material and a growing collection of foreign spiders including many specimens from mainland Europe, Uganda, Malaysia and Madeira. The collections are maintained by a part-time curator of arachnids, traditionally nominated by the British Arachnological Society, the post being currently held by Stanley Dobson who succeeded David Mackie in 1984. The spider collections housed at the museum may be examined on site by prior arrangement, or in exceptional circumstances specimens may be posted, subject to status of the borrower. Laboratory facilities, including the use of a microscope and associated equipment are available. To view the collection, contact The Keeper of Zoology, Manchester Museum, The University, Oxford Road, Manchester M13 9PL.

It should be pointed out here, for the benefit of anyone who may be involved with Leicestershire Voucher Material that the Reference Collection species numbers differ from the current species index number given in chapter two. Ideally, these numbers should be identical but the original numbers were based on an earlier checklist and it is too big a task to change these now. After 30 years a revision has been necessary, and the current British Arachnological Society checklist has generated new species numbers which are used in this book and incorporated into the *SPIREC* computer program. For information the obsolete numbering system is given here. The species checklist in *British Spiders* Vol.II (Locket & Millidge, 1953 pp.423-428), lists 581 species in 24 families; these are numbered in increments of 10 from 0010 – *Atypus affinis* to 5810 – *Mengea warburtoni*. The following 'new' species have been added to bring the list represented by the collections up to date. Nomenclatural changes retain their original numbers and the names shown here are the names on the specimen labels.

0982	*Zora silvestris*	3854	*Pocadicnemis juncea*
1285	*Philodromus praedatus*	3952	*Trichopterna cito*
1308	*Philodromus collinus*	4102	*Trichoncus hackmani*
1345	*Philodromus albidus*	4315	*Carorita paludosa*
1576	*Euophrys browningi*	4372	*Micrargus apertus*
2672	*Theridion mystaceum*	4412	*Glyphesis servulus*
2674	*Theridion melanurum*	4622	*Diplocentria saxetorum*
3592	*Dicymbium brevisitosum*	4746	*Praestigia duffeyi*
3615	*Entelecara congenera*	5212	*Centromerus albidus*
3794	*Gonatium corallipes*	5750	*Pityohyphantes phrygianus*
3812	*Minyrioloides maritimus*	6000	*Lasiargus gowerensis*

Living Fossils

Charnwood Forest is well-known for its Pre-Cambrian fossil coelenterate *Charnia masoni*, from 600 million-year-old rocks at Woodhouse Eaves, but it is also important for its 'living fossil' spider *Mastigusa macrophthalma* - a species with an almost identical relative *Mastigusa acuminata* preserved in 30 million-year-old Baltic amber. When this spider was entrapped there was a substantial land-bridge between south-west England and the north European continent which enjoyed a tropical climate. In the Baltic region of Prussia there were forests of pine trees whose resins, which trapped invertebrates, hardened into amber, preserving a perfect fossil fauna. The fossil record includes many species of extinct animals representing unsuccessful experiments in structural innovation, but there are also examples of plants and animals very close to organisms living today. The phenomenon of persistent types is seen, for example, among plants in the maidenhair tree *Ginkgo*, and among arthropods in the king-crab *Limulus*. Representatives of these groups have changed little since the Jurassic period, 150 million years ago. Recognisable ancestors of the 'living fossil' coelacanth *Latimeria chalumnae* flourished in the Devonian seas 350 million years ago, and fossils of this fish have been found in more recent Cretaceous rocks. Likewise, 100 million-year-old lithographic limestone of this period, in south-west Europe, has yielded fossil spiders which are clearly recognisable as members of modern living genera.

Various stages in the evolutionary cycle are evident in the life strategies of species today (see table 14, page 195) and commence with **divergence**, or the splitting of a species into two or more lineages. This is followed by a period of **improvement** where the new taxa adapt to environmental conditions, and then exploit available niches by adaptive radiation, sometimes referred to as '**explosive evolution**'. Often a new cycle of evolution can begin again during this phase. Survivors of lineages in the **stability** phase, where no further evolution takes place, give rise to persistent types or 'living fossils'. Finally, all lineages unable to adapt to environmental changes enter the final phase of **extinction**.

With changing climate, even tolerant organisms succumb to the new pressures to which they are subjected. In the early Pliocene 10 million years ago, the *Ginkgo*, now found wild only in China, was living in Germany, but 9 million years later had disappeared from Europe. Unlike plants, however, invertebrates can take advantage of subterranean habitats to survive the cold, and this may have happened with *Mastigusa* during the Pleistocene period.

During the coldest periods the ice-sheet spread across most of lowland Britain; only the present south-coastal strip of England was free of permanent ice, and tundra conditions extended over northern Europe, to which Britain was joined at that time. Between the main glaciations the climate allowed the re-establishment of open grassland and adventitious woodland, developing into luxuriant forest at the climatic optimum when, for example, hippopotamus lived as far north as Leeds. After the retreat of the last glaciers, 10 000 years ago, the pine and birch forest returned to the south of England, then spread to the Scottish Highlands. This was followed by the return of temperate mixed deciduous forest to lowland Britain, with dense oak/ash forest (*Quercus robur* + *Fraxinus excelsior*) on the damp clays, sessile oakwood (*Quercus petraea*) in siliceous well-drained, often hilly regions, and beech-forest (*Fagus sylvatica*) on and near the chalk. In south-eastern England hornbeam (*Carpinus betulus*) is associated with oak, and like beech, was a late post-glacial immigrant. This climax woodland survived in Britain until early historical times and many great forests such as Windsor Forest, the Forests of Essex, Arden (Warwickshire), Dean (Monmouthshire), Wychwood (Oxfordshire), Wyre (Worcestershire) and Sherwood Forest remained intact throughout the Middle Ages.

Mastigusa macrophthalma m. (x10)

Mastigusa in Baltic Amber

It was of great interest therefore when it was discovered that a species of spider, almost identical to *Mastigusa macrophthalma*, was living in Prussian pine forests 30 million years ago (Wunderlich, 1986). The fossil spider *Mastigusa acuminata* was perfectly preserved in amber and first described by Menge in 1854. However, he mistakenly assigned this hitherto unknown genus and species to the family Salticidae, a fact that was overlooked when, in 1871, Thorell described the living species *Cryphoeca arietinus* (in the family Agelenidae) as new to science. This name is derived from the Latin *aries* (a ram), after the shape of the male palpal lamella and stylus which are coiled like a ram's horn. Simon also overlooked the fossil spider when, in 1884, he erected a new agelenid genus, *Tetrilus* and transferred to it the Palearctic species *arietinus*.

Then, in 1897, Kulczynski described a large-eyed form (*macrophthalma*) of *arietinus* from Hungary. The size and disposition of the eyes of both forms are discussed and illustrated by Bristowe (1939, pp.147 & 167), who states "this difference in the eyes is the only distinguishing feature between the two forms", and suggests that inter-breeding experiments would make an interesting study. Simon (1937 p.1022-1023) asserts that these are separate species, and Locket & Millidge (1955 p.23) preferred to follow Simon in regarding them as such "at least until intermediates are recorded". Roberts (1995, pp.250-251) also follows this argument. Jackson, however, in writing to Donisthorpe (Donisthorpe, 1927) seemed quite certain that *macrophthalma* was only a form of *arietina*, and not a separate species. An examination of Jackson's '*Tetrilus arietinus*' specimens has provided a further piece of evidence: tube 5 labelled "Sherwood *T.recisa* male x" contained 4 males, 3 of which were typical *Mastigusa macrophthalma*, the fourth had an irregularity in the posterior median eyes, which were of different size (right p.m. =*arietina*; left p.m. =*macrophthalma*). Lehtinen (1964) also, has recorded intermediate forms from Finland, from the inner parts of the nests of the ant *Formica rufa*. After examining spirit material at the British Museum (N.H.), and looking at living specimens from Charnwood Forest, Wanless (pers.comm. 17.10.1972) was of the opinion that there does appear to be inter-specific variation. Wunderlich (pers.comm. 28.5.1985) was more assertive after examining Charnwood specimens of '*macrophthalma*' and comparing them with continental specimens of *arietina*: he states "the species name is not *macrophthalma* but *arietina* (*macrophthalma* is a species from the south-east of Europe only!)" i.e. Hungary and Yugoslavia. Certainly the two forms which occur in Britain are distinctive and easily separated on the arrangement and size of the eyes, therefore it is considered useful to retain the individual names for the time being, in order to differentiate field records, until the matter of specific integrity is conclusively resolved. This will involve examination of all the types and intermediates.

The difficulty in studying these extremely 'rare' spiders has been in obtaining living material. *Mastigusa arietina* is recorded in Britain only from Surrey (Oxshott, Weybridge & Woking), Berkshire (Windsor Forest), Somerset (Wellington College), Glamorgan (Craig-yr-Eglwys) and Cumberland (near Carlisle). These are very old records of isolated occurrences and the actual sites are not known. Much more is known about *Mastigusa macropthalma*, which is associated with ancient forest sites at Sherwood Forest, Donington Park and Charnwood Forest, one-time continuous pre-historic climax woodland. There are also old records from Surrey (Oxshott, Weybridge & Woking) and Somerset (Porlock & Wellington College), but not from Glamorgan (Jackson's record is of *arietina*, not *macrophthalma*).

The early records of *Mastigusa*, in Britain, were mostly from ants' nests, but Donisthorpe (1927) also includes records of 3 females collected by others around the turn of the century: *arietina* on railings near Carlisle, and under a stone in Glamorgan; and *macrophthalma* swept off heather in Sherwood Forest. There is doubt about the Carlisle specimen which was collected in 1892 (before Kulzynski's description of *macrophthalma*), but the other two are Jackson's determinations and are sound. More recently Roberts (1985) has collected *macrophthalma* from grass tussocks in Sherwood Forest; and on Charnwood Forest it has been taken in pitfall traps in acid grassland, and amongst grass roots and leaf litter at various sites. Otherwise, the typical micro-habitat is amongst dead wood and litter inside dead and dying oak trees; also under rocks and stones, often embedded in peaty soil (Crocker, 1973). Only 15 of the 45 records for *macrophthalma* on Charnwood Forest are from ants' nests (*Formica fusca*), though foraging ants are often present with free-living *macrophthalma*. With the exception of the Carlisle record, all British *arietina* have been from deep inside ants' nests, and this suggests a caverniculous affinity rather more strongly than in the case of *macrophthalma*. *Mastigusa arietina* is known to be a cave dweller on the continent (Bristowe, 1939 p.147). A further observation on the free-living habits of *macrophthalma* is of interest. At a new site for this species at the edge of Charnwood Forest, a female was taken in a pitfall trap in rough, unmanaged, rocky acid grassland adjacent to a former granite quarry at Markfield. The site is typical of many on Charnwood with rocky outcrops, large granite boulders and oak scrub, and with no known history of land use other than rough grazing. The quarry, known locally as 'Hill Hole', was abandoned in the 1920s and was quarried at two levels; the first a wide area of shallow workings forming a plateau surrounded by a sheltered and vegetated quarry face on three sides. Some time around the end of the nineteenth century a deeper excavation was commenced at the open end of the plateau, extending to about a third of the plateau area. This deeper quarry has now filled with water, and part of the plateau is used as a tailings (quarry waste) dump, forming smooth sparsely vegetated mounds. The remainder of the plateau now has a varied but sparse calcifuge flora with gorse/bramble scrub, and a large number of various sized rocks around the perimeter, which have fallen from the old quarry face. On 13.5.95 two females and a male *macrophthalma* were found under some of the smaller of these rocks which were embedded in grass. The significance of this is twofold, in that the species was obviously established in this relatively small relict of ancient Charnwood, but more so in that it had recolonised a new site which is only 100 years old.

Crocker (1973) records *Mastigusa macrophthalma* from five locations within Charnwood Forest: at Bradgate Park, Bardon Hill, High Sharpley & Gun Hill, Ives Head and Beacon Hill, and a further site at Donington Park. Since then this 'living fossil' spider has been discovered at a further three sites, all relicts of the ancient climax forest: at Buddon Wood, Hill Hole and Altar Stones. It may be of some future significance to record here that the plates (Roberts, 1985 Vol.1 plate A p.167 and Vol.3 plates 95 & 96a) and line drawings of the epigyne and palp (Roberts, 1995 p.251) of *Mastigusa macrophthalma* are of Charnwood Forest specimens.

Lepthyphantes midas

Lepthyphantes midas (Simon) is an extremely rare ancient woodland species with very few records, and never found in any numbers. It is known only from Europe. Until the synonymy was established in 1971, with the discovery of both sexes and subadults at Donington Park, Leicestershire, the species was known from the syntype females *L.midas* (Simon, 1884) from Seine-et-Marne, Fontainebleau Forest, ancient climax forest in France, and the holotype male *L.carri* (Jackson, 1913) from Sherwood Forest. Other early records are: a female (*midas*) collected by J.Braendegaard from Fortuens Indelukke State Forest, Denmark, in May 1923 (Braendegaard, 1932); and a male (*carri*) collected in Windsor Forest, November 1928, from a jackdaw's nest by H.St.J.Donisthorpe (Jackson, 1932). A male (*carri*) was collected by E.C. Rosenberg from a bird's nest in Jaegersborg Dyrehave (deer park), Denmark, October 1944 and identified by S.Langemark. No further details are available, but this specimen was not necessarily from Fortuens Indelukke (pers.comm. O. Boggild 9.7.1976). Fortuens Indelukke is an enclosure within Jaegersborg Dyrehave, which was established as a royal hunting park in 1670, and is now a state forest. It is an ancient beech/oak wood with open commons having groups of trees and single trees. The trees were allowed to stand until they fell of old age, there being no economic management of the woodland before state control, however there was (and is now) a large population of fallow, red and sika deer which browse on any seedling trees, resulting in open grassland between isolated old trees. However, modern forestry management has removed nearly all trace of the past, as all trees in decay or bad health are immediately removed (pers.comm. B.O. Nielsen 1.11.1977; S. Toft 4.11.1977). The next record is of a female *midas* and two juveniles from the Swietokrzskie National Park, ancient fir/beech forest (*Abletetum polonicum + Fagetum carpathicum*) in Poland, collected by S. Pilawski, July 1960 (Pilawski, 1966). The juveniles were not taken with the female, and were from an atypical site, so should be discounted.

On August 8th 1971 two females, a subadult female and a subadult male *Lepthyphantes midas* were found on pieces of dead wood inside the hollow trunk of a very old, decaying pedunculate oak tree at Donington Park, Leicestershire (SK4126) by J.& M.Crocker. A male *Lepthyphantes carri* was also taken in litter under a piece of dead bark lying on the ground close to the base of the same tree. Further visits were made to this tree in July 1972 and additional specimens obtained (a male and a subadult male under dead wood on ground against tree), June 1973 (subadult female from leaf litter under piece of dead wood on ground near tree), August 1977 (female in sheet web inside cavity of rotten root limb of the same tree), September 1978 (female on the underside of a piece of dead wood inside the hollow trunk of this tree, and a subadult male in loose dry litter around the piece of wood, also a suspected but unconfirmed immature male in the same place), and finally in September 1979 (subadult male in a mesh cage containing loose oak leaf litter and dead wood from inside the hollow trunk, also an adult female on underside of piece of dead wood lying against this cage and completely covered with debris. The female was in a small sheet web spun over a depression in the wood). Living subadults of both sexes are clearly recognisable; the male by excrescences on the outside of the papal envelope, also by their reddish colour - though not as rich as in the adults. Subadults have been reared through to maturity in captivity. The colour of living adults is a rich reddish-brown with clearly defined sooty grey-black abdominal markings, but the spiders are well camouflaged against the red rot-wood of decaying oaks. Synonymy is established on the basis of the Donington Park population, where both sexes and subadults have been taken together in a long-established colony, and by comparison with Simon's syntypes and Jackson's holotype. Specimens figured by Roberts (1987 Vol.II, text figures 81b, p.155 - *L.midas* epigyne and palp; and plate A, p.156 of the female) were drawn from Donington specimens. A male and a female *Lepthyphantes midas* from Donington Park are lodged in the British Museum (N.H.) collections [BMNH 1977.9.5.2/3], together with a further female [BMNH 1977.9.5.4] collected by A. Russell-Smith in Epping Forest.

Several weeks after the discovery of *L.midas* in Leicestershire, Russell-Smith (pers.comm.) collected the female, referred to above, from the remains of a squirrel's old drey in a pollarded hornbeam at Epping Forest, near Loughton (TQ4196). A further female was found by Russell-Smith in a similar situation, September 1972; and in 1980 between May and July, he carried out a survey in Epping Forest and at Hainault Forest, Essex, yielding a further 6 males and 13 females from Epping and a male from Hainault. Subadults collected in May and June all matured within 4–16 days. Samples (squirrels' dreys, birds' nests and leaf litter from holes and crowns of trees) were taken from 30 hornbeam, 22 oak and 6 beech trees, yielding specimens of *L.midas* from 7, 4 and 1 of the samples respectively. In all of these samples collembola were common, and probably the main prey. The 12 samples containing *midas* were from the following micro-habitats:

Squirrels' dreys	6
Pigeons'/Jays'? nests	3
Thrushes nest	1
Leaf litter	2

Four other records of *Lepthyphantes midas* complete this summary of known occurrences. In May 1978, during a British Arachnological Society survey in Windsor Forest, a female *midas* was collected by P.E. Jones from rotten wood, beech leaves and moist humus inside a hollow (living) beech tree at SU927748. Then later the same year, in September, a female was taken at Sherwood Forest in an artificial nest made from a plastic 'string' bag filled with sterile wood shavings, previously placed inside a hollow oak tree by L. Bee during May (Crocker, 1979). A further female was collected at Sherwood the following year by Bee from a jackdaw's nest in one of the old oak trees. Since then three female *midas* have been recorded from Czechoslovakia by Vlastimil Ruzicka (pers.comm. 26.2.1988), from the Trebon Basin in South Bohemia - an area of old waterlogged forest with oak/lime communities on drier ground. The first specimen was collected between May and August 1986 in a pitfall trap placed inside a hollow linden tree *Tilia* sp., and two more specimens between May and September 1987 from pitfall traps in hollow pedunculate oak trees *Quercus robur*. All of the above sites are relicts of ancient forest and have an unbroken succession of vegetation, albeit much altered during the last 500 years.

Donington Park is an ancient woodland site in the final stages of decline; even in the eleventh century this was an area of decaying wood pasture, out of which the Park was later created. The centuries of close grazing by deer and rabbits is evident, in that there has been no natural regeneration of the old trees. There is a lack of young tree succession and no protective shrub layer. Occasional replacement trees have been planted close to dead trees within recent years, but this does not redress the need for continuity of all stages of tree cover to ensure cross-over habitats for invertebrates. There are still many venerable old pedunculate oak trees *Quercus robur,* but most are in the final stages of decay. The '*midas*' tree is still intact with sufficient living wood to support a small spread of leaves, but the danger of complete collapse in heavy winds or winter snow storms is very real. All these ancient trees stand in isolation, and despite positive endeavour to discover *midas* populations away from this single tree, no further evidence of its survival elsewhere has been found.

In Sherwood and Windsor Forests *L.midas* has been equally elusive with only single specimens turning up after concentrated effort by experienced survey teams. In Epping Forest, however, the species would appear to be well established and more mobile.

Sherwood Forest is an inheritance of the Forest Law of the Normans, possibly with direct connections with the great oak forests which extended throughout Britain in post-glacial times. Its exact bounds, including most of the central part of Nottinghamshire, were laid down in 1232, and as late as the sixteenth century covered an area around 50 000 ha (24 miles by about 8 miles) between Nottingham, Mansfield and Worksop. Today the private and public parks of Clumber, Thoresby, Welbeck, Rufford and Bestwood occupy some of the original forest land, and the old forest is now reduced to 350 ha. Here, the oldest stag-headed giant oaks are over 300 years old. The next generation are 150-200 year old replacements for the timber felled during the 18th century. These are followed in succession by 80-130 year old trees naturally regenerated after pannage ceased, whilst the youngest generation are seedlings regenerated during the dearth of rabbits thirty years ago caused by myxomatosis. The over-mature and dead trees are of particular importance for many specialist invertebrates such as the spiders *Mastigusa macrophthalma*, *Lepthyphantes midas* and the pseudoscorpion *Dendrochernes cyrneus*.

Windsor Forest is largely beech and oak, growing much as they have done since at least Norman times, whilst Epping Forest - part of the original Waltham Forest - is regenerating beech/oak wood with hornbeam, which have been pollarded for centuries. Although the trees in the Loughton area of Epping Forest are rather small and probably of no great age, the forest has had a continuous history of woodland cover and centuries of commonage. Originally a Royal Forest, it passed, in 1878, to the London Corporation who have managed the woodland as a public open space for over 100 years.

Lepthyphantes beckeri

Unlike the previous two species which are 'living fossils' *Lepthyphantes beckeri* would appear to be a new 'divergent' species in the *tenuis* group, closely related to *L.flavipes* and *L.mengei*, or it could possibly be an aberrant *L.cristatus*.

Only two specimens are known to science, both females collected in March, in open areas of ancient deer parks, and both carrying a well developed ectoparasitic larva. In each case, diligent searching after the discoveries failed to produce further specimens. The first female was collected by Professor K.Becker from an old open-woodland nature reserve on the outskirts of Berlin, in a BARBER-trap put down in a small clearing in the Park during March 1953. Despite further searches throughout 1972 no more specimens were encountered. A description of this female was published (Wunderlich, 1973) with figures of the epigyne and vulval structure, and named as a new species after Becker, the male being unknown. On 22nd March 1970, J.& M.Crocker collected an unknown female *Lepthyphantes* from leaf litter around the base of a low brick wall surrounding a sunken lawn, within the Bradgate House ruins (SK535102), near Leicester.

The abdominal pattern of this Bradgate spider is similar to that of *Lepthyphantes mengei*, but much less distinct; it has the same basic pattern but with less pigmentation, so that the 'chevrons' are not as well defined. The taxonomic position is difficult to establish without the male, and follow-up collections of litter, from the original and adjacent sites, were made over a period of several years in an attempt to secure more specimens, but none were found. On the basis of external morphological characters and the vulval anatomy (Wanless, 1973) of this single female, it is considered to be closely related to *Lepthyphantes mengei* Kulcz, 1887 and *Lepthyphantes flavipes* (Blk., 1854), and is temporarily placed between the two in the systematic list. *Lepthyphantes mengei* and *Lepthyphantes flavipes* are both recorded from Bradgate Park, the latter often in litter around the base of old oak trees, the former more commonly, and at the same site as the new species, and in the same micro-habitat.

With Wunderlich's description of *Lepthyphantes beckeri* it became clear that the specimens from Berlin and Bradgate Park were the same species. Although there are small differences in the positions of trichobothria and tibial spines, these are within the normal range of variability.

Tm I Berlin: position = 0.22
Tm I Bradgate: position = 0.31
Tib II Berlin: 1 retrolateral and 1 prolateral spine
Tib II Bradgate: 1 retrolateral spine only, approximately level with the dorsodistal spine.

The specific integrity of this taxon will be in question until males are found and breeding populations discovered. Possibilities exist of these females being hybrid freaks or subject to biochemical modification by the parasite. Even if no more specimens are found and the questions remain unanswered, these are important and intriguing scientific discoveries.

Description of *Lepthyphantes beckeri* Wunderlich, 1973 - new to Britain.

Female: Bradgate Park, Leicestershire, SK535102. Collected 22.3.1970 from leaf litter around base of brick wall of Bradgate House ruins, J. & M.G. Crocker. Specimen with ectoparasitic larva. British Museum accession number: 1977.9.5.1

Overall length: 2.3 mm over spinners

Abdomen: Shape typical of genus, tapered posteriorly, overall length 1.8 mm. Very pale yellowish white with ill-defined grey-brown markings (vestigial chevrons) on dorsal surface, and dark flanks (which carry whitish patches in the medial and posterior half of the abdomen); similar to light coloured specimens of *L.mengei*.

Carapace: length 0.98 mm. Pale yellow-brown with darker fovea and feint striae. Darker inverted triangle anterior to fovea, apex touching fovea, base anteriorly, base line drawn out medially towards posterior median eyes, and feint lines from ends of base line to posterior lateral eyes (*cf. L.flavipes*), markings grey-brown. Carapace with blackish margin dorsally.

Cleypus: slightly concave, a.m. eyes projecting over cleypus.

Eyes: large, on black spots (*cf. L.cristatus*), anterior row strongly recurved, posterior row much less so. Posterior median eyes largest, 0.5 d apart and about same from posterior laterals. Anterior median eyes smaller than anterior laterals, c.0.5 d apart and 0.75 d from laterals. Laterals eyes almost touching and about equal in size.

Sternum: broader than length, light brown evenly suffused with black and having a blackish margin.

Chelicerae: Outer margin with three teeth; inner margin with one small tooth and four very small teeth, all in a closely grouped row towards the fang base.

Legs and palps: yellow-brown

Legs long and slender, suffused with black at joints.

Femora I with prolateral spine position = 0.55, all patellae with one long distal spine and one short proximal spine. All Mt with one dorsal spine, Mt I spine position = 0.31. Tm IV absent, Tm I position = 0.22. Tib I-IV with two dorsal spines, Tib I proximal spine position = 0.29, apical spine position = 0.66. Tib I with a further pair of lateral spines (1 retrolateral & 1 prolateral approx level with dorso-distal spine). Tib II with retrolateral spine only, approx level with dorso-distal spine. Tib III-IV without lateral spines. Spinal armature similar to *L.flavipes*.

Epigyne and vulva: Scape parallel, longer than width, with deep incisure distally. Stretcher long and prominent. Seminal receptacles widely separated (*cf. L.flavipes*) with seminal ducts lying roughly parallel to scape and disposed similarly to those of *L.flavipes*. Photographs and drawings of genitalia prepared by F.R.Wanless.

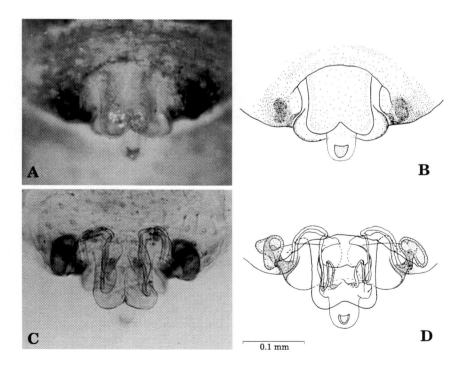

Epigyne (A, B), and vulva (C, D) of *Lepthyphantes beckeri*

CHAPTER FIVE

Atlas of Leicestershire and Rutland Spiders

The distribution maps are laid out three to a page in check list order (see chapter two), and include several blank maps at the end of the chapter, for the addition of future 'new' species. A map with an open circle is included for *Araneus marmoreus*, for which there is only one historical record (Rowley, 1897) localised to Owston Wood, and a blank map for *Micrommata virescens* for a record from the same source with no location, since both these species could yet be re-discovered in east Leicestershire woodland.

Not all records of Leicestershire spiders can be plotted on the distribution maps (see table 5, page 55) because many have insufficient localised detail, but all records in table 6 (page 57) are included in the atlas. Old records between 1948 and 1959 are shown as open circles, and all post-1959 records are shown as solid circles. These are entered as six or four figure national grid references (i.e. SK494134 or SK4913) and are plotted on a 2 km square (tetrad) grid within the national framework of 10 km squares. Summaries of all first occurrences in the county are listed in chapter three, tables 5 and 6.

Spider Biographies

A short biography is given against each distribution map relating to the status of each species in Leicestershire. Space does not permit descriptions of the species, and the reader is referred to dedicated books on identification such as Roberts (1995) *etc*. Where helpful, a comment may be made on a spider's wider status, but generally, comments are restricted to the distribution of the species within the county. Likewise, the reader is referred to chapter two for the national status, since a nationally common species may be local or even rare in Leicestershire. Habitat details and phenology are also based on Leicestershire records and, where space permits, locations are given for the less common spiders. Since tables 5 and 6 in chapter three list new county records in date order, these are repeated here in species order for ease of reference.

The Maps

Field records have been entered into the computer via SPIREC software (see page 68) and transferred to the main database in blocks of 200-300 records. SPIREC generates all current species names, authorities, B.R.C. reference numbers, checklist index numbers and sequential record numbers, and there are facilities for copying the whole or part of the previous record into a new record. The main database is used for data manipulation and information retrieval such as mapping files. The mapping file consists of lists of grid references for each species, headed by the species index number, and is compiled as an ASCII file which can be read and updated using the standard DOS editor or any word processor. Distribution maps and text figures in chapters four, five and six were generated by the mapping software program DMAP for Windows[1] in the form of Encapsulated PostScript (EPS) files. These were incorporated into the relevant DTP pages, for output at 600 dpi. The outlines and isolines for maps in chapter one were produced using the GRUTIL facility of DMAP. Firstly, the isolines were digitised from original maps using a graphics (digitising) tablet. These were then combined with the digitised county boundary and displayed in DMAP. Finally the resulting maps were imported into a graphics program where the different sections of the maps were filled with various infill patterns and map text added.

Rare Spiders

The 32 species listed in tables 7 and 8 can be considered to be very rare in Leicestershire since they are known only from single records, or from a single site. Whilst this is understandable for the local, notable and Red Data Book species, it is less obvious why most of the 10 nationally 'common' species (marked with an asterisk) are so under-represented in the county. Future records for these species will be viewed with assiduity.

---

[1] Further details of DMAP can be obtained from Dr Alan J. Morton, Blackthorne Cottage, Chawridge Lane, Winkfield, Windsor, SL4 4QR

Table 7: List of species only known in the county from a single record.

Clubiona subtilis	Seaton Meadow	L	f	1994
Micrommata virescens	old record (Rowley, 1897)	L		
Xysticus lanio	Sheet Hedges Wood	C*	f	1978
Ozyptila scabricula	Geeston Quarry	Nb	f	1995
Philodromus albidus	Essendine churchyard	Nb	m	1994
Pardosa monticola	Ives Head, Shepshed	C*	4m	1976
Trochosa robusta	Ketton Quarry	Nb	m	1993
Tegenaria saeva	Leicester	S*	f	1989
Araneus marmoreus	old record, Owston Wood (Rowley, 1897)	L		
Agalenatea redii	Ketton Quarry	L	fm	1995
Walckenaeria capito	C.L.N.R.	L	f	1964
Moebelia penicillata	Outwoods, Loughborough	C*	m	1963
Baryphyma trifrons	Ulverscroft N.R.	L	m	1967
Minyriolus pusillus	Norris Hill, Ashby Woulds	C*	f	1993
Thyreosthenius biovatus	old record, Buddon Wood (Donisthorpe, 1927)	L	ffmm	
Porrhomma egeria	Bardon Hill	L	f	1970
Meioneta innotabilis	Burbage Wood	C*	m	1994
Lepthyphantes beckeri	Bradgate Park	RDBK	f	1970
Pityohyphantes phrygianus	Pickworth Great Wood	L	f	1994

Table 8: List of species for which there are several records from a single site.

Psilochorus simoni	Coleorton Hall	S*	established colony	1965/1993
Haplodrassus silvestris	Buddon Wood	Nb	fm+subf	1975/1977
Heliophanus cupreus	The Brand & Swithland Wood, adjacent sites	C*	ff	1966
Ballus chalybeius	Buddon Wood	L	ff	1963/1995
Alopecosa barbipes	Geeston Quarry	C*	ffmm	1994/1995
Tegenaria parietina	Leicester (old records)	S*	mm	Mott, 1890 *etc.*)
Tetragnatha striata	Groby Pool	Nb	ffm	1971/1975
Pelecopsis nemoralis	Swithland Wood	L	ff	1967/1968
Monocephalus castaneipes	C.L.N.R.	L	ff	1963/1964
Gongylidiellum latebricola	Ketton Quarry	L	mm	1994
Porrhomma convexum	Swithland Wood	L	fm	1967/1968
Agyneta cauta	Bardon Hill	L	ffm	1962
Lepthyphantes midas	Donington Park	RDB2	ffmm	1971/1979

Whilst on this subject it is perhaps permissible to conjecture upon the overall number of species which may be expected to occur in Leicestershire. In the long term a further 16 non-linyphiids and 14 linyphiid spiders could be added to the county list. Merrett (1990), in his review of the British spider fauna, comments upon the ratio of linyphiids to non-linyphiids, and quotes figures for some selected well-worked counties in different regions of England and Scotland. In essence, these show an inverse progression from large numbers of species and lower percentages of linyphiids in the south, to lower numbers of species and higher percentages of linyphiids in the north. It is interesting to see that Leicestershire produces a 'good fit' in these statistics for both the known spider fauna and the conjectured optimum.

	Species	Linyphiids	% Linyphiids
South of England	416	168	40.3
Leicestershire (actual)	326	156	47.8
Leicestershire (optimum)	356	170	47.7
North of England	345	182	52.7
Scotland	213	126	59.1

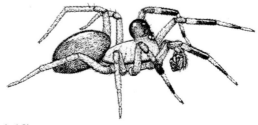

Peponocranium ludicrum m. (x16)

CHAPTER FIVE

Atlas of Leicestershire and Rutland Spiders
AMAUROBIIDAE

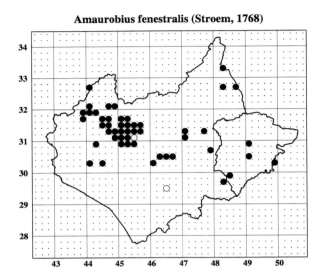

Amaurobius fenestralis (Stroem, 1768)

Amaurobius fenestralis is widespread and common throughout Leicestershire and Rutland particularly in rural areas, where it occurs under bark – often associated with *Oonops pulcher*, under stones and rocks, in rural artifacts, piles of building materials, leaf litter and rubbish, walls, fences, *etc.*, also in dense bushes such as gorse. Adults have been taken in every month of the year. The species is more common than the map suggests and will be found in many of the unrecorded squares by careful searching. It needs to be separated from *A.similis* with which it can be superficially confused, as the two species are sometimes found together. First recorded by Chalcraft (Rowley, 1897).

Distributed all over the British Isles, **Amaurobius similis** is similar in appearance to the above species and is likewise widespread and common, but seems to have a greater association with human activity. In the vice-county it is found throughout the year in domestic and other buildings, churchyards, old ruins and man-made structures, also under stones, in walls, crevices, behind bark, amongst rubbish and on shrubbery. It is often found in company with *A.fenestralis* in rural situations, and *A.ferox* in houses. *A.similis* probably occurs in every building in Leicestershire and Rutland but actual records do not satisfactorily reflect this distribution pattern. First record by Bristowe (1939).

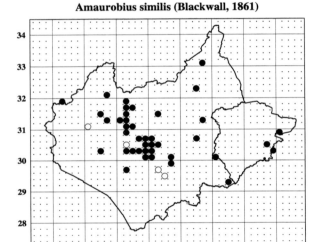

Amaurobius similis (Blackwall, 1861)

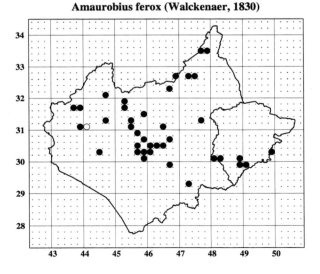

Amaurobius ferox (Walckenaer, 1830)

This spider is larger and darker than the two preceding species and is found in similar situations to *A.similis*. **Amaurobius ferox**, first recorded by Chalcraft (Rowley, 1897), is encountered less frequently than *A.similis* but is just as widely distributed, with most records in the spring and summer months; females occur throughout the year. Amongst recorded localities are: Belton, Bisbrooke, Glaston, Heather, Leicester, Loughborough, Little Dalby, Mountsorrel, Redmile, Seaton, Shepshed, Swithland and Wardley Wood. All three species construct similar retreats in holes or crevices with untidy webs around the entrance of the retreat. The flocculated silk has a bluish tint when fresh.

Atlas of Leicestershire and Rutland Spiders
DICTYNIDAE

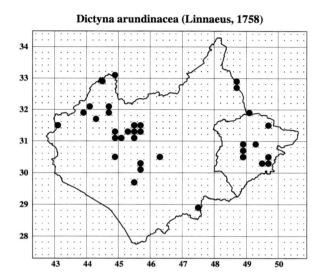

Dictyna arundinacea (Linnaeus, 1758)

Dictyna arundinacea is common and widespread. Adults May – September with peak activity noted in May and June. It is found in woodland, rough grassland, marshes and reed beds in the field and shrub layers where the webs are constructed on heads of dead and dry vegetation, such as tall herbs, rough grass, heather, small shrubs and gorse. Large numbers occur in the early summer. The species was first recorded in the county by H.A.B. Clements in 1962 at Shepshed, and has since been found at typical sites such as Aylestone Meadows, Botcheston Bog, Buddon Wood, Cademan Moor, Ketton Quarry, Lockington Marsh, The Drift and Ulverscroft Nature Reserve.

Widespread and common, but not as frequent as the previous species, ***Dictyna uncinata*** is found in similar situations, including gardens, on shrubs, hedgerows and young trees. Occasionally *D.uncinata* has been noted on prostrate rockery plants where they have been spun up in the dead flower heads of the previous year. Peak activity, with greatest numbers of both sexes, occurs in May and June. Adult males are active in April and females have been taken as late as October. First recorded for Leicestershire by Bristowe (1939) the species has also been taken at Burbage Common, Geeston Quarry, Luffenham Heath, Narborough Bog, Sheet Hedges Wood, Swithland Reservoir *etc*.

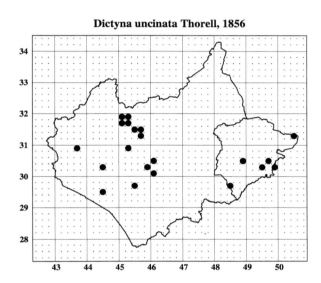

Dictyna uncinata Thorell, 1856

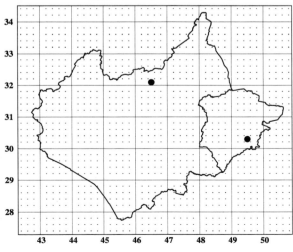

Dictyna latens (Fabricius, 1775)

Dictyna latens is established in Leicestershire on the basis of five records. It is found in the same situations as the previous two species, particularly on heather, gorse and other low vegetation, but is much less common. Rare in the north of England and recorded from only two sites in the county – Luffenham Heath and Six Hills – *D.latens* was first collected in 1994 by J. Daws at the former site. It is adult in spring and summer but only females have been taken (June & July) in Leicestershire. The specimens were obtained by sweeping stable rough grassland with tall herbs, sweeping field layer vegetation in herb-rich heathy grassland with scrub oak/gorse, and beating oak/gorse scrub.

CHAPTER FIVE

Atlas of Leicestershire and Rutland Spiders
DICTYNIDAE, OONOPIDAE

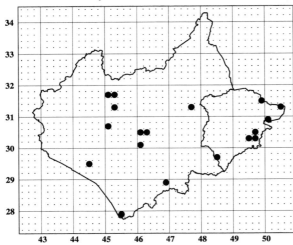

Lathys humilis (Blackwall, 1855)

Though widespread, this species is very local in England and Wales. In Leicestershire ***Lathys humilis*** is known from 20 sites in 18 tetrads. First recorded by J. Crocker in 1967 from Swithland Wood, *L.humilis* has since been found at Burbage Wood, Burrough Hill, Eyebrook Reservoir, five churchyards, three Leicester gardens, Ketton Quarry, Knighton Spinney, Luffenham Heath, Martinshaw Wood, Pickworth Great Wood, *etc*. All specimens were obtained by beating trees and bushes – oak, hawthorn, gorse, garden shrubbery, woodland rides, yew, blackthorn and pines. Reaching maturity in spring and summer; females taken in May–July, males in May when large numbers were noted.

Oonops pulcher is widespread and common, particularly in the Charnwood Forest area. The species was first record by E. Duffey in 1953 at Beacon Hill. Adults of both sexes have been taken in all months of the year except January, with peak numbers in March and August–September. Found under bark, in dry-stone walls, under piles of tiles, bricks and stones, in ants' nests under embedded rocks, in oak leaf and pine needle litter and in old *Amaurobius*, *Segestria* and *Coelotes* webs. Habitats range from deciduous and mixed woodland, conifer plantations, farmyards and farm buildings, derelict industrial sites, rocky heathland, old deer parks, rubbish dumps and gardens.

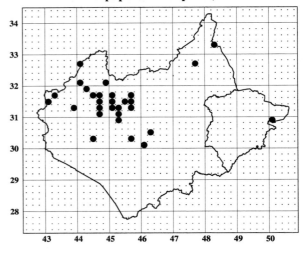

Oonops pulcher Templeton, 1835

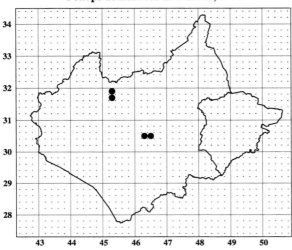

Oonops domesticus de Dalmas, 1916

Oonops domesticus is associated with man, inside houses and other buildings. Adults have been recorded throughout the year with males between August and January. A male was found outdoors in January, inside a pile of plant pots stacked against the side of a house. First recorded for Leicestershire by Bristowe (1963) and subsequently from only four separate houses. This small, pink, elusive spider is more active at night than in the daytime and is more widespread and common than present records suggest. It has been recorded inside houses (bathrooms, toilets and kitchens) on door panels and walls, inside cupboards, on window ledges, on wall mirrors, in hand basins and behind tiling.

Atlas of Leicestershire and Rutland Spiders

DYSDERIDAE

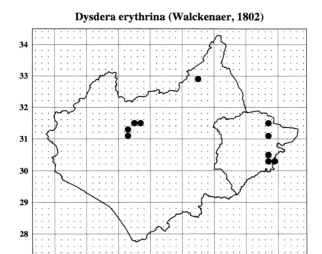

Dysdera erythrina (Walckenaer, 1802)

A distinctive species which, together with the closely related *D.crocata*, is adapted to feed on woodlice. ***Dysdera erythrina*** is widespread but local in Leicestershire, occurring in short permanent grassland, often with ant hills, rabbit-grazed short turf, disused railway tracks and in ancient woodland with an open canopy and grassy floor. Found under stones in grass and amongst grass roots, particularly favouring sparse vegetation on otherwise bare ground (Crocker, 1966). Adult males have been taken April–November and females April–October with peak activity noted May–June. First recorded from Bradgate Park (Crocker, 1962). Also from Buddon Wood, Swithland Wood and Pickworth Great Wood.

Whereas the previous species seems to prefer natural habitats, ***Dysdera crocata*** is mainly synanthropic in Leicestershire, being found in domestic, industrial and farm buildings, on stony waste ground, refuse tips, in disused quarries and churchyards. It has been taken inside houses and on the outside walls of buildings, in silk cells under stones in short calcareous grassland and in 'colonies' under rubbish. Though widespread it is nevertheless met with infrequently. Adults have been taken March–November. *D.crocata* is more robust than the previous species and in direct competition could displace *D.erythrina*, however they currently co-exist in Ketton and Geeston quarries. First record Mayes (1934).

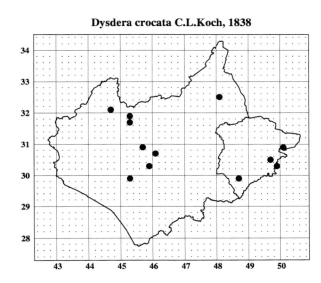

Dysdera crocata C.L.Koch, 1838

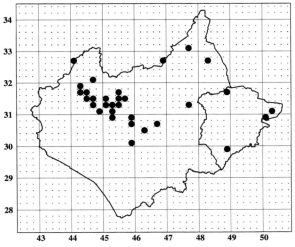

Harpactea hombergi (Scopoli, 1763)

Harpactea hombergi is widespread within the county but local, except on Charnwood Forest where it is common. Adults recorded throughout the year: males January–October and females March–December. Found in woodland, parkland, rocky acid grassland, heathland, moorland and reservoir margins, also disused quarries, buildings, stone walls, rubbish tips, farm yards, gardens and churchyards; usually in silk cells under bark of old trees and in crevices, under stones and in holes in walls. Also found under dead wood on the ground, from litter inside hollow trees and from bark traps; from grass roots between rocks, in leaf litter, under rubbish and inside buildings. First record, Lowe (1912).

Atlas of Leicestershire and Rutland Spiders
SEGESTRIIDAE, PHOLCIDAE

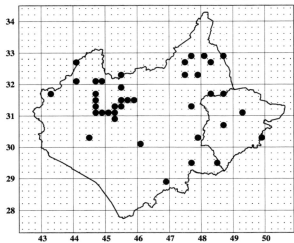

Segestria senoculata (Linnaeus, 1758)

This distinctive and unmistakable spider is common and widespread in Britain, and is certainly under-recorded in Leicestershire. First recorded by Wild (1952), ***Segestria senoculata*** is found in similar situations to the previous species. Adults of both sexes have been taken March–November, with main activity of males May–September. The characteristic retreat in holes in walls and bark can be identified by the radiating surface trap-lines. *S.senoculata* is associated mainly with old trees, stone walls and piles of loose materials such as tiles, rocks, timber *etc*. It is often found in leaf litter inside and around old trees and occasionally beaten from oak, hawthorn and willow.

Pholcus phalangioides is a synanthropic species. Though it appears to be rare in Leicestershire, it is more widespread than records suggest. First recorded for the county by J. Woodhead in 1990, from Leicester, the species is now known to be established also in Enderby, Whetstone, Saddington and Loughborough. Transported by man, it is found inside heated buildings where it can be quite conspicuous hanging upside down in its tangled web, often in the corners of ceilings. Males will wait in the female's web for several days before mating, and females carry a flimsy bundle of eggs in their chelicerae. Females and immature specimens have been observed all year, males May–October.

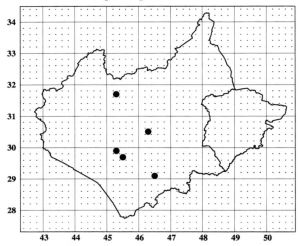

Pholcus phalangioides (Fuesslin, 1775)

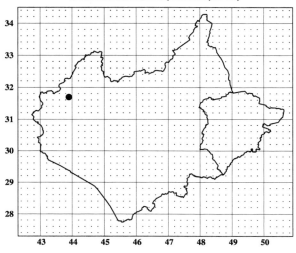

Psilochorus simoni (Berland, 1911)

Psilochorus simoni is known from only a single site in the county – Coleorton Hall SK391173 – where it is well established in the cellars, but probably occurs elsewhere in similar situations. First recorded in 1965, J. Crocker, this 'colony' has been monitored on two occasions, February 1965 and July 1993, when specimens were numerous with both sexes at all stages of development, including females with ova. Similar in its habits to *Pholcus*, *Psilochorus* is a smaller spider and was almost certainly introduced in imported wine consignments some time after the cellars were renovated and extended in 1808. The webs are found in corners of walls and roof, at floor level and on shelving.

Atlas of Leicestershire and Rutland Spiders
GNAPHOSIDAE

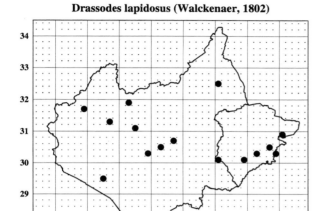

Drassodes lapidosus (Walckenaer, 1802)

Widespread but infrequent in Leicestershire, **Drassodes lapidosus** is probably under-recorded. Often confused with *D.cupreus*. Females can be found all year but are more active May–September. Males May–August with peak activity noted in June and July. More generally associated with human activity than *D.cupreus*; in buildings, gardens, farmyards. churchyards, urban derelict land, disused quarries and railway sidings, stony wasteland, rubbish tips and on recreational scrubland. Found inside buildings, under stacked building materials, under stones, in walls, behind gravestones, beating scrub *etc.* and spun up in a cell on an old tractor. First recorded for the county by Lowe (1912).

Drassodes cupreus is widespread and common; more so than *D.lapidosus*. There is a wide variation in size, larger females being quite aggressive. Though females probably occur all year we have only taken them May–November and males May–July with highest numbers in June. *D.cupreus* is recorded from damp acid grassland, open moorland, wet heathland, rocky *Calluna/Vaccinium* heath, dry-stone walls, disused quarries and mature colliery spoil heaps, herb-rich limestone grassland, woodland rides, wood pasture and scrub gorse. Spun up in silk cells under stones, in gorse bushes, in walls and amongst short grass, also in leaf litter and *sphagnum* moss. First record Chalcraft (Rowley, 1897).

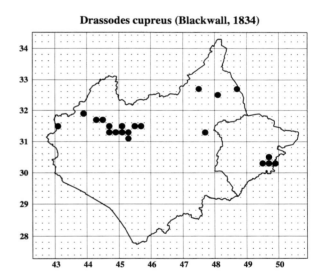

Drassodes cupreus (Blackwall, 1834)

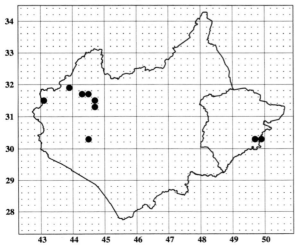

Haplodrassus signifer (C.L.Koch, 1839)

Though widespread **Haplodrassus signifer** is local, occurring mainly on Charnwood Forest and in the east of the county. It was first recorded by M.J. Roberts in 1969 at Newbold Verdon, and has since been taken at Bardon Hill, Cademan Moor, C.L.N.R., Geeston Quarry, High Sharpley, Coleorton, Moira, Newfield Heath and North Luffenham Quarry. Adults of both sexes have been taken April–September. Recorded from disused ironstone and limestone quarries, damp heathland, heath grassland, *Calluna* heath, neutral grassland marsh and *sphagnum* bog, generally in short vegetation or on bare dry ground, under stones and among heather roots, also in leaf litter and grass detritus.

CHAPTER FIVE

Atlas of Leicestershire and Rutland Spiders
GNAPHOSIDAE

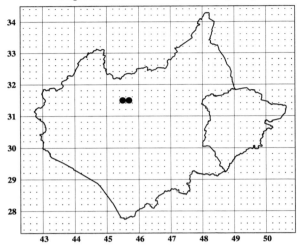
Haplodrassus silvestris (Blackwall, 1833)

Haplodrassus silvestris is very rare in Leicestershire, known only from Buddon Wood, Quorn, where it was once widespread but due to quarrying is now threatened. A nationally notable species, it is mature in spring and autumn, and was first recorded in 1975 by M.G. Crocker when a male and sub-adult females were collected in May amongst grass roots on a rocky slope in regenerating birch/oak woodland. The females were reared through to maturity in early June. In September 1977 a mature female was taken in the same type of habitat but at a different site in the wood. Since then, despite extensive pitfall trapping in what is left of the site, no further specimens have been taken.

This synanthropic species is widespread and common but is under-recorded in the county. ***Scotophaeus blackwalli*** has been recorded mainly from domestic buildings but also from gardens on bushes, fences and under debris, farm yards, stables, rural industrial buildings, and also from reservoir margins under debris. Outdoor specimens are generally much larger than those from inside houses. They can be quite aggressive and can inflict a painful bite. *Scotophaeus* is a nocturnal wanderer and will be found on house walls and ceilings after dark. Females occur in every month of the year, males have been taken July–January with peak numbers of each sex noted in July and August. First record Mayes (1934).

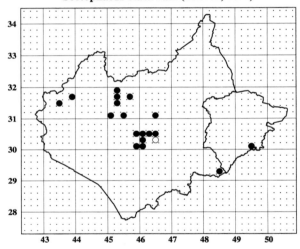
Scotophaeus blackwalli (Thorell, 1871)

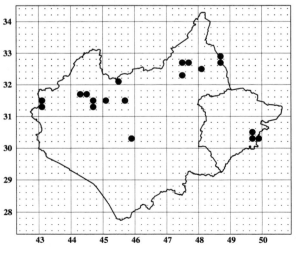
Zelotes latreillei (Simon, 1878)

Zelotes latreillei is widespread, but uncommon in Leicestershire. Its local status is, however, being reviewed nationally. First recorded by T.A. Walden in 1948 from Waltham-on-the-Wolds, *Z. latreillei* has since been taken at a further 22 sites in 19 tetrads, including Acresford Sandpit, Bardon Hill, Beacon Hill, Buddon Wood, Cademan Moor, Geeston Quarry, Harby Hills, Stonesby Quarry and The Drift. It has been found in calcareous and neutral herb-rich grassland, damp acid grassland, *Vaccinium/Calluna* heath, rocky grass heathland, *Molinia* moorland, ancient woodland and disused quarries, mostly under stones and amongst grass roots. Adults, both sexes April–November.

Atlas of Leicestershire and Rutland Spiders

GNAPHOSIDAE

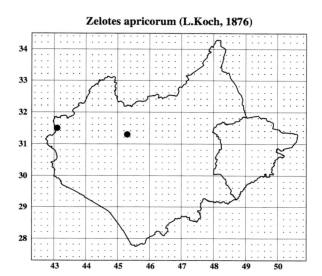
Zelotes apricorum (L.Koch, 1876)

Zelotes apricorum. A very local spider, rare in Leicestershire, recorded from only three sites: The Brand (first record E. Duffey, 1952), Swithland Wood and Newfield Heath. The former sites are landscaped slate quarries in ancient woodland within the same tetrad, whilst Newfield Heath is a naturalised 19th century closed colliery site, with a well developed *Calluna* heathland flora. Females April, June and September, and a male in May, under stones and pieces of slate amongst sparse vegetation on scree, in grass and leaf litter, and from a pitfall trap amongst heather. Nationally, the spider is generally distributed, but is rather local, being found under stones and amongst debris, heather *etc*.

This very rare spider **Urozelotes rusticus** is known in the county from three individual males 1961, 1962 and 1965. A synanthropic species, it was first recorded from a refuse tip at Shepshed where the wandering male was collected by torchlight at night amongst sparse vegetation (Crocker, 1962a). The other two specimens were from the basement of the County Museum, New Walk, Leicester where they were found in a jar and on a shelf three years apart. *U.rusticus* is nationally rare and has been recorded from under stones and detritus often around houses. It is thought to have been accidentally imported from the U.S.A. The Leicestershire specimens were taken in June, September and December.

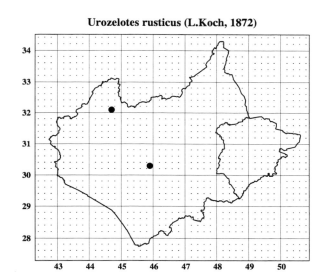
Urozelotes rusticus (L.Koch, 1872)

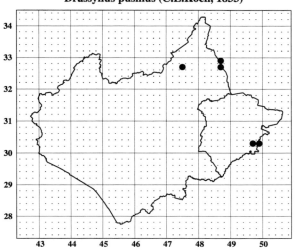
Drassyllus pusillus (C.L.Koch, 1833)

Drassyllus pusillus, a rare local spider, is recorded from only five calcareous grassland sites in the east of the county. First recorded by J. Daws in 1992 from The Drift, it is also found at Geeston Quarry, Harby Hills, King Lud's Entrenchments and North Luffenham Quarry. Females have been taken May–August and males June–August with the highest number of individuals recorded for June. The species prefers dry situations and has been collected in Leicestershire from short grass and amongst sparse vegetation in rough calcareous grassland, and in disused limestone and ironstone quarries. It has been taken elsewhere, outside the county, under stones and among rubbish on heathland.

Atlas of Leicestershire and Rutland Spiders
GNAPHOSIDAE, CLUBIONIDAE

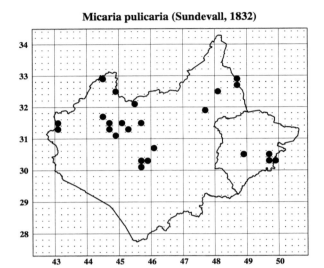

Micaria pulicaria (Sundevall, 1832)

Widespread and common, in open sunny situations, **Micaria pulicaria** is a ground hunting spider, first recorded in 1962 by H.A.B. Clements from Charnwood Lodge N.R. It has since been taken at a variety of sites including Acresford Sandpit, Bardon Hill, Beacon Hill, Geeston Quarry, Humberstone Quarry, Leicester Cattle Market, Stonesby Quarry *etc*. Adults of both sexes April–November. Woodland, calcareous and heath grassland, vegetated sand banks, urban derelict land, rocky *Vaccinium* and *Calluna* heath, shingle bank at edge of river, sedge bed with bare areas, disused quarries and railway sidings. The species is caught readily in pitfall traps especially on open ground.

Clubiona corticalis, first recorded by Cooke and Lampel (1953), is a widespread and common species under-recorded in Leicestershire, for which we have only 31 records. Distinctive and easily recognised, both sexes have been found March–August but immatures and sub-adults commonly encountered throughout the year, under bark, beating shrubs and trees, under stones, under garden rubbish and in damp litter. Recorded from Bradgate Park, Buddon Wood, Charnwood Lodge N.R., Coleorton, Cropston Reservoir, Donington Park, Knighton Spinney, Leicester and Loughborough gardens, Nether Broughton, Piper Wood, Queniborough, Scraptoft and Thorpe Langton.

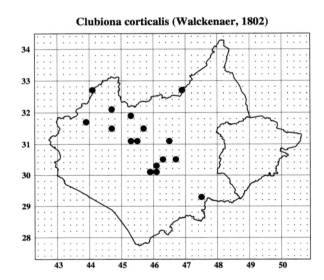

Clubiona corticalis (Walckenaer, 1802)

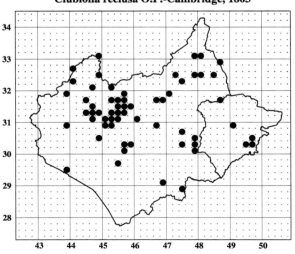

Clubiona reclusa O.P.-Cambridge, 1863

Clubiona reclusa. A widespread and common species, represented in the county from a wide range of sites, often in damp or marshy situations, where it is frequently found inside retreats constructed from curled leaves on a variety of field layer plants, also in ground zone litter. The habitat of this species also includes woodland with herb-rich ground and shrub layers, rough scrub grassland, nettle beds, rubbish tips and heathland, but most records are from wet areas with sedges, meadowsweet, reeds, *juncus etc*. C.reclusa adult females have been taken January–October and adult males May–August with peak activity May–June. First record I.M. Evans, 1961, G.U. Canal, Gumley.

Atlas of Leicestershire and Rutland Spiders

CLUBIONIDAE

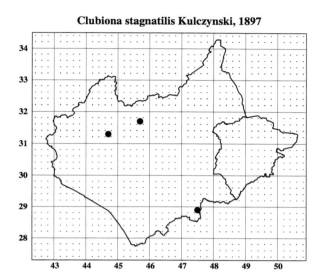

Clubiona stagnatilis Kulczynski, 1897

Clubiona stagnatilis is rare in the county, recorded only from three widely separated wetland sites. First recorded by Lowe (1912) without details of location, the three subsequent records are from a small *Phragmites* marsh on the embryo River Sence brook below Bardon Hill, at Great Bowden Pit from a *sphagnum* bog with *juncus* and cotton grass, and from a damp meadow beside the River Soar at Quorn. In each case the spiders were obtained from wet ground zone vegetation, females in the summer and October, and males in May and July. Generally accepted as maturing in the spring and summer *C.stagnatilis* is considered to be common and widespread throughout Britain.

Clubiona pallidula. Widespread but infrequent, recorded from only twelve sites – five of these urban or suburban gardens. First recorded for Leicestershire in 1963 by P.H. Gamble from Buddon Wood and since then from six other natural or semi-natural habitats: Barkby Holt Wood, Knighton Spinney, Luffenham Heath, Newton Burgoland Marsh, and two sites at Rutland Water. Usually found on bushes and trees and under bark, the county material has been obtained by beating birch, oak, lime, willow, hawthorn, gorse and garden shrubs; also from inside houses and outhouses, under stones and artifacts, and sweeping grass. Adult females May–October, males in May and June.

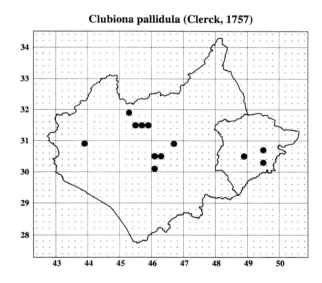

Clubiona pallidula (Clerck, 1757)

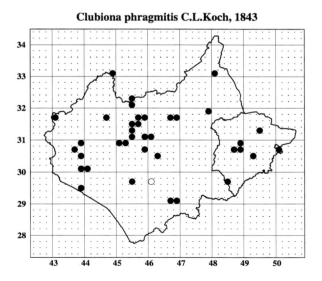

Clubiona phragmitis C.L.Koch, 1843

Clubiona phragmitis is a widespread wetland spider and often abundant in its natural habitat. It has been recorded from marshes, swamps and marginal vegetation associated with small ponds, streams, rivers, canals, lakes, reservoirs and gravel pits. Females April–November, males April–October, with highest numbers May–July. From dry, damp and wet ground zone litter, on wet mud and sweeping marginal vegetation. In standing water *C.phragmitis* can be observed stalking prey on reed stems just above the water-line, and silk retreats are often found on reed stems and leaves. The spider is characterised by its dark ocular area and black swollen chelicerae. First record (Bristowe, 1939).

CHAPTER FIVE

Atlas of Leicestershire and Rutland Spiders

CLUBIONIDAE

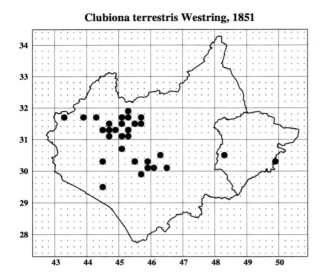

Clubiona terrestris Westring, 1851

This common widespread, species is frequently met with in and around the house and in the garden. **Clubiona terrestris** is under-recorded in certain parts of the county where it will be found on trees, shrubs and other low vegetation. First record Cooke and Lampel, 1953. Established in deciduous and mixed woodland, *C.terrestris* also occurs in scrub land, wooded parkland and gardens with plenty of shrubs. The spider likes dry sunny situations and is extremely agile, particularly the male. Specimens have been collected by beating the shrub and canopy layers, sweeping grassy clearings and grubbed from ground zone litter; also from walls. Adults, both sexes, every month of the year.

Classified nationally as a local species, **Clubiona neglecta** is rare in Leicestershire. Recorded for the first time in the county by I.M. Evans in 1961 from the Grand Union Canal, Gumley, *C.neglecta* has been subsequently taken at two sites in Coleorton, Acresford Sandpit, Humberstone Quarry, Egleton and Rise Rocks (Markfield), by beating young trees and shrubs, sweeping field layer, and in leaf litter and under stones. Females June–November and males May–July. From canalside vegetation, disused sand and brick pits, a re-profiled colliery dirt tip, a mound of fly ash, unmanaged acid grassland with gorse scrub, unimproved hay meadow and a young deciduous plantation.

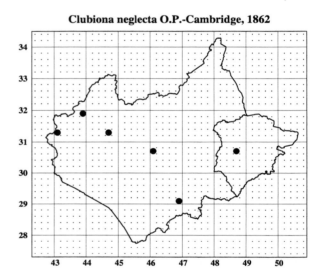

Clubiona neglecta O.P.-Cambridge, 1862

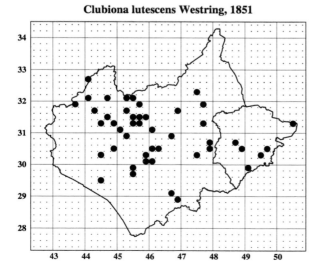

Clubiona lutescens Westring, 1851

Clubiona lutescens. Widespread and common in the county, mostly from damp rather than wet places. First record 1962 (J. Crocker), since then from deciduous dry and wet woodland, ancient parkland, calcareous and neutral grassland, churchyards, gardens, allotments and refuse tips, also from marshes, wet meadows, reed and sedge beds, and a variety of waterside habitats ranging from closed plant communities, emergent vegetation and matted litter, to bare mud. Spiders have been obtained by beating oak, birch, hawthorn and gorse, sweeping field layer vegetation, grubbing and sifting litter. Females February–December, males May–July; highest numbers May and June.

Atlas of Leicestershire and Rutland Spiders
CLUBIONIDAE

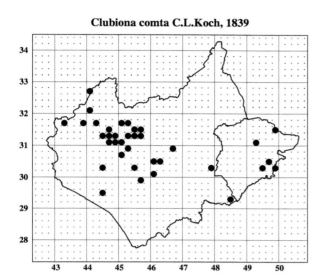

Clubiona comta C.L.Koch, 1839

Clubiona comta is widespread and common, more so than records indicate. This species is arboreal, living on the leaves of trees and shrubs and in retreats under bark and in crevices. Collected by beating oak, hazel, sycamore, lime, birch, hawthorn, ivy, rhododendron, gorse and young spruce, also sweeping the field layer under trees, and in ground zone litter. Females have been taken throughout the year and males March–July, with main activity period May and June. Immatures are common in the ground and field layers in the late autumn and winter; they migrate to the canopy during the spring. Bristowe (1939) records *C.comta* for the first time in Leicestershire.

Not as common as the previous species, ***Clubiona brevipes*** is equally widespread in Leicestershire and certainly under-recorded. The reason these two species have been overlooked is probably due to the less frequent application of beating techniques, since their main habitat is on leaves of trees and bushes. First record P.H. Gamble, 1962, Buddon Wood. Elsewhere from woodland and scrub, open parkland with trees, hedgerows and occasionally rough grassland. The majority of records are from beating oak trees. Also from other trees (birch, hawthorn, spruce and gorse), stone walls and leaf litter under trees. Adult females April–November, males May–July; main activity June.

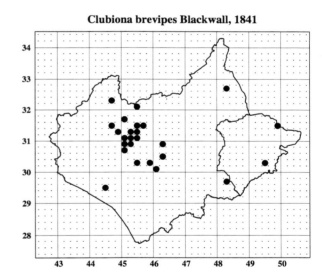

Clubiona brevipes Blackwall, 1841

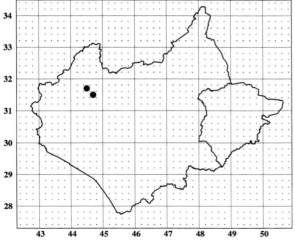

Clubiona trivialis C.L.Koch, 1843

Though widespread in Britain and frequent in mountainous, heathery places, especially in the north, ***Clubiona trivialis*** is very rare in Leicestershire, known only from three single specimens taken at two moorland sites on Charnwood Forest. First record J. Crocker, 1962, C.L.N.R. (SK467156), a female under heather at side of brook. Another female, 1992, High Sharpley (SK446170) *Calluna* heath, from pitfall trap in heather, and then a male, December 1995, again from C.L.N.R., this time from a pitfall trap in *sphagnum* under heather on *Vaccinium* heathland (SK473148). At 240m this is one of the highest points in Leicestershire. The two females were taken in July and September.

Atlas of Leicestershire and Rutland Spiders
CLUBIONIDAE

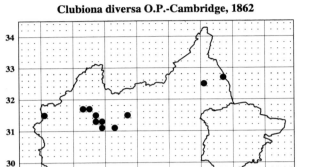

Clubiona diversa O.P.-Cambridge, 1862

Clubiona diversa. Widespread but local in Leicestershire, mostly on or around Charnwood Forest; but also at Stonesby Quarry and King Lud's Entrenchments in calcareous grassland which is typical of its southern habitats. In the west of the county the spider occurs on rocky heathland, acid grassland and marshy grassland – habitats associated with the species' northern distribution. *C.diversa* is primarily a field layer dweller but most of the records are from pitfall traps and grubbing amongst grass roots. Occasionally it has been taken in vole runs, under rocks and stones, and beaten from young oaks. First record 1962, C.L.N.R., J. Crocker. Females April– August, males June–November.

A southern wetland species, also occurring in coastal sandhills, **Clubiona subtilis** is very rare in the county, represented by a single female collected by J. Daws in May 1994 at Seaton Meadow (SP916979). The specimen was taken in a pitfall trap in damp grassland with tall herbs within this herb-rich hay meadow on the River Welland flood plain. The site is a SSSI and is characterised by herbs such as the great burnet, meadowsweet and meadow vetchling. It is subject only to traditional low intensity agricultural management and is cut for hay in September. The ancient meadow is flooded periodically during winter snow-melt and summer flash floods (Barfield, 1995).

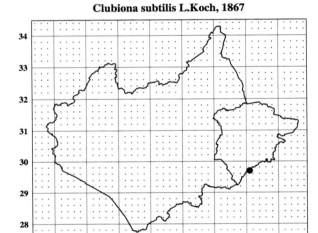

Clubiona subtilis L.Koch, 1867

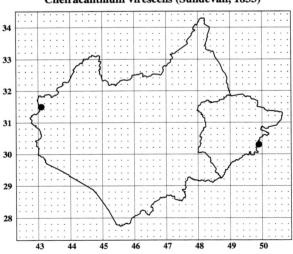

Cheiracanthium virescens (Sundevall, 1833)

Cheiracanthium virescens is very rare and localised in Leicestershire, being recorded from only two sites at opposite sides of the county. First recorded by J. Daws; a female, November 1992, Moira Junction (SK306159), in a heathy area with sparse *Calluna*, from a pitfall trap in bare ground amongst meagre grassy vegetation. In April 1995 a further female was taken at Geeston Quarry (SK981037) and the following month another five females and two males were caught in the same place. This site is a disused ironstone quarry with scant vegetation and pitfall traps were set in the bare clay substrate amongst small patches of grass, similar to the placement of the Moira trap.

Atlas of Leicestershire and Rutland Spiders
LIOCRANIDAE

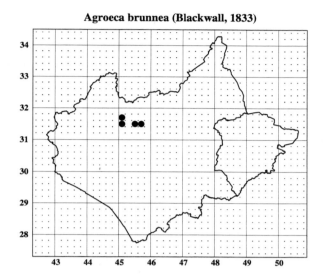

Agroeca brunnea (Blackwall, 1833)

Agroeca brunnea is considered to be a widespread but uncommon woodland species. It is rare and very localised in Leicestershire, having been found – since Agar's 1890 record – in only two ancient woodland sites. (Buddon Wood and Outwoods). The 19th century record is based on the distinctive shape of the egg cocoon (Neilsen, 1932) and is likely also to have been from Buddon Wood. Females have been taken in April, July and November, a male in April, and a well developed sub-adult female in September. From birch/oak and alder woodland, a larch plantation, and a marshy woodland clearing; in grass/bracken litter, wet litter at the base of alders and in larch needle litter.

Widespread and much more common than the previous species, ***Agroeca proxima*** is found (infrequently) in more open habitats; rocky open heathland with and without heather, damp acid grassland with invading scrub, deciduous woodland, a former colliery dirt tip, wet neutral grassland, calcareous grassland and a disused ironstone quarry. First record 1967, Ulverscroft N.R., J. Crocker, and subsequently from Bardon Hill, Buddon Wood, Cademan Moor, Coleorton, C.L.N.R., Cooper's Plantation, High Sharpley, King Lud's, Newfield Heath and N. Luffenham Quarry. Amongst grass roots, pitfall traps, and litter. Females all year, males August–November. Large numbers in November.

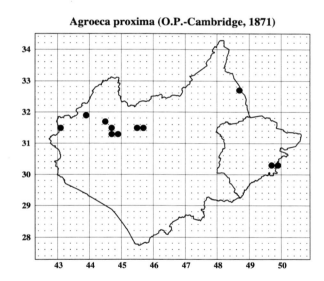

Agroeca proxima (O.P.-Cambridge, 1871)

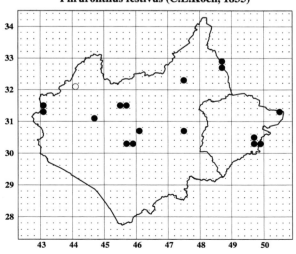

Phrurolithus festivus (C.L.Koch, 1835)

Phrurolithus festivus. Widespread and common, typically from open clay and sand pits, limestone and ironstone quarries, urban derelict land *etc.*, mostly from bare open ground in dry conditions with sparse vegetation, but occasionally from grass and bracken litter. The majority of records are from pitfall traps. *P. festivus* is an active hunter on open ground where it can be found running about in sunshine, sometimes with ants with which it can be confused. In Leicestershire, females May–November, males May–September with very large numbers in July. Sub-adults are noted September–November after which activity seems to cease. First record 1959, I.M. Evans, Breedon Cloud Quarry,

CHAPTER FIVE

Atlas of Leicestershire and Rutland Spiders
ZORIDAE, ANYPHAENIDAE, HETEROPODIDAE

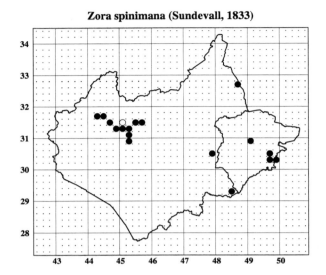

Zora spinimana (Sundevall, 1833)

Zora spinimana is an attractive grassland hunting spider, usually encountered in the ground zone, amongst grass roots and in leaf litter but also under stones, pieces of damp litter *etc.*, and occasionally by sweeping. Widespread but local in Leicestershire, it was first recorded in 1953 by E. Duffey from Beacon Hill, and since then found at relatively few sites: Barnsdale Wood, Benscliffe Wood, Bradgate Park, Buddon Wood, C.L.N.R., Geeston Quarry, Grace Dieu Wood, Great Easton, High Sharpley, Ketton Quarry, King Lud's, Launde Big Wood, Luffenham Heath, North Luffenham Quarry, Swithland Wood, The Brand and Ulverscroft N.R. Adults of both sexes March–December.

The 'buzzing spider', so named from the male's habit of drumming its abdomen on a leaf surface during courtship, **Anyphaena accentuata** is widespread but local in the county. It is an arboreal species taken mostly by beating, but sometimes whilst sweeping scrubby grassland and from leaf litter, never far from trees or bushes. Records are from fourteen ancient woodland sites, one churchyard, Ketton Quarry and Luffenham Heath – where there is a lot of oak/gorse scrub, as also at Ketton. No males taken, and females only in May and June. Sub-adults of both sexes have been numerous in March and April; immatures noted throughout the year. First record Chalcraft (Rowley, 1897).

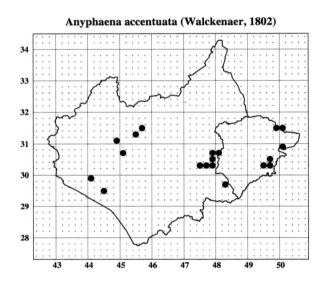

Anyphaena accentuata (Walckenaer, 1802)

Micrommata virescens (Clerck, 1757)

Micrommata virescens. A large and distinctive spider, the adult males not easily overlooked; widespread throughout Britain, rare in the north and uncommon in the south. It is known only in Leicestershire from Chalcraft's 19th century record (Rowley, 1897), which was quite probably from Owston Wood where the species could still be established. The spider prefers sheltered woodland clearings and has a short maturity period (May–June). Immatures and sub-adults are cryptically coloured and are inconspicuous. *Micrommata* is generally found in ancient oak woodland with a well structured shrub layer, a rich ground flora and having a good history of continuity.

Atlas of Leicestershire and Rutland Spiders

THOMISIDAE

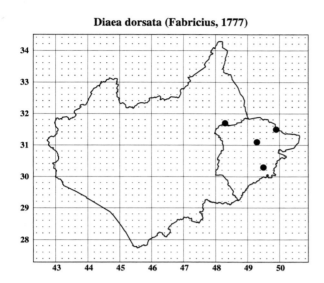

Diaea dorsata (Fabricius, 1777)

Diaea dorsata appears to be a declining species, commoner in the south of England it is rare in the county, being restricted to east Leicestershire. At present known from only four sites: Exton churchyard, Luffenham Heath, Pickworth Great Wood and Wymondham Rough – where it was first recorded by I.M. Evans in 1984. Living mainly on the foliage of trees and shrubs it is occasionally encountered in the field layer of undisturbed scrubby grassland; in this habitat its cryptic coloration makes it difficult to see. The maturity season is May–June; both sexes recorded in the county in June by beating broadleaf trees and shrubs, and a sub-adult female in February under bark of Scots Pine.

As with *D.dorsata*, **Misumena vatia** is a southern species, rare in Leicestershire and known only by a few individual occurrences. First recorded by Chalcraft (Rowley, 1897) from Owston Wood on a valerian flower, and later, almost certainly introduced, from a Leicester suburban garden on a *Chrysanthemum frutescens* flower; it was also taken at Pickworth Great Wood in mixed woodland with good shrub and ground layers, where a single female was collected by grubbing along the side of an open ride. Female May, male July. A male was also swept from roadside vegetation against ancient woodland at Bedford Purlieus, Northamptonshire – close to the Rutland county boundary – July 1962.

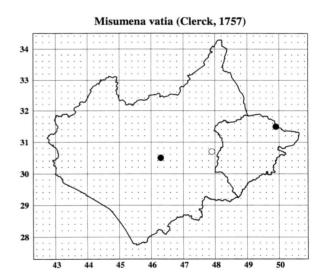

Misumena vatia (Clerck, 1757)

Xysticus cristatus (Clerck, 1757)

One of Leicestershire's commonest spiders and the most widespread of its genus, ***Xysticus cristatus*** is a field layer species of the leaf and stem zone, but frequently found at ground level on bare ground, amongst sparse prostrate vegetation and in grass tufts, also under stones and in leaf litter. It is a regular aeronaut in the immature stages and all instars appear commonly in pitfall trap catches. The map shows only where specimens have been collected but *X.cristatus* must occur in every terrestrial tetrad in the county. Adults of both sexes have been taken April–November with peak activity May–July and very large numbers in June. First record Chalcraft (Rowley, 1897).

Atlas of Leicestershire and Rutland Spiders

THOMISIDAE

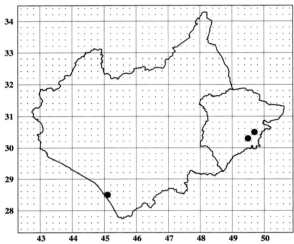

Xysticus audax (Schrank, 1803)

Another species generally affirmed as widespread and common throughout Britain; but **Xysticus audax** is rare in the county. No males have been taken; and females in April, August and October represent the five records for Leicestershire from three sites: Bitteswell Aerodrome (SP5085), Ketton Quarry (SK9705) and Luffenham Heath (SK9502) which produced the first record in 1994 (J. Daws). These sites are featured by rough scrubby grassland where the spiders were beaten from gorse and tall herbs. In appearance *X. audax* is similar to *X. cristatus* but Roberts' (1985 p.101) figures are helpful in separating the two species. It may be that *X. audax* has been overlooked in the past.

As with the previous species, **Xysticus erraticus** is considered to be common, but in Leicestershire it is very local, with only seven records from three widely separated sites: High Sharpley (SK4417), Bloody Oaks Quarry (SK9710) and Ketton Quarry (SK9705), the former unmanaged grassy *Calluna* heathland and the latter two disused limestone quarries with herb-rich calcareous grassland. Females August–October, males June–August and October. Except for one male obtained by sweeping field layer vegetation, all other specimens were from pitfall traps in both acid and limestone grass. High numbers of males are recorded in August and October. First record J. Daws, 1992, High Sharpley.

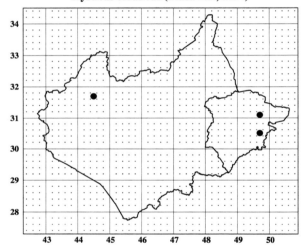

Xysticus erraticus (Blackwall, 1834)

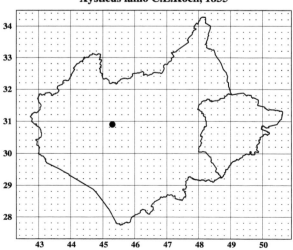

Xysticus lanio C.L. Koch, 1835

Xysticus lanio. Only one record 1978 (J. Crocker), a female from Sheet Hedges Wood (SK530085), an area of open recently felled woodland on an ancient woodland site. The specimen was swept from dog's mercury and yellow rattle under Alders in June. The species is commoner in the south than in the north, is adult May–July and generally found in woodland in the shrub layer and lower canopy. It would seem that beating would be the most likely collecting technique to capture *X. lanio*, and whilst it is a fact that this is the least often used method, sufficient beating has been done in a wide variety of ancient woodland sites to have found this species elsewhere had it been present.

Atlas of Leicestershire and Rutland Spiders
THOMISIDAE

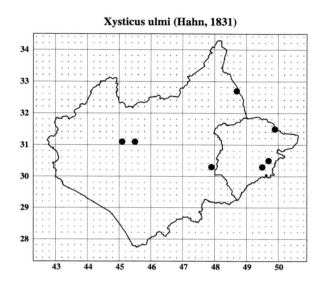

Xysticus ulmi (Hahn, 1831)

Widespread but rare in the county, **Xysticus ulmi** favours wet habitats where it has a preference for field layer vegetation. First recorded in 1962 by I.M. Evans from a typical site at Cropston Reservoir, where it was collected from amongst sparse vegetation at the edge of a filter bed. Subsequent records from Ketton Quarry, King Lud's Entrenchments, Launde Big Wood, Lea Wood and Luffenham Heath may appear to be less typical habitats, but do contain wet areas which may not have been differentiated in the collections. Females June and October, males May and June. Sweeping nettles in woodland, sweeping limestone grassland and from pitfall traps in rough grass.

Ozyptila scabricula. A single female from Geeston Quarry (SK981037) April 1995, collected by I. Phillips in a pitfall trap set in the bare clay substrate on a well drained slope at the edge of the disused ironstone quarry. The site was very open with sparse vegetation. This nationally notable species is very rare. Reaching maturity in spring and summer, it prefers dry light soils and the presence of bare ground would appear to be important. Elsewhere it has been found under stones and amongst low vegetation in sandy areas such as breckland, also on coastal dunes. It is tempting to speculate that *O.scabricula*, a notable (Nb) species, is extending its range westwards from East Anglia.

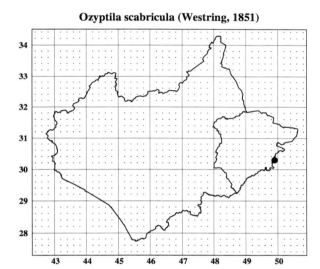

Ozyptila scabricula (Westring, 1851)

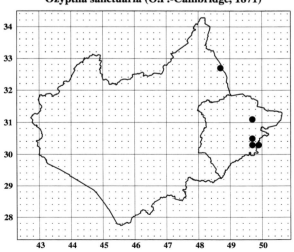

Ozyptila sanctuaria (O.P.-Cambridge, 1871)

A local spider restricted to east Leicestershire with records from only five sites, all on the oolitic limestones and sandstones: Bloody Oaks Quarry, Cooper's Plantation, Geeston Quarry, Ketton Quarry and North Luffenham Quarry. **Ozyptila sanctuaria** is a southern species generally rare but often abundant on chalk grassland. In the county mostly single specimens have been taken in pitfall traps placed in short calcareous grassland with a certain amount of bare open ground, and also (a male) by beating gorse. Females March–June and September–October, males August–November. First record produced by Leicestershire Museums' survey team in 1978 from Cooper's Plantation.

CHAPTER FIVE

Atlas of Leicestershire and Rutland Spiders
THOMISIDAE

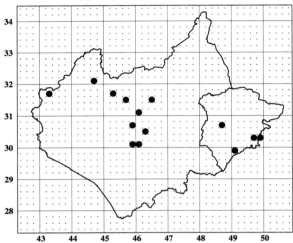

Ozyptila praticola (C.L.Koch, 1837)

Ozyptila praticola is widespread but infrequent in the county mostly from semi-natural habitats such as gardens, parkland, a cemetery, quarries and gravel pits *etc*. First recorded in 1963 by H.A.B. Clements from a brookside marsh at Shepshed, *O.praticola* has subsequently been found in a wide variety of situations: by beating shrubs and sweeping field layer vegetation, under stones, amongst prostrate plants and in damp detritus, but most records are from pitfall traps in grass, deep litter and on bare ground with sparse vegetation. Adult females April–October, males May–November, with highest numbers of both sexes in June. More common in the south of England than in the north.

Recorded from only five sites in four tetrads, ***Ozyptila trux***, which is widespread and common in Britain, must be considered rare in Leicestershire. Hand collected from under a damp log in oak/hazel coppice (Swithland Wood), and by grubbing in marsh vegetation (Newton Burgoland). Elsewhere it has been taken in pitfall traps in calcareous grassland (King Lud's and The Drift) and under a species-rich hedgerow (Coalfield West). Females May–November, males June and July. The literature indicates that this species is adult throughout the year and that it moves higher up in the vegetation at night, as do many daytime ground zone dwellers. First record 1966, Swithland Wood, J. Crocker.

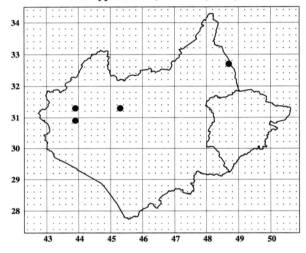

Ozyptila trux (Blackwall, 1846)

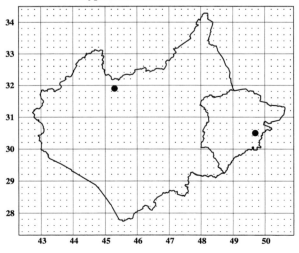

Ozyptila atomaria (Panzer, 1801)

Ozyptila atomaria. Very rare in the county, recorded from only a single 'natural' site in east Leicestershire and a solitary male from a garden at Loughborough where it was probably introduced. At Ketton Quarry (SK9705): a female under a stone on open ground among the ground flora, May 1993 (J. Daws), and another female in August 1994 by beating oak/gorse scrub some distance from the original site. The Loughborough adult male was found in February 1970, living under a rockery plant (*Dianthus sp.*) in association with an active colony of springtails. A single female, brought home from Box Hill, Surrey, for photography was released on the rockery in September 1968.

Atlas of Leicestershire and Rutland Spiders
THOMISIDAE, PHILODROMIDAE

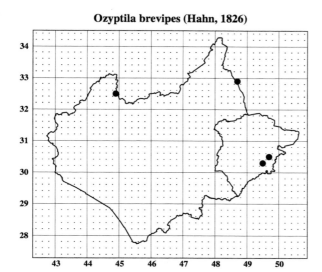

Ozyptila brevipes (Hahn, 1826)

Widespread but rare in Leicestershire with only ten records, **Ozyptila brevipes** was first recorded by Bristowe (1939) but no site details are given. A southern species having a fairly wide distribution throughout England, where it is reported from damp marshy areas in summer. From Long Whatton brookside marsh (SK4824) grubbed from ground zone, Luffenham Heath (SK9502) by beating scrub gorse, males swept from wide calcareous grass verges along 'The Viking Way' in May (The Drift, SK8628), and from a number of different sites at Ketton Quarry (SK9705) where the species is well established in limestone grassland and on gorse. Females June and July, males May, June and July.

Philodromus dispar. Widespread and common; under-recorded within the county. First recorded by Lowe (1912), this southern arboreal spider is frequently seen indoors on window frames and house plants during the early summer months, particularly the distinctive black males which are extremely agile. Mature in spring and summer, females have been taken May–August and males April–July, with highest numbers recorded for May and June. Many records are from domestic buildings and gardens but elsewhere it is found in woodland, scrubby marsh and grassland and on hedgerows, where it is collected by beating and sweeping. Females can be mistaken superficially for *P.cespitum*.

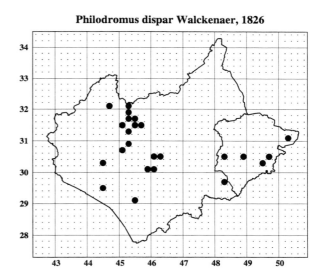

Philodromus dispar Walckenaer, 1826

Philodromus aureolus (Clerck, 1757)

Widespread and common this species is particularly numerous on gorse, hawthorn and oak scrub; also on pine trees and low vegetation. Generally obtained by beating trees and shrubs, **Philodromus aureolus** is also encountered occasionally sweeping field layer vegetation along hedgerows and woodland edge. It is found in a wide variety of woodland and scrubby sites, including parkland and commons, churchyards, gardens, reservoir margins, gorse scrub and deciduous and mixed woodland. As with most canopy invertebrates *P.aureolus* reaches maturity in spring and summer at the onset of a regular food supply. Females May–August, males May–July. First record Lowe (1912).

CHAPTER FIVE

Atlas of Leicestershire and Rutland Spiders
PHILODROMIDAE

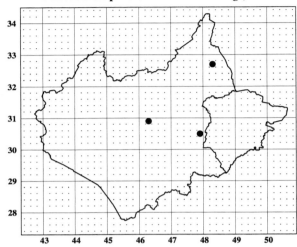
Philodromus praedatus O.P.-Cambridge, 1871

Philodromus praedatus is an uncommon spider and is probably much more widespread than hitherto suspected. It has been recorded from many new sites in England in recent years. In Leicestershire it is known from three widely separated locations, all grazed parkland with old oak trees (Barkby Thorpe Park, Croxton Park and Launde Park). It has been beaten from the lower foliage of isolated oak trees. First recorded for Leicestershire by J. Daws in 1994 at Croxton Park, the species is very local. Females have been taken in July and August. males in July and immatures in August, probably missing the main period of activity (May to July).

Widespread and common, **Philodromus cespitum** was first recorded for Leicestershire and Rutland by Bristowe (1939). Since then, the species has been taken at 34 sites ranging from urban gardens to ancient woodland, mostly on scrubby vegetation, particularly gorse and hawthorn, but also from nettles, ivy and on walls of buildings. Females have been collected in every month from May to November, and males May to July with late specimens in October. The main activity period is in June and July. This spider is very similar in appearance to *Philodromus aureolus* and occurs in the same situations but is easily separated on genitalia.

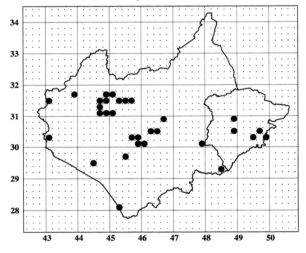
Philodromus cespitum (Walckenaer, 1802)

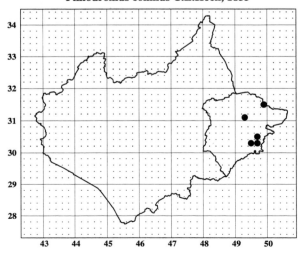
Philodromus collinus C.L.Koch, 1835

Philodromus collinus is a nationally notable species (Nb), mainly from the south of England and East Anglia, often on conifers, but also on Yew trees. The first record for Leicestershire was from Luffenham Heath by J. Daws in 1994, and subsequently from Exton churchyard, Ketton Quarry and Pickworth Great Wood, all from shrubby situations – beating gorse/oak scrub, beating box bushes in an open mown churchyard, beating conifers in a mixed plantation and beating shrubs in rough limestone grassland. Never taken in any numbers, females are recorded for June and males in June and July, but the seasonal maturity period is likely to extend from May to August.

Atlas of Leicestershire and Rutland Spiders
PHILODROMIDAE, SALTICIDAE

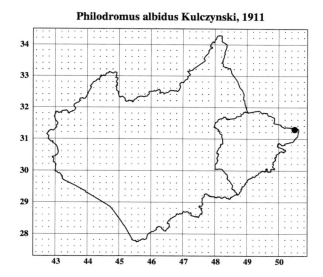

Philodromus albidus Kulczynski, 1911

Philodromus albidus is a rare species (Nb) of the more southern and eastern counties of England. It is known in Leicestershire from only a single male, beaten from shrubs at the edge of an open, mown churchyard at Essendine (TF049128). This churchyard is adjacent to a scrubby spinney on the site of an ancient Motte and Bailey. The specimen was collected by J. Daws in May 1994. Even in Hampshire, where *P.albidus* is widespread, only individual specimens have been encountered and this seems generally to be the case elsewhere. In Suffolk, females have been beaten from hawthorn and churchyard Yews in May, June and August. The spiders are very pale with greenish legs.

Tibellus oblongus. Widespread but local throughout Leicestershire in a range of grassy sites, including rough calcareous grassland, moorland, heath and neutral grassland – often in damp situations. This is a field-layer species obtained by sweeping long grass, but it turns up occasionally in pitfall traps in rank vegetation. Peak numbers occur in June and July, with adult females being recorded from March through to September, and males May, June and July. The first record for the county was published by M.B. Wild (1952). *Tibellus* is a hunting spider and, typical of the family, spends much of its time cryptically concealed on grass stems, waiting to pounce on unsuspecting prey.

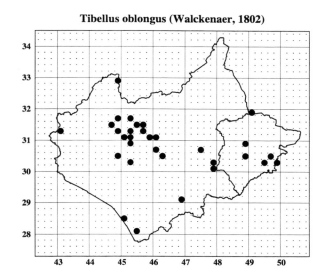

Tibellus oblongus (Walckenaer, 1802)

Salticus scenicus (Clerck, 1757)

Salticus scenicus. The 'zebra spider' of sunlit window ledges and house walls. Widespread and very common, it is certainly under-recorded in the county, and will be found on buildings, fences, tree trunks and rock faces throughout the summer. As with other salticids, it hunts in bright sunshine and usually retires to its retreat in dull weather. In Leicestershire it has been found away from buildings in woodland, quarries, grassland and stony wasteland. First recorded for Leicestershire by Chalcraft (Rowley, 1897); there are only 62 records for the county and it is suspected that it has been generally overlooked. Females May to September and males May to July, with sub-adults in April.

Atlas of Leicestershire and Rutland Spiders
SALTICIDAE

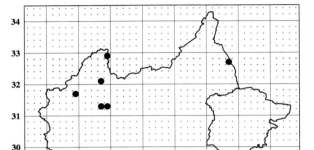

Salticus cingulatus (Panzer, 1797)

Salticus cingulatus is widespread but uncommon in Leicestershire. It prefers damp situations, but occurs on fences and tree bark in drier conditions. The spider is well camouflaged on silver birch trunks where it forms retreats in the typical curled bark features of this tree. Similar in appearance to *S.scenicus*, the impression of the living spider is of white stripes on a black abdomen, whereas in *scenicus* the impression is of black stripes on a white abdomen. Adults of both sexes May to July. First record by I.M. Evans, 1962, from King Lud's Entrenchments. Also from Ulverscroft N.R., Coleorton, Blackbrook, Bardon Hill, Lockington Meadows and Saddington Reservoir.

Heliophanus cupreus. Very rare in Leicestershire, recorded from a single divided site where it appears to be well established. The site comprises landscaped disused slate quarries in mixed woodland – The Brand (SK535132) and Swithland Wood (SK538130) – with a metalled road and high dry-stone wall boundaries between the two. First recorded in 1966 by M.G. Crocker from grass and leaf litter in mixed woodland, and the same year in silk cells under pieces of loose slate on an open slate tip at the side of a water-filled quarry. Adult females in May and subadults and immatures in September. A black spider with yellow legs and palps, easily confused in the field with *H.flavipes*.

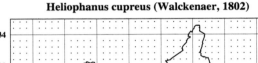

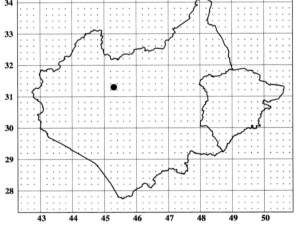

Heliophanus cupreus (Walckenaer, 1802)

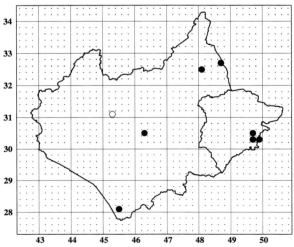

Heliophanus flavipes (Hahn, 1832)

Similar in appearance to the previous species, **Heliophanus flavipes** is more commonly encountered, and has been recorded from a Leicester garden, Bradgate Park, King Lud's Entrenchments, Stonesby, North Luffenham, Geeston and Ketton Quarries, and from Shawell Gravel Pits. Females have been taken each month from April to August, and males in June and July. The first county record was established by Lowe (1912), but no details of location, date or habitat were given. In Leicester, the species was caught in a Malaise trap which indicates aerial dispersal, and elsewhere under stones, in leaf litter, beating gorse, and sweeping rough grassland with tall herbs.

Atlas of Leicestershire and Rutland Spiders

SALTICIDAE

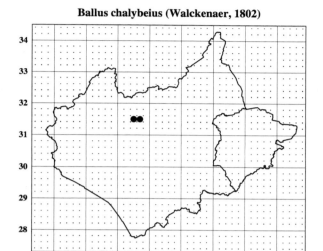

Ballus chalybeius (Walckenaer, 1802)

Ballus chalybeius. A very rare spider in Leicestershire, only known from a single site (Buddon Wood) where it is under threat from quarrying activities. This location appears to be the species' northernmost station in Britain, and extensive searching throughout Sherwood Forest has failed to yield specimens. There is, however, a doubtful record from Yorkshire. In the more southern parts of the country *Ballus* is fairly common, but local, in wooded areas, particularly on young oak trees where it spins a retreat on the upper side of leaves. At Buddon, females have been beaten from Durmast Oak in May, June and July. No males have been taken. First record P.H. Gamble, 1963.

Common on Charnwood Forest where it is widespread; elsewhere **Neon reticulatus** is recorded only from Donington Park and Launde Big Wood, but because of its small size may have been overlooked. It has been found in damp leaf litter, under stones, and from sphagnum moss, also occasionally by beating gorse and heather. All the locations for the 38 records are from ancient undisturbed sites, mostly from deciduous woodland. The first county record was from Swithland Wood, collected by P.H. Gamble in 1961. Females have been taken in each month from February to October and are probably mature throughout the year. Males are recorded for April, May and June.

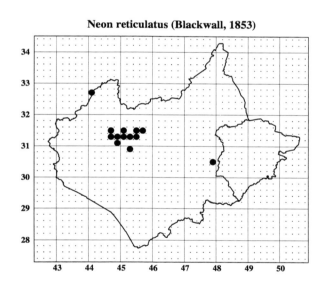

Neon reticulatus (Blackwall, 1853)

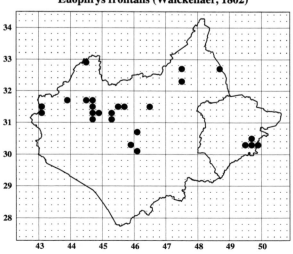

Euophrys frontalis (Walckenaer, 1802)

Euophrys frontalis. Widespread and common throughout Leicestershire, and first recorded by J. Crocker in 1961 from Coleorton Churchyard, it is the commonest species of the genus. It is found in woodland, acid, neutral and limestone grassland, sparsely vegetated stony wasteland, gardens, sandy areas, quarries *etc.*, often in silk cells under loose stones, rocks and litter. The species has generally been collected by grubbing amongst grass roots and from pitfall traps in closed grassland communities and on open sparsely vegetated ground; also by sweeping low vegetation. Females are recorded from May to November, and males May to September, with high numbers in June.

Atlas of Leicestershire and Rutland Spiders

SALTICIDAE

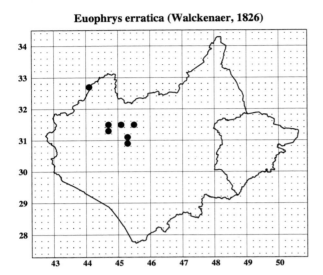

Euophrys erratica (Walckenaer, 1826)

Euophrys erratica. First recorded in 1962 by A.E. Squires from Bradgate Park, and subsequently taken at Bardon Hill, Beacon Hill, Buddon Wood, Charnwood Lodge N.R. and Donington Park. The species is very local, with only 18 records for the county – all from rugged heathland, scrubby acid grassland and ancient deer parks with over-mature oak trees. In each case, bracken is a dominant feature of the habitat. The spiders have been observed stalking prey on rock surfaces and on an oak stump in bright sunshine, they have also been collected from dead wood and under bark of old oak trees, under stones and in leaf litter. Females are recorded from May to November, and males in May, June and July.

Generally considered to be a southern species, **Euophrys aequipes** occurs on hot sunny banks and dry places. There are 24 records for Leicestershire, mostly from short vegetation in open disturbed areas and on stony wasteland. The first county record was from Cropston Reservoir in 1965, where it was swept from horsetails in wet mud by J. Crocker. This small jumping spider has also been collected from *Calluna* heathland, herb-rich limestone grassland, a vegetated sand bank, a disused ironstone quarry, in an abandoned brickpit, and on urban derelict land. Within the county it is widespread but local and would appear to be a successful aeronaut. Females April to November, males May, June and July.

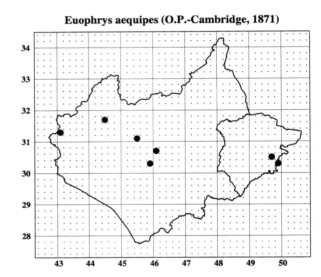

Euophrys aequipes (O.P.-Cambridge, 1871)

Euophrys lanigera (Simon, 1871)

Euophrys lanigera. A synanthropic species extending its range in Britain. Very local in Leicestershire with only 9 records from 5 sites, but certainly under-recorded. First record, J. Crocker, 1979, Ashby-de-la-Zouch and subsequently from buildings in Leicester, Kirby Muxloe and Loughborough. Inside and outside buildings, on walls, window ledges, roofs and ceilings; active in full sunshine. Adult females have been recorded from March to August and males April to October. *E. lanigera* was first discovered in Devon as a new species to Britain in 1930, and soon became common and widespread along the south coast. It is still spreading northwards through the midland counties of England.

Atlas of Leicestershire and Rutland Spiders
SALTICIDAE, LYCOSIDAE

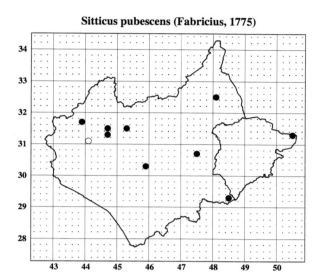

Sitticus pubescens (Fabricius, 1775)

Sitticus pubescens. Widespread but infrequent in the county, with only 13 records from 10 sites. First record by M.A.H. Watts in 1959 from Ibstock and since then from Great Easton, Essendine, Leicester, Bardon Hill, Charnwood Lodge N.R., Stonesby Quarry, Coleorton Hall, John-o'-Gaunt Fox Covert and Woodhouse Eaves. On walls, window frames *etc*. of buildings (rural, suburban and urban), often inside, but mostly outside in full sunshine. Also on tree trunks, stumps and fences, and in heather. Females are recorded for May, June, August and November and males in June and September. A marked decline has been noted nationally in this species since the early 1970s.

A very attractive jumping spider, this woodland species is widespread but very local in Leicestershire. ***Evarcha falcata*** was first recorded by Lowe (1912) but without details of where it was taken. It has since only been found at Buddon Wood and Swithland Reservoir, and at Ketton Quarry where it occurs on oak/gorse scrub on calcareous grassland. Most of the 20 records for the county are from the Buddon/Swithland Reservoir woodland, where specimens were beaten from a range of deciduous trees, and birch, oak and heather scrub. Adults of both sexes have been taken in each month from May to September. The male is particularly handsome and is very active in full sunshine.

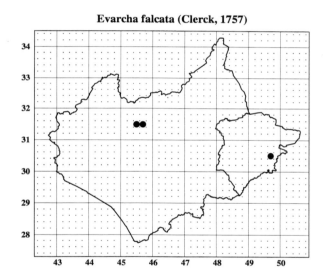

Evarcha falcata (Clerck, 1757)

Pardosa agrestis (Westring, 1861)

Pardosa agrestis is a very local (Nb) 'wolf spider' of the southern East Midlands down to London. It is associated with open clayey ground, disused quarries and stony calcareous areas. First recorded for Leicestershire by J. Daws in July 1994 from Seaton Meadow, a herb-rich hay meadow and grazed marsh on the River Welland flood plain. Here, pitfall traps in tall dry grass (SP916979) and damp marsh vegetation (SP915980) have yielded high numbers of both sexes, so the species is obviously well established. Elsewhere, at Geeston (SK981037), a disused ironstone quarry adjacent to the River Welland, a single male was taken in June 1965, on bare ground with sparse vegetation.

CHAPTER FIVE

Atlas of Leicestershire and Rutland Spiders

LYCOSIDAE

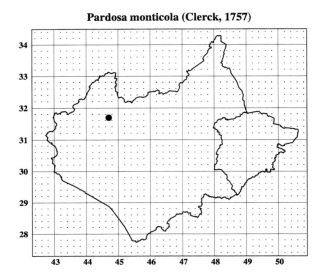
Pardosa monticola (Clerck, 1757)

Pardosa monticola. Although widespread throughout Britain where it is considered to be fairly common, this spider is very rare in Leicestershire, recorded only from a single site. It is an open ground species with a restricted distribution, confined to dry areas including inland heaths, and preferring habitats with open short vegetation. Adults are generally encountered between May and August, with females as late as November. The only record for the county is from Ives Head (SK477170), where 4 females were collected by J. Crocker in June 1966, on acidic rock-strewn short rabbit-cropped grassland. These spiders were running on open ground around the exposed summit area.

Pardosa palustris. Widespread and common in Leicestershire, often locally abundant and a successful aeronaut. First recorded for the county by J. Crocker, 1964, from Charnwood Lodge N.R., and since then from a wide range of open sites including Acresford Sandpit, Bardon Hill, Burbage Common, Kendall's Meadow, Ketton Quarry, Loughborough Meadows, Seaton Meadow and Stonesby Quarry N.R. On bare ground along river and reservoir margins, open arable ground, ley, heath grassland, limestone grassland, moorland and river flood meadows. Mostly from pitfall traps which have yielded huge numbers of specimens. Adults of both sexes May to November, with activity peak in June–July.

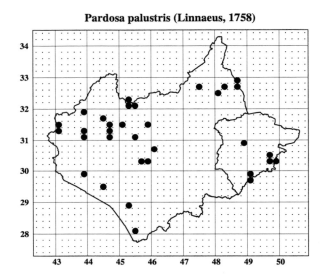
Pardosa palustris (Linnaeus, 1758)

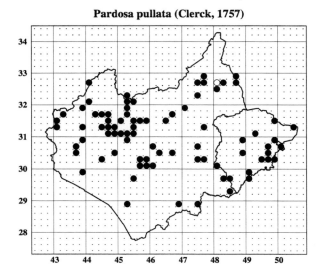
Pardosa pullata (Clerck, 1757)

Pardosa pullata. One of Leicestershire's most common spiders, first recorded by Mayes, 1934. Widespread throughout the county in a very wide range of habitats and sites. Mainly an open ground and ground zone species, found in both dry and wet grassland, it occurs in woodland glades and along rides, but always at ground level. In the garden *Pardosa pullata* females can be observed to establish their own territorial zones, especially on rockeries with short vegetation and will defend this territory until the eggs hatch. Adults of both sexes recorded April to November with main activity in June. Females with egg sacs July to November. The late cocoons are often parasitised.

Atlas of Leicestershire and Rutland Spiders
LYCOSIDAE

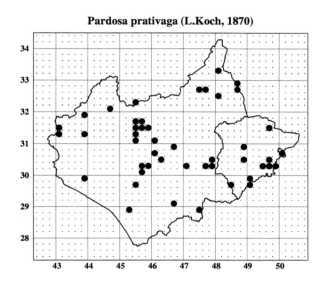

Pardosa prativaga (L.Koch, 1870)

Pardosa prativaga. Similar to the previous species, with which it often occurs. Much less frequent than *P.pullata*, *P.prativaga* is nevertheless widespread and common throughout Leicestershire, where it is recorded from 45 different sites, generally in the same habitats as *pullata*, but more often from wetter situations. These include wet flushes in woodland, flood meadows, damp hollows, sedge-beds, reed-swamp, sphagnum bog and *Glyceria* swamp. Otherwise in a wide variety of grassland and derelict sites. First recorded for the county by P.H.Gamble in 1961 from Swithland Wood. Large numbers of adults of both sexes are recorded for June and July with males and females occurring May to November.

Pardosa amentata. Widespread, and one of the county's most commonly encountered spiders, this ubiquitous 'wolf spider' is found in the same habitats as *Pardosa pullata*, often the two species occurring together. Easily separated in the field, *Pardosa amentata* is a larger spider and has clearly annulated legs. It favours open sunny situations but prefers damp, even wet places. The Leicestershire data show adults of both sexes occurring between April and November, with highest numbers in May, June and July, but males are scarce after July. Females with egg sacs are recorded from May to November. The species was first recorded for Leicestershire by Chalcraft (Rowley,1897).

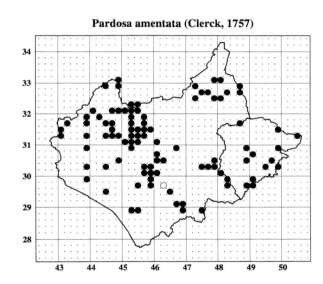

Pardosa amentata (Clerck, 1757)

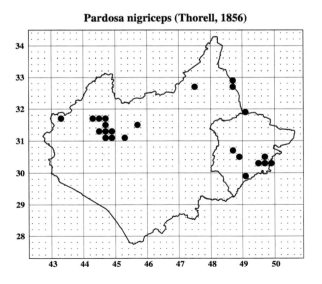

Pardosa nigriceps (Thorell, 1856)

Pardosa nigriceps. Widespread and common in open field layer vegetation. Readily climbs tall vegetation when hunting and is particularly active in sunshine. Though recorded from two woodland sites (Buddon Wood and Cooper's Plantation) these records were from open areas not closed in by the canopy. First recorded by J. Crocker (1962) from Bradgate Park. The 69 subsequent records are from moorland, heathland and a wide range of scrubby grassland, obtained by beating, sweeping and pitfall trapping. As with other lycosids maturity and sexual activity is phased to late spring and early summer, with peak numbers in June. Females April to November; males April to July.

CHAPTER FIVE

Atlas of Leicestershire and Rutland Spiders
LYCOSIDAE

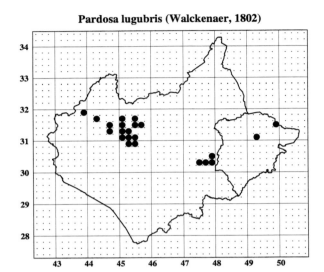

Pardosa lugubris (Walckenaer, 1802)

Pardosa lugubris. Widespread and often locally abundant in woodland, but with few records away from Charnwood Forest, where it is probably under-recorded. It occurs in or near woodland in sunlit clearings and at the woodland edge, on the ground and in the field layer. Away from Charnwood *Pardosa lugubris* has been collected from Exton churchyard, Lount, Launde Big Wood, Loddington Reddish, Pickworth Great Wood and Skeffington Wood. Females have been collected during May, June, July and September, and males during April, May and June, with highest numbers in May. Females with egg sacs – June to September. First record, P.C. Jerrard, 1962, Bradgate Park.

Alopecosa pulverulenta. Widespread but local. This is a well marked and distinctive field layer species of the leaf and stem zone (15-50cm), occurring on open areas of heathland, grassland, and occasionally gardens and stony wasteland. First record for the county – T.A. Walden, 1948, Waltham-on-the-Wolds. Most of the 143 records for Leicestershire are from pitfall traps, with some additional hand collected specimens from the ground zone, occasionally from under stones, and by sweeping field layer vegetation. Adults occur from April onwards, females being recorded through until November and males until September, with very large numbers in June.

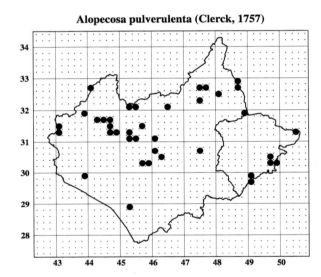

Alopecosa pulverulenta (Clerck, 1757)

Alopecosa barbipes (Sundevall, 1833)

Alopecosa barbipes. Very rare in Leicestershire, recorded from only a single site. First record E.Caradine, 1994, Geeston Quarry, and subsequently from the same site by I. Philips during a ten-month pitfall trapping project. This handsome spider prefers dry open habitat where the vegetation is very short, typically found at Geeston. The male is easily recognised in the field by its elongated swollen tibiae on the first pair of legs. Specimens were obtained by hand collecting amongst sparse vegetation adjacent to scrubby grassland, otherwise from pitfall traps on the bare coarse substrate and in short turf. The main activity period here is March to May, with the first appearance of both sexes in February.

Atlas of Leicestershire and Rutland Spiders
LYCOSIDAE

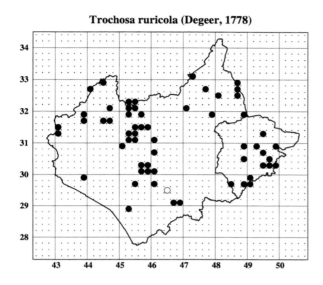

Trochosa ruricola (Degeer, 1778)

Trochosa ruricola. Widespread and common, this nocturnal 'wolf spider' is often found under stones, pieces of wood, logs *etc*. It prefers damp situations and will form retreats under piles of garden rubbish. First recorded for Leicestershire by Chalcraft (Rowley,1897), the species is now shown to be well distributed throughout the county in damp grassland, river margins, marshes, disturbed ground, gardens, allotments, manure heaps *etc*., also in deciduous woodland. Some females can be quite aggressive, particularly when guarding eggs. Adults of both sexes are recorded from March to November (probably mature all year). Females with ova noted May to September.

Trochosa robusta. Very rare in Leicestershire, this nationally notable (Nb) spider is recorded from a single site at Ketton Quarry (SK976053), where one male was collected by J. Daws on 21st March 1993, amongst stones and rough grass within limestone grassland. Despite follow-up hand searches, no further specimens have been found. It may be necessary to carry out a series of pitfall trap experiments to ascertain the status of this species and its relationship with other members of the genus. *Trochosa robusta* is a Breckland species and is mostly restricted to chalk grassland, conditions which are generally replicated at Ketton, where it is assumed this spider is established.

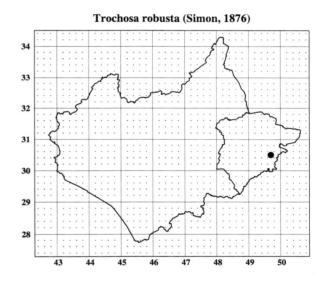

Trochosa robusta (Simon, 1876)

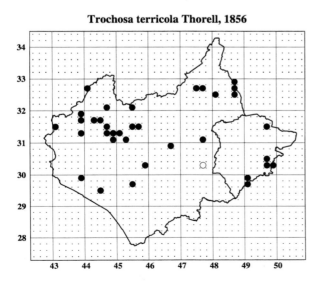

Trochosa terricola Thorell, 1856

Trochosa terricola is widespread and common throughout the county, and often found in the same situations as *T.ruricola* but with fewer records. It does, however, seem to prefer the drier heathy conditions of Charnwood Forest and north-west Leicestershire, where it has been recorded typically from ungrazed *Nardus* grassland, rocky *Vaccinium/ Calluna* heathland, heath birch/oak scrub and bracken dominated pasture woodland. Elsewhere it is recorded from deciduous woodland, limestone grassland, rough neutral grassland and derelict industrial sites. The phenology of this species mirrors that of *T.ruricola*, with highest numbers in June. First record E. Duffey, 1955, Loddington.

CHAPTER FIVE

Atlas of Leicestershire and Rutland Spiders

LYCOSIDAE

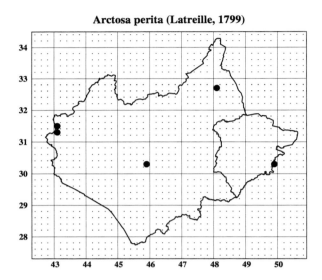

Arctosa perita (Latreille, 1799)

Arctosa perita. A local species restricted to sandy places and dry heaths where the soils are of light texture, enabling the spider to construct its retreat burrow in the substrate. First recorded by Lowe, 1912, the species has been recorded from 7 sites, 4 of these quite close together – Acresford Sandpit, Donisthorpe Colliery spoil tip, Newfield Heath and Moira Junction. Elsewhere, from the derelict C.E.G.B. site (wasteland in the centre of Leicester), and from Geeston and Waltham disused quarries (on an ash tip and sandy rabbit-grazed turf). See Arnold and Crocker (1967). Females have been taken from May to November and males March to June and November, suggesting adults may overwinter.

Pirata piraticus. Widespread and common, this attractive water-loving spider is found mostly in wet situations, but occasionally in damp depressions on flood meadows, well vegetated marshes and seasonally dried out ponds. The majority of the 80 records for Leicestershire are from margins of established water bodies from small ponds to rivers, where it is found in fen, reedswamp and on bare mud. First recorded for the county by T.A. Walden in 1948 from Scraptoft and now known from 43 sites, it is certain, however, to be found in many more locations. Adult females are recorded May to October and males May to August. Females with ova June to October. Highest numbers June and July.

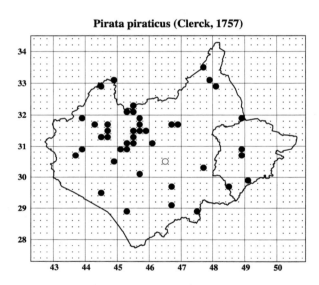

Pirata piraticus (Clerck, 1757)

Found in the same wet habitats as the previous species, ***Pirata hygrophilus*** is more tolerant of drier conditions and also occurs in rough grassland, scrub and wooded areas. It is a much darker spider than *Pirata piraticus* and, judging by Leicestershire records, seems to have a longer maturity period, females occurring from April to November and males April to July, with peak activity in June and July. This species was first recorded for the county by J. Crocker in 1961, with specimens from Wardley Wood. It is widespread and locally abundant in the county, being recorded from 24 sites, mostly waterside and marsh habitats, and often found with *P. piraticus*.

Pirata hygrophilus Thorell, 1872

Atlas of Leicestershire and Rutland Spiders
LYCOSIDAE, PISAURIDAE

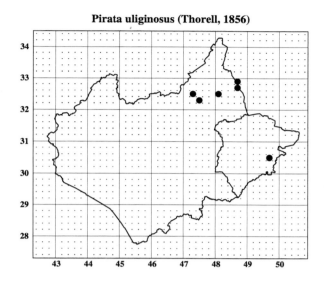

Pirata uliginosus is a local species, first recorded for Britain in 1951 from limestone grassland at Wytham, Berkshire (Duffey, 1953). First discovered in Leicestershire by H.A.B. Clements in 1963 at Holwell Mouth, it has since been taken at Brown's Hill, Holwell North, Ketton, and Stonesby Quarries, and at King Lud's Entrenchments and The Drift – all ironstone and limestone grassland sites. These habitats are much drier than those in which *P.piraticus* and *P.latitans* are found. Mature in spring and summer, the peak activity of *P.uliginosus* at King Lud's occurred in June and July, with highest numbers in July (36f 160m in one pitfall trap placed in rough calcareous grassland).

Very localised in Leicestershire, but widespread, **Pirata latitans** is recorded for the county from only four sites: Botcheston Bog, Bradgate Park, Holwell Mouth and Ulverscroft N.R. It is a spring and summer maturing species of wet habitats, locally occurring around ponds, in a valley bottom marsh, in wet flushes in marshy meadows and in an area of wet fen at the side of a hay meadow. In each case, specimens were from the ground zone (0–8cm) in wet depressions or from juncus and moss, again in very wet conditions. Females have been taken in June and July, and males in May and July. First recorded for Leicestershire in 1961 by I.M. Evans, from Bradgate Park.

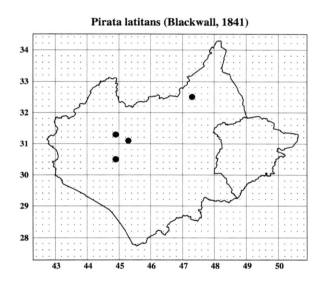

Pisaura mirabilis. Common and widespread, this distinctive spider is well established in the county, but would appear to be under-recorded. It is a field-layer species of rough grassland, heather and open woods, and seems equally at home in gardens, on wasteland and roadside verges as in woodland clearings, scrubby grassland and herb-rich meadows. *Pisaura* is essentially a summer spider with adult females active during May to September, and the occasional late specimen in October, often with parasitised ova. Males occur in May, June and July, and high numbers of sub-adults in April and May. Tents are common in July and August. First record Chalcraft (Rowley, 1897).

CHAPTER FIVE

Atlas of Leicestershire and Rutland Spiders
ARGYRONETIDAE, AGELENIDAE

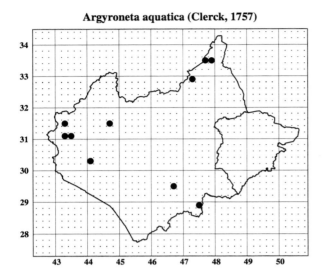

Argyroneta aquatica (Clerck, 1757)

The 'water spider' *Argyroneta aquatica* is widespread but local in Leicestershire, probably decreasing as a result of water pollution and loss of suitable habitat. This aquatic spider requires clean, well vegetated water with very little current, and is well established in the Grantham Canal, Ashby Canal and the Grand Union Canal at Kibworth. Elsewhere, *Argyroneta* has been recorded in ponds and lakes at Market Bosworth, Newfield Heath and Charnwood Lodge N.R., also at Great Bowden Pit. First recorded for the county by I.M. Evans in 1960 from the Grand Union Canal. Adult throughout the year, but apparently more active between April and September.

Agelena labyrinthica. This common southern species is rare in Leicestershire and recorded locally from only 5 sites in the east of the county, all in Rutland and mostly on calcareous grassland (Duddington roadside verge, Geeston Quarry, Ketton Quarry, Seaton Meadow and Tixover Quarry). The large sheet webs constructed by this spider are very conspicuous and, where they occur, often numerous. Open, herb-rich grassland is preferred as this habitat carries a larger number of prey species such as grasshoppers and crickets. Webs have been found on short rabbit-grazed limestone turf, tall dry scrub grassland and herb-rich hay meadow. Adult in summer. First record Lowe, 1912.

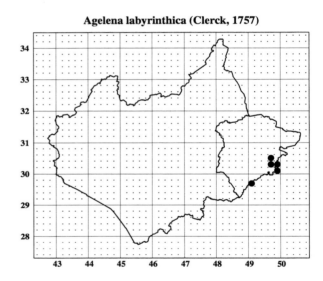

Agelena labyrinthica (Clerck, 1757)

Widespread but local, ***Textrix denticulata*** is common on Charnwood Forest, where it is found in dry-stone walls, under stones and rocks, and inside houses. Elsewhere it has been recorded in churchyards, abandoned quarries, scrub grassland and woodland. The spider is occasionally encountered on bare ground amongst sparse vegetation, when it can be mistaken for a 'wolf spider'. It is a sheet-web builder and will sometimes be found on bushes and low vegetation, where the webs are constructed. Females are adult throughout the year and overwinter in silken retreats in crevices. Males are recorded in May, June and July. First county record, Chalcraft (Rowley, 1897).

Textrix denticulata (Olivier, 1789)

Atlas of Leicestershire and Rutland Spiders
AGELENIDAE

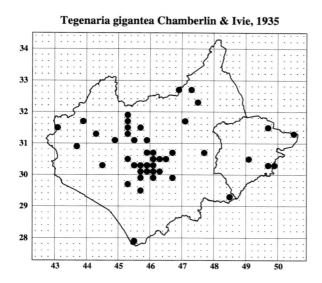

Tegenaria gigantea Chamberlin & Ivie, 1935

Tegenaria gigantea. This common 'house spider' is widespread in Leicestershire and well established in buildings and gardens throughout the county, apparently to the exclusion of competing species such as *T.saeva* and *T.atrica*. Considering the successful exploitation of suitable habitats (houses, garages, outbuildings and garden shrubs and structures) by *T.gigantea* it is surprising that there are so few records for the county, and the species must be significantly under-recorded. Adult females are recorded for every month of the year and males from April to November, with a single record for January. First county record Chalcraft (Rowley, 1897).

Very similar in appearance to the foregoing species, ***Tegenaria saeva*** is an invasive spider, exploiting the same habitat as *T.gigantea*, and elsewhere (Yorkshire) is known to be interbreeding with and replacing the former resident *gigantea*. Great care is needed in differentiating these species, and a microscope is necessary. *T.saeva* is recorded for Leicestershire from a single female handed in to Leicester Museum in 1989 by B.E. Wills. This was taken inside a Leicester house, where it was found in a bowl of fruit. There is also an unconfirmed report of this species from Ashby-de-la-Zouch, but no specimen has been seen. A detailed survey of the *Tegenaria* species in Leicester would be worthwhile.

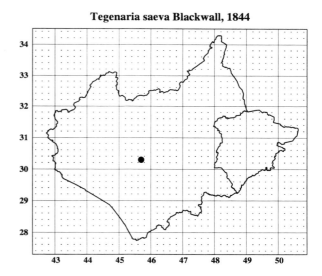

Tegenaria saeva Blackwall, 1844

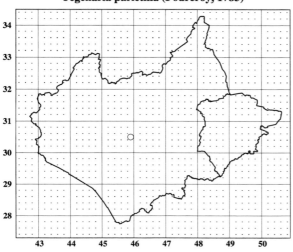

Tegenaria parietina (Fourcroy, 1785)

Tegenaria parietina. The status of this spider as a Leicestershire species has been discussed on pages 37 and 48. Now very rare in the county it is established by old records. Probably quite frequent at one time in old buildings, warehouses *etc.*, but likely to have been reduced in numbers following the post-war rebuilding of much of old Leicester. However, this handsome but intimidating spider is still being recorded from new sites outside the county – transported by the agency of man – so could still turn up again within the Shire. First record Headly (Mott, 1890); second reference Mayes, 1935 (see p.40); last reference Foxwell, 1961 (in Evans, 1963), all from the city of Leicester.

Atlas of Leicestershire and Rutland Spiders
AGELENIDAE

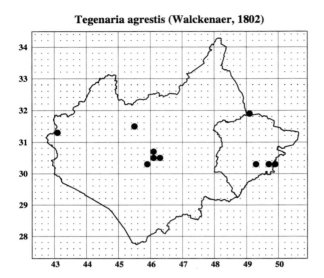

Tegenaria agrestis (Walckenaer, 1802)

Tegenaria agrestis is a local, free-living species recorded from 10 sites in the county: Acresford Sandpit, Buddon Wood Quarry, C.E.G.B. Rawdykes, Geeston Quarry, Humberstone Quarry, Leicester Cattle Market site, Morcott Gullet, North Luffenham Quarry, Thistleton Gullet and an urban garden in Leicester – all disturbed habitats. From amongst vegetation, under stones and debris and in dry-stone walls. First record J. Clarke, 1960, from an old fishtank lying in his garden (Leicester). Adult females recorded throughout the year, with males in September and November. A foreign species new to Britain in 1949 and since then extending its range northwards from Dorset.

The ubiquitous 'house spider' of domestic dwellings and outbuildings, **Tegenaria domestica** is widespread and common throughout Leicestershire. Transported by man in boxes, furnishings and artefacts it does not take this species long to colonise new houses. With only 44 records from 30 sites in 23 tetrads, *T.domestica* is grossly under-recorded. Being smaller than *T.gigantea* and much less conspicuous, its presence is likely to be noticed more by their webs behind pictures on a wall or books on a bookshelf, than by the spiders themselves. Adults of both sexes occur throughout the year. First recorded for Leicestershire by Chalcraft (Rowley,1897).

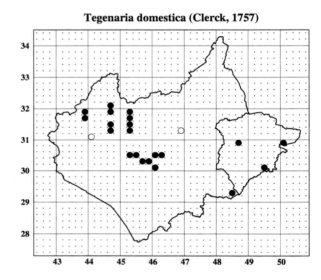

Tegenaria domestica (Clerck, 1757)

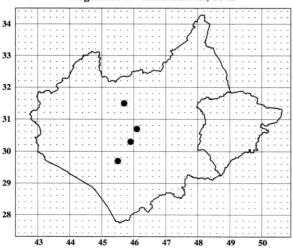

Tegenaria silvestris L.Koch, 1872

Tegenaria silvestris is a local, free-living spider, rare in Leicestershire and known from only 4 sites: Buddon Wood, C.E.G.B. Rawdykes, Gypsy Lane wasteland and Narborough Bog. In woodland it occurs generally in damp situations with good structure where sheet webs can be constructed, such as low banks and ditches, overhanging grass, tree stumps, natural debris and in open vegetation along rides. *Tegenaria silvestris* will take advantage of rubbish and coarse vegetation to support its web. The spider will retire behind bark, and under stones and logs. Adult females May, June and December, and a male in June. First recorded by Lowe, 1912.

Atlas of Leicestershire and Rutland Spiders

AGELENIDAE

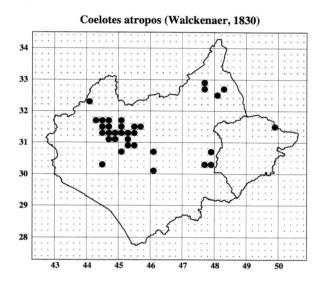

Coelotes atropos (Walckenaer, 1830)

Coelotes atropos. Widespread and common, particularly so on Charnwood Forest where it is found under rocks, stones, bark, logs and discarded litter, preferring damp situations. The majority of records are from woodland and rocky heathland, but the species is also recorded from grassland, reservoir environs, brickpits and quarries, churchyards and farmyards. *C.atropos* is a powerful spider and the remains of quite large beetles can often be seen scattered around the entrance of its retreat. Probably adult all year – mature females recorded February to November, and males March to November, with high numbers of both in November. First record Chalcraft (Rowley, 1897).

Cicurina cicur. Local, occurring in three distinct and widely separated localities within the county, all having a limestone connection. The first locality comprises Geeston, Ketton and North Luffenham Quarries, with open stony ground and short calcareous turf. The second site, Thistleton Gullet, is a former ironstone quarry in Jurassic limestone, and the third location, in west Leicestershire at Lount, is on the site of early nineteenth century lime kilns. Specimens were mostly collected on bare, and sparsely vegetated, disturbed ground in open situations, also in woodland litter. Adult females October to April, and males November and January. First county record D.A. Lott, 1991, Lount.

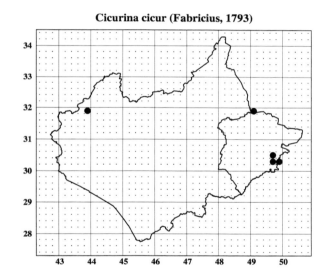

Cicurina cicur (Fabricius, 1793)

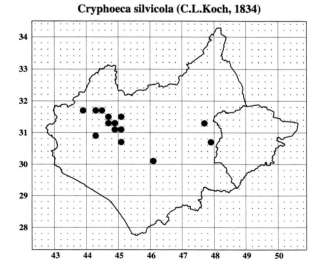

Cryphoeca silvicola (C.L.Koch, 1834)

Widespread but infrequent, except on Charnwood Forest where it is well established, ***Cryphoeca silvicola*** is confined to woodland and open ancient woodland sites. Away from the Forest it is recorded from Burrough Hill, Coleorton Hall woods, Knighton Spinney, Martinshaw Wood, Nailstone Wiggs and Owston Wood. *Cryphoeca* is found mostly in leaf litter and twiggy growths around the base of trees, but also under bark, under stones, amongst moss and in grass tussocks. Females are recorded from February to November and males February to May and again in September and October, with high activity levels in the spring. First record J. Crocker, 1961, Coleorton.

CHAPTER FIVE

Atlas of Leicestershire and Rutland Spiders
AGELENIDAE, HAHNIIDAE

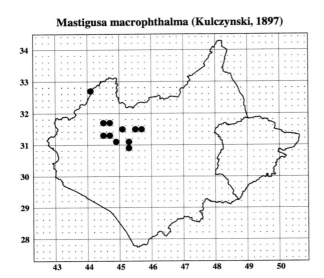

Mastigusa macrophthalma (Kulczynski, 1897)

Mastigusa macrophthalma. This distinctive and vulnerable spider is a national rarity (RDB3) with its stronghold on Charnwood Forest, where it is recorded from ten tetrads. There are two main habitats (Crocker, 1973), both associated with ancient oak forest. As at Bradgate Park, where *Mastigusa* was first discovered (Donisthorpe, 1927), and Donington Park, it has been found in rotten wood and beetle larvae borings behind dead oak bark on dead and over-mature trees. At Altar Stones, Bardon Hill, Beacon Hill, Buddon Wood, High Sharpley, Hill Hole and Ives Head, under large rocks and stones in peaty soil, usually in ants' nests. Adults of both sexes and sub-adults are recorded April–November; probably mature all year.

Antistea elegans. Widespread in north-west Leicestershire but local, usually in very wet situations: canal, reservoir and river environs, woodland ponds, marshes, and wet grassland. At the first mentioned sites specimens were collected from wet mud around the base of emergent vegetation and in rotting litter in reedswamps and *Phragmites* beds. Similarly at the other sites from wet litter, and around the wet basal region of *juncus*, grasses, moss and other marsh plants. Mature females from February to October and males in August and September only. Sub-adult males are recorded for April, June and October suggesting summer maturity. First county record I.M. Evans, 1962, Holwell Mouth.

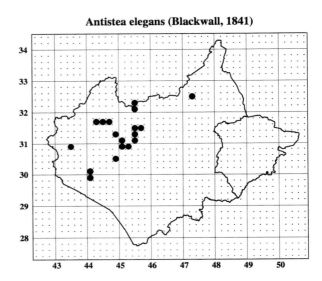

Antistea elegans (Blackwall, 1841)

Hahnia montana (Blackwall, 1841)

Hahnia montana. Widespread and common, particularly on Charnwood Forest, with 63 records from 18 sites. Prefers damp situations but not too wet. Taken in woodland leaf litter, pine needles, moss and at grass roots; also from ground zone detritus in calcareous and heathy grassland, *Calluna/Vaccinium* heath, and under stones in a derelict quarry. In Leicestershire, adult females are recorded for every month of the year, but males only in July, August, September and November. Since highest numbers of females occurred in April this would indicate that males also are mature throughout the year, but overlooked. First record H.A.B. Clements, 1962, C.L.N.R.

Atlas of Leicestershire and Rutland Spiders

HAHNIIDAE, MIMETIDAE

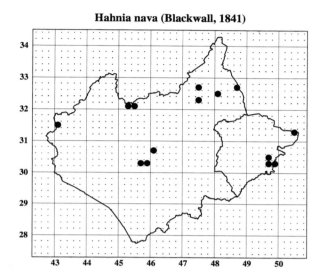

Hahnia nava (Blackwall, 1841)

A local spider but well represented in Leicestershire, with 78 records from 14 widespread sites, ***Hahnia nava*** occurs in grassland, heathland and open wasteland. First recorded in 1992 by J. Daws from St. Mary's Allotments, Leicester, it has since been found at Brown's Hill Quarry, C.E.G.B. Rawdykes, Essendine churchyard, Evesham Rd. Allotments, Leicester, Geeston Quarry, Harby Hills, Humberstone Quarry, Ketton Quarry, King Lud's, Loughborough Meadows, Moira Junction, North Luffenham and Stonesby Quarries. From under stones, grass tussocks, cleared ground, short turf, rank grass and tall herbs. Adult females throughout the year, males January–July, with high numbers of males May to early July.

Hahnia helveola. A local, woodland species only recorded from the Charnwood Forest area, firstly by J. Crocker in 1962, at Loughborough Outwoods, and subsequently from Blackbrook Reservoir, Buddon Wood, Stoneywell Wood and Swithland Wood. Most of the 16 records are from Buddon Wood, where it was widespread in leaf litter around oak and hazel, amongst grass roots between rocks, and in clumps of woodrush. Elsewhere it has been taken in pine needle litter, in acidic grassland dominated by bracken, under heather, and in damp vegetation at side of reservoir. Possibly mature throughout the year, but the small samples have yielded only adult females January to October, with no males.

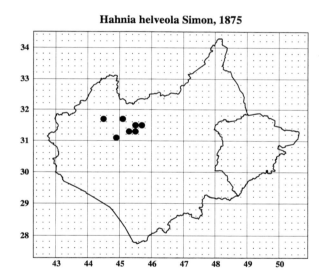

Hahnia helveola Simon, 1875

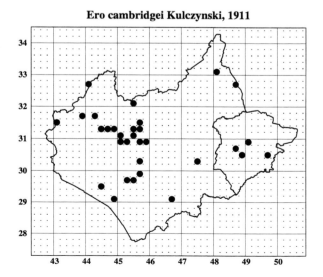

Ero cambridgei Kulczynski, 1911

Ero cambridgei. Widespread and fairly common, amongst grass, low vegetation and shrubs. First record for Leicestershire, J. Crocker, 1968, Coleorton rough grassland, and since then from a further 31 woodland, grassland, marsh, fen and reedswamp sites. The spider has been found under logs, in vole runs, leaf litter and under heather by grubbing and sieve & sort techniques. It has been shaken from sedge/grass overhanging a stream and found in a dense *Carex paniculata* tussock, also beaten from gorse, low vegetation and young deciduous trees. Adult females have been taken every month of the year except December, and mature males February to April and August to October.

Atlas of Leicestershire and Rutland Spiders

MIMETIDAE, THERIDIIDAE

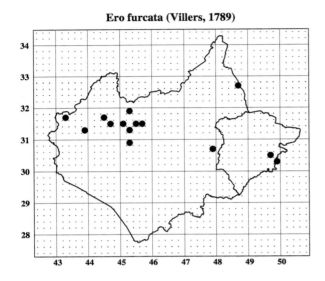

Ero furcata (Villers, 1789)

Ero furcata. In similar situations to *E. cambridgei*, but less common, and sometimes found together. There are 33 Leicestershire records from 14 sites: deciduous and mixed woodland, hedgerows, pasture woodland, wet heath and moorland, grassy scrub heathland, *Vaccinium/Calluna* heath, rough grassland (neutral and calcareous), disused ironstone and limestone quarries and a suburban garden. From damp situations amongst grass roots, in *sphagnum*, under stones, bark and pieces of wood, and sweeping long grass and low vegetation. First record (Wild, 1952). Adult females recorded from March to November and males every month from June to November, also February.

Episinus angulatus. Very local and rare in Leicestershire, recorded from two widely separated localities: Buddon Wood and The Brand on Charnwood Forest, and Geeston and Ketton Quarries in Rutland. The species was first recorded for the county by Lowe, 1912, but his location is unknown. In Charnwood this distinctive spider is associated with heather among scrub oaks, and undisturbed grass along woodland rides, obtained by sweeping at ground level. The Buddon site is outside the permitted quarrying area. Elsewhere *Episinus* has been taken in herb-rich grazed limestone grassland and rough grass with tall herbs by pitfall trapping. Adults May, June and July.

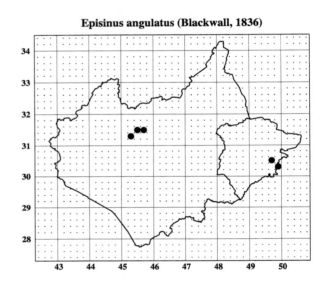

Episinus angulatus (Blackwall, 1836)

Euryopis flavomaculata (C.L.Koch, 1836)

Euryopis flavomaculata. Very local, and rare in Leicestershire, this spider is known only from two localities in west Leicestershire, material from both sites turning up in pitfall traps. First recorded for the county from Moira Junction marshy grassland adjacent to the dismantled railway line, by J. Daws in 1992, *Euryopis* has also been taken there in pitfall traps in cleared ground with a little heather. At the other site – Buddon Wood – the pitfall traps were placed amongst grass on a steeply sloping rocky ridge within the open birch/oak woodland. Adult males occurred in June, July and August with highest numbers in June, whilst the only mature females appeared in July.

Atlas of Leicestershire and Rutland Spiders
THERIDIIDAE

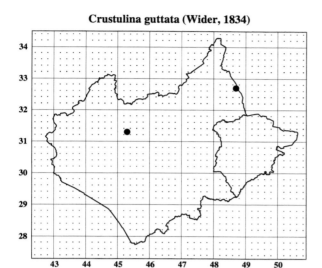

Crustulina guttata (Wider, 1834)

Crustulina guttata. Rare, only males, from two sites: The Brand and King Lud's Entrenchments. First recorded for Leicestershire by Chalcraft (Rowley, 1897), this early record is suspect but stands on the basis of subsequent records. The Brand site is a series of landscaped old slate quarries, surrounded by mixed woodland, with an acidic substrate. Specimens were hand collected from bracken and grass litter at the base of a rock face. At King Lud's Entrenchments, *Crustulina guttata* males were caught in pitfall traps placed in rough calcareous grassland at the edge of scrubby woodland. This is a summer maturing species and all 5 records are of males in June and July.

Steatoda bipunctata. Widespread and common, this ubiquitous species is grossly under-recorded in the county. With only 88 records from 48 sites in 43 tetrads there is plenty of scope for a massive improvement in distributional coverage of the county, for *Steatoda bipunctata* must occur in every terrestrial tetrad where there are buildings, old trees and rubbish dumps. First recorded for Leicestershire by Chalcraft (Rowley, 1897) this distinctive spider is easily recognised, even in its immature stages. It is one of the common inhabitants of old and dead trees, where it occurs under bark and amongst dry litter inside hollow trunks and branches. Adults of both sexes throughout the year.

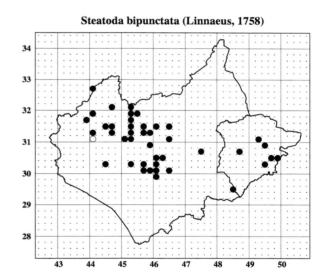

Steatoda bipunctata (Linnaeus, 1758)

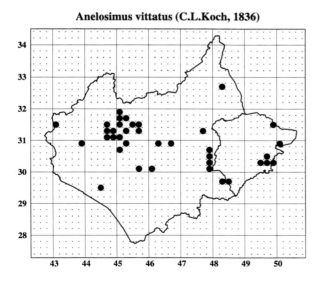

Anelosimus vittatus (C.L.Koch, 1836)

Anelosimus vittatus. Widespread and common, obtained mostly by beating. First county record P.H. Gamble, 1962, Buddon Wood, subsequently recorded from a further 36 sites. Old and young deciduous and mixed woodland, hedgerows and gorse scrub, also rough grassland – particularly with tall herbs and scrub. Additionally from reservoir margins, disused quarries, railway embankments, churchyards and gardens. Mainly by beating foliage of trees (oak, birch, hawthorn, spruce) and shrubs (particularly gorse). Also sweeping (grass, tall herbs, heather) and occasionally in leaf litter and on tree trunks. Adult females May–November, males May–July, high numbers in June.

CHAPTER FIVE

Atlas of Leicestershire and Rutland Spiders

THERIDIIDAE

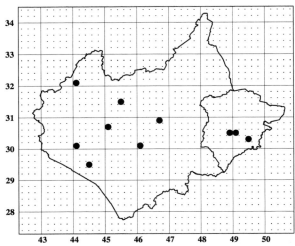

Achaearanea lunata (Clerck, 1757)

Achaearanea lunata. A local southern spider, widespread in Leicestershire but uncommon, with 13 records from 10 sites. First county record J. Crocker, 1977, Buddon Wood and since then from Barkby Holt Wood, Bosworth Duckery, Burbage Wood, Cloud Wood, Knighton Spinney, Luffenham Heath, Lyndon Wood, Martinshaw Wood and Rutland Water. From woodland sites, on the lower branches of trees and shrubs, especially in dark or shady situations, obtained mostly by beating oak and gorse, but also sweeping heather, and from inside an old summerhouse and inside a newly constructed bird-watching hide. Mature females April to August, and males May, June and July.

Rare in Britain, this nationally notable (Nb) species is very local in Leicestershire with only 6 records from 5 sites. ***Achaearanea simulans*** was first recorded for the county by J. Daws, 1992, from Moira Junction, where a female was taken in a pitfall trap on cleared ground with a little heather. The remaining specimens, occurring in similar situations to *Achaearanea lunata* were obtained by beating oak/ash at Knighton Spinney, yew at Martinshaw Wood and (oak?) at Lyndon Wood, also by sweeping field layer vegetation around Shawell churchyard, which is surrounded by tall hedges and trees. Only one female has been taken (in September); males are recorded in May, June and July.

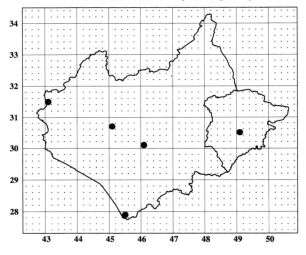

Achaearanea simulans (Thorell, 1875)

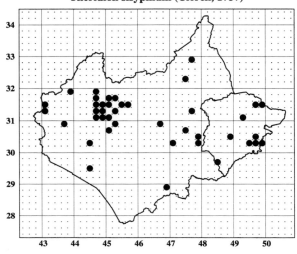

Theridion sisyphium (Clerck, 1757)

Theridion sisyphium. Widespread and common, especially on gorse. First record Chalcraft (Rowley, 1897). 89 records from 50 sites in 41 tetrads. Webs on gorse bushes and other shrubs, also on tall herbs especially along hedgerows. From deciduous, mixed and conifer woodland, along rides and woodland edge, gorse scrub, hedgerows, roadside verges and rough scrubby grassland. Also from disused sandpits, brickpits and quarries, railway sidings and churchyards. Beating gorse, oak, birch, conifers, hawthorn and blackthorn, and sweeping tall herbs, grass, nettles and heather. Females May to September, males May, June and July. Main activity in June with females and ova in June and July.

Atlas of Leicestershire and Rutland Spiders
THERIDIIDAE

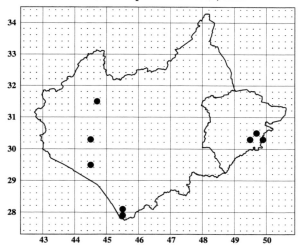

Theridion impressum L. Koch, 1881

Very similar in appearance and habits to the previous species **Theridion impressum** is widespread and local in Leicestershire, with 14 records from 8 sites. First recorded from Newbold Verdon in 1969 by M.J. Roberts, this species has also been taken at Burbage Common, C.L.N.R., Geeston and Ketton Quarries, Luffenham Heath, Shawell churchyard and Shawell Gravel Pits. The habitat is the same as for *T.sisyphium* and specimens were collected by beating shrubs, mostly gorse, and yew trees, also sweeping heath grassland with heather and bilberry. This species has a short summer season, females maturing in June, with males active in July. No records after August.

Theridion pictum. Another local species, widespread throughout the county but uncommon. Found mainly in damp situations, on shrubs and low vegetation: on gorse, oak, blackthorn and bilberry scrub, nettles, *juncus* and waterside vegetation, also from *glyceria* litter and amongst *sphagnum* and cotton grass. First record D.B. Forgham, 1973, Buddon Wood, and subsequently from Ashby Canal, Great Bowden Pit, River Soar and River Welland environs, Saddington Reservoir, Sharnford Meadows and Shawell working gravel pits. Adult females are recorded for June, August and September, and adult males in May, June and July. Sub-adults have been recorded June and September.

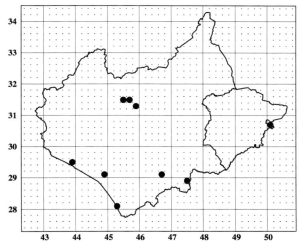

Theridion pictum (Walckenaer, 1802)

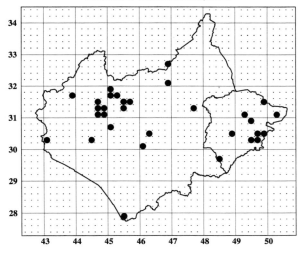

Theridion varians Hahn, 1833

Theridion varians is widespread and common in Leicestershire with 63 records from 32 sites in 31 tetrads. Obtained almost exclusively by beating. Very common on gorse and oak scrub in a wide variety of situations; rough grassland – particularly with tall herbs and scrub, heathland, hedgerows, woodland including conifer plantations, wet carr and thorn scrub. Also from structures and artefacts, gardens – on buildings and vegetation; and frequent in churchyards – many records from yew trees and box, on gravestones and walls. Adult females recorded for each month from May to October, and males May through to August, with activity peak in June. First record (Bristowe, 1939).

Atlas of Leicestershire and Rutland Spiders

THERIDIIDAE

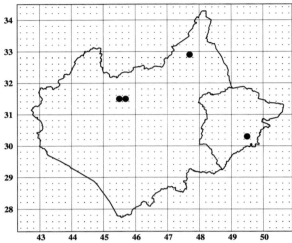
Theridion simile C.L.Koch, 1836

Theridion simile. A southern species, on bushes and low vegetation; in Leicestershire almost exclusively on gorse and heather. Rather rare in the county and very localised, it is known from only four sites. First recorded by J. Crocker in 1975 from 75cm high heather bushes on rocky ground in Buddon Wood. Since then, also recorded from nearby Rothley Common (SK5614) on gorse, Stathern (SK7629) on gorse in pasture land, and at Luffenham Heath Golf Course where the species was found to be well established on gorse. Adults of both sexes have been taken, by beating and sweeping gorse and heather, in May, June and July, with highest numbers in May and June.

Theridion melanurum is widespread and common, usually on or around buildings but also on vegetation well away from houses. Indoors it is often found on window ledges and in the corners of window frames, ceilings and walls. This is a very common species, and the few records for Leicestershire belie its true distribution. First recorded for the county by G. Smith, 1960 from Scraptoft, and subsequently from houses, stables, offices, churches and outbuildings. Also from under pieces of rock in a working quarry, on garden shrubs and under the bonnet of a car! Adult females are recorded from March to August, and females with eggs in June and July. Mature males February to July. See notes in the next species.

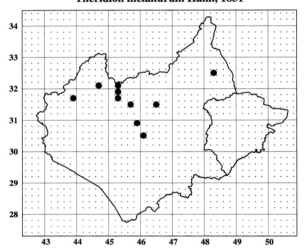
Theridion melanurum Hahn, 1831

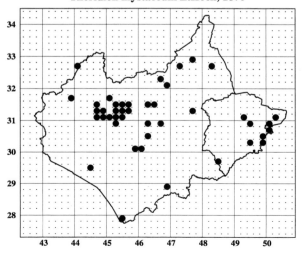
Theridion mystaceum L.Koch, 1870

Similar in appearance to the previous species ***Theridion mystaceum*** is widespread and common outdoors on trees and bushes and is sometimes found overwintering indoors. Though each species has its preference, there is an overlap in habitat, and specimens must be examined under a microscope for identification. The phenological evidence from Leicestershire suggests that *mystaceum* has a slightly later maturity period than *melanurum*, with adult females April to October, males May to August and ova up to August. Mostly on tree trunks, also on foliage, under stones and on walls and structures, such as gravestones. First record J. Crocker, 1961, Coleorton Hall Woods.

Atlas of Leicestershire and Rutland Spiders
THERIDIIDAE

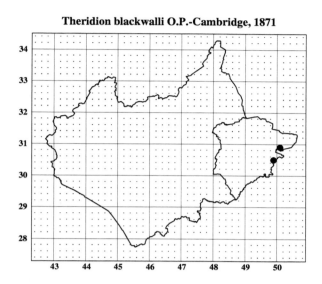

Theridion blackwalli O.P.-Cambridge, 1871

Theridion blackwalli. Very rare in Leicestershire, recorded from only two sites in the extreme east of the county at Great Casterton Church (TF001087) and Ketton Church (SK981043). A male was discovered, new to the county, by J. Daws at the former site in May 1994, at the base of the old stone boundary wall. The second specimen, a female, was collected from Ketton Church on an outside wall, in July 1995. *T.blackwalli* is said to have a very short maturity period, from the end of May to early July, and the Leicestershire specimens fit in with this concept. Mainly of south-eastern distribution, this species is usually found amongst low plants and on tree trunks and buildings.

Theridion tinctum. Widespread but local. Obtained mainly by beating gorse, oak, hawthorn, blackthorn, yew and other evergreens, and sweeping field and shrub layers along open woodland rides. Also on walls, fences and artefacts. Typical habitats in Leicestershire comprise parkland trees and bushes, woodland and scrub gorse, grassland with scrub oak/gorse, churchyards with yew trees and other evergreens, rural structures, domestic buildings and outhouses, gardens and garden structures. Adult females mainly April to November, but also one January record. Males, May to July, with high activity in June. Females with ova are recorded for July and August. First record (Bristowe, 1939).

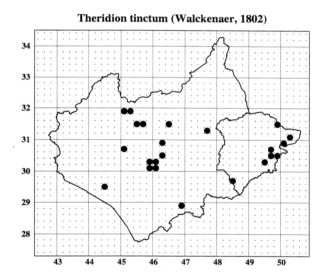

Theridion tinctum (Walckenaer, 1802)

Theridion bimaculatum (Linnaeus, 1767)

Theridion bimaculatum. Widespread and common, this small theridiid is easily recognised; the males are quite distinctive in colour and form, and the female carries her large white egg cocoon, attached to her spinners, on perambulations through grass and low herbage. First recorded by Lowe, 1912, and more recently from a wide variety of sites, often in damp habitats. Woodland, frequently beaten off gorse and scrub oaks; grassland, under stones, amongst grass and from wet flushes, marshes, sedge fen and reed beds; heathland, on heather; and gardens on rockeries and flower beds. Females are recorded for all months from May to October, and males during May, June, July and August.

Atlas of Leicestershire and Rutland Spiders
THERIDIIDAE

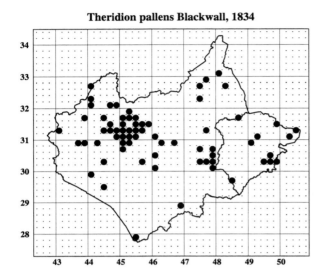

Theridion pallens Blackwall, 1834

Theridion pallens. Widespread and common, in a wide variety of habitats, but mainly females with ova on underside of broad-leaf foliage. Woodland, parkland and wayside trees, hedgerows, bushes and low vegetation; rural and urban, including gardens. Collected by beating and sweeping, and occasionally in leaf litter. On oak, ash, beech, sycamore, birch, willow, hazel, field maple, gorse, hawthorn, blackthorn, alder, sweet chestnut, yew, spruce, pine, larch, ivy and rhododendron; also nettles, heather and marsh herbs. Adult females are recorded for all months April–October, males similarly April–August, and females with ova June–July. Highest numbers May. First record Mayes, 1934.

Enoplognatha ovata. Widespread in Leicestershire, and one of the county's most commonly encountered species, it is easily confused with *E.latimana*, not yet found in Leicestershire. *E.ovata* occurs in three colour forms, often in mixed communities. The female constructs a loose retreat in a curled leaf, in which the bluish egg cocoon is guarded. In a wide range of habitats on low vegetation and bushes, particularly nettles and brambles, and in litter. Woodland, hedgerows, roadside verges, grassland, marshes, waterside vegetation, wasteland, farmyards, allotments and gardens. Females May–October, males June–August, ova July and August. First record Lowe, 1912.

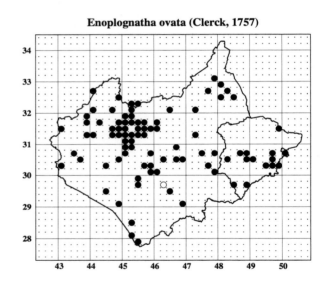

Enoplognatha ovata (Clerck, 1757)

Enoplognatha thoracica (Hahn, 1833)

Enoplognatha thoracica. Widespread but local, from 14 sites in 12 tetrads. First record M.G. Crocker, 1975, Buddon Wood. Also from Brown's Hill Quarry, Cademan Moor, Geeston Quarry, Harby Hills, Ketton Quarry, King Lud's, Lount Colliery reprofiled dirt tip, Luffenham Heath, Moira Junction, Newfield Heath, Saltby disused wartime camp site, Stoke Dry Churchyard and Stonesby Quarry. Generally in damp situations, in grass, under stones and pieces of rubbish, woodland, disused quarries and wasteland with bare ground and sparse vegetation, grassland, heathland and churchyards. Adult females recorded May through to November, and males – April to July, with main activity period in June.

Atlas of Leicestershire and Rutland Spiders
THERIDIIDAE

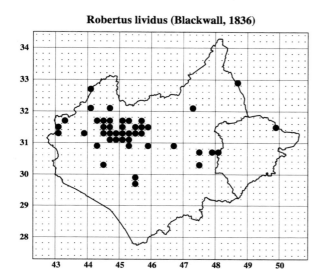

Robertus lividus. Widespread and common, especially on Charnwood Forest. Under stones and rocks, amongst grass roots and moss, leaf and pine needle litter, and in heather, bilberry and bracken detritus. Also under bark, damp rubbish, in pitfall traps and occasionally swept from field-layer vegetation. From moorland, rocky heathland and woodland, also grassland, grassy heath, hedgerows, edges of cultivated fields, river and open water environs, reed-beds, peat bog, bracken dominated parkland, railway embankments, vegetated sand bank, old quarries, refuse tips and gardens. Adults of both sexes recorded every month of the year. First county record Lowe, 1912.

Robertus neglectus. Very rare in Leicestershire, only known from three sites. First recorded by Leicestershire Museums, Arts and Records Service (LMARS) in 1973 from Welby Osier Beds (wet woodland), and subsequently from Luffenham Heath Golf Course (calcareous grassland with large banks of rough grass and scrub) and Owston Wood (ancient woodland site). In each case amongst grass and moss – two males were collected from the ground zone and the third obtained by sweeping. These adult males were taken in March, September and October. Nationally, this species, which is very similar to *R.lividus*, is widespread throughout Britain but uncommon and local.

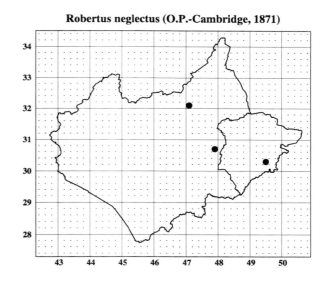

Pholcomma gibbum. Widespread and fairly common in north-west Leicestershire. First county record J. Crocker, 1962, C.L.N.R., and since then from a further 17 sites, only 5 of which are away from Charnwood Forest. Under stones and rocks, and in dry-stone walls, in the ground zone amongst grass roots, bracken and leaf litter, and occasionally swept from grass and heather. From rocky heathland, moorland, heath grassland, bracken dominated furze heath and parkland, woodland, hedgerows, reservoir environs and marshy ground, generally in damp situations. Adult females are recorded for all months – February to November, and males similarly March to November.

CHAPTER FIVE

Atlas of Leicestershire and Rutland Spiders
THERIDIIDAE, NESTICIDAE, TETRAGNATHIDAE

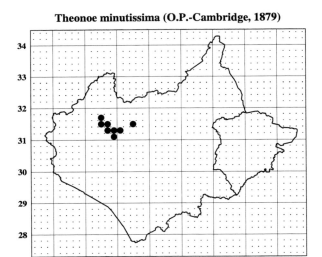

Theonoe minutissima (O.P.-Cambridge, 1879)

Theonoe minutissima. Recorded only from Charnwood Forest where it is very local. Under stones and rocks, amongst heather, bilberry, leaf, pine-needle and bracken litter, *sphagnum* and grass roots, in vole runs and occasionally swept from grass. From rocky *Vaccinium/Calluna* heath, *Molinia* grassland, neutral and acidic scrubby grassland, and deciduous and coniferous woodland. First recorded for Leicestershire by J. Crocker, 1963, from Charnwood Lodge N.R., and since then from Bardon Hill, Benscliffe Wood, Buddon Wood, High Sharpley, Stoneywell Wood and Ulverscroft N.R. Adult females each month February to October and males over the same period, except May and July.

Nesticus cellulanus. Widespread but local, from 15 sites but certainly under-recorded; always in damp, often wet and dark situations. In drains, culverts and on walls of dark, damp tunnels, under manhole covers, inside hollow trees, deep down in rock piles, under corrugated iron sheets and wet fibre board, in dark recesses under tussocky grass and in wet vegetation at water level. From waterside habitats, marshes, rocky and grassy heathland, damp and wet woodland, buildings, gardens, farmyards and churchyards. A large and distinctive 'cave spider', often found in cellars and sewers. Adults of both sexes March to November. First record J. Crocker, 1964, Beacon Hill.

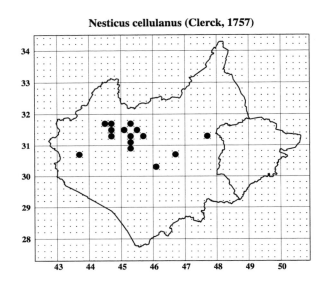

Nesticus cellulanus (Clerck, 1757)

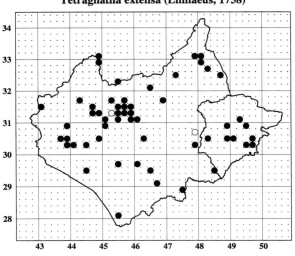

Tetragnatha extensa (Linnaeus, 1758)

Tetragnatha extensa. Widespread and common, locally abundant in reeds and waterside vegetation; a frequent aeronaut, occasionally found in quite dry situations. Obtained by sweeping field-layer vegetation and beating shrubs. Recorded from 56 sites in 57 tetrads, mostly in waterside and marshy habitats. From the sides of rivers, canals, reservoirs, flooded gravel pits and ponds, reedswamps and marshes, flood meadows, wet moorland, grassland, deciduous woodland, shrubby parkland, disused quarries and rough wasteland, rural hedgerows, churchyards and gardens. Adults of both sexes recorded each month May to October. First record Chalcraft (Rowley, 1897).

Atlas of Leicestershire and Rutland Spiders
TETRAGNATHIDAE

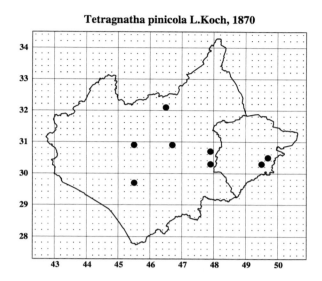

Tetragnatha pinicola. Widespread but rare, usually only single specimens taken. A nationally notable (Nb) species established in the county by records from 8 separated locations. First recorded by O.H. Black in 1972 from Anstey (Near Bradgate Park), and since then from Barkby Holt Wood, Ketton Quarry, Launde Big Wood, Luffenham Heath Golf Course, Narborough Bog, Owston Wood and Twenty Acre Piece (Six Hills). On shrubs, rough grass and tall herbs, obtained by beating and sweeping. From woodland edge, open canopy deciduous woodland with a diverse shrub layer, rough scrubby grassland with tall herbs and wet hay meadow. Adults of both sexes June and July.

Tetragnatha montana. Widespread and very common, in the same situations as *T.extensa* but more often in drier habitats well away from water. On low vegetation, shrubs and trees, obtained by beating and sweeping, occasionally in pitfall traps and ground zone vegetation. From woodland, especially with good shrub layer, woodland edge, open glades and rides, hedgerows and roadside verges, rough grassland with scrub and tall herbs, fen, peat bog, marsh, riverside reedbeds, reservoir and canal margins, osier beds, gardens, churchyards, *etc*. Adult females recorded from May to September, males May to August, with high numbers in June. First record I.M. Evans, 1960, Eyebrook Reservoir.

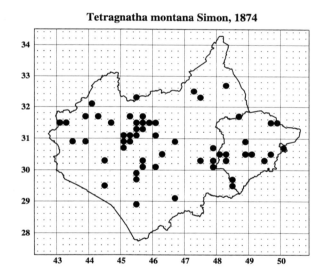

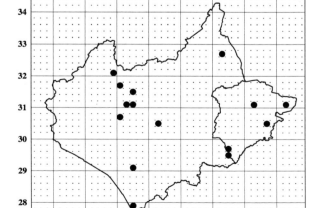

Tetragnatha obtusa. Widespread but local, with records from 15 sites, mostly on trees. Obtained by beating and sweeping, also by hand collecting at night when the spiders are more active. On various trees, including oak, willow, hawthorn, cedar, yew *etc.*, also on stone structures (reservoir dam) and field-layer vegetation, including horsetails. From open woodland, over-mature oak trees in grazed parkland, hedgerows, churchyards with yew trees and box bushes, some with rough unkempt grass, others mown and tidy, reservoir margins and a garden with trees and plenty of cover. Adult females recorded in June, July and August, males in April, June and July. First record (Bristowe, 1939).

CHAPTER FIVE

Atlas of Leicestershire and Rutland Spiders

TETRAGNATHIDAE

Tetragnatha nigrita Lendl, 1886

Tetragnatha nigrita. Widespread but local in Leicestershire. A southern species and rather rare, established in the county on the basis of 18 records from 12 sites in 15 tetrads. On trees and field layer vegetation, often near water but also in drier situations. Collected by beating trees and shrubs (oak, ash, birch, hawthorn, willow, yew, pine *etc*), sweeping grassland, heather and waterside vegetation (including *glyceria/urtica*). From woodland including damp ash wood, hedgerows, rough grassland with tall herbs and scrub, marsh, open water margins, and on railings and a fence. Adult females June, July and August, males June and July. First record I.M. Evans, 1960, Eyebrook Reservoir.

Tetragnatha striata. Very rare in Leicestershire, recorded from a single site, but probably under-recorded due to inaccessibility of habitat. A nationally notable (Nb) species with local populations scattered throughout England, Wales and Ireland, and found recently at two sites in Derbyshire. First recorded in 1971 by J. Crocker from a lakeside *Phragmites* bed at Groby Pool, and later shown to be a stable population. The spiders were found only along the open-water edge of the reedbed, on vegetation in deep water (75cm), in small orb-webs about 10cm above the water, at night. During daytime the spiders lie along the reedstems in cryptic attitude. Adult females June and July, males in July.

Tetragnatha striata L.Koch, 1862

Pachygnatha clercki Sundevall, 1823

Pachygnatha clercki. Widespread and one of the county's commonest spiders, in damp and wet situations amongst grass, leaf litter and low vegetation. Collected mostly in the ground zone and by pitfall trapping, also sweeping the field layer, occasionally beating shrubs, and inside buildings. In riverside, canal and open water marginal vegetation, reedswamp, reed beds, sedge fen, *sphagnum* bog, marshes, wet meadows, herb-rich hay meadows, rough grassland, rocky heathland, alder carr and woodland with wet flushes, osier beds, garden and churchyards. Adults of both sexes recorded every month of the year except January (an underworked month for fieldwork). First record (Bristowe, 1939).

Atlas of Leicestershire and Rutland Spiders
TETRAGNATHIDAE, METIDAE

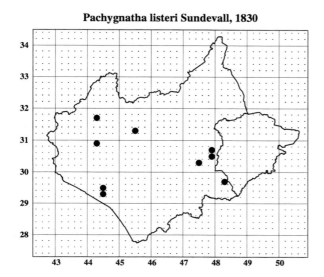

Pachygnatha listeri Sundevall, 1830

Pachygnatha listeri. Widespread but very local in the county, with 14 records from the following 8 ancient woodland sites in 9 tetrads (Burbage, Grace Dieu, Gt.Merrible, Launde Big Wood, Nailstone Wiggs, Owston, Skeffington and Swithland), and considered to be an indicator of woodland with a long history of continuous tree cover. It is associated with deciduous woodland with rich herb and shrub layers, coppiced areas and wide rides. From grass, woodrush and litter, especially around the base of hazel, under oaks, brambles and birch scrub, also swept from sedge and dog's mercury. Adult females February – July, males April, May and June. First record Lowe, 1912.

Pachygnatha degeeri. Widespread and one of the county's commonest spiders, mostly in wet and damp grassland: – marshes, wet pasture, waterside vegetation, sedge beds, *sphagnum* bog, rocky heathland, furze heath, rough grassland, hay meadows, wheat fields, ley, open bare ground with weeds, disused allotments, urban derelict land, farmyards and gardens. Also occasionally in woodland and woodland edge, hedgerows *etc*. In the ground zone amongst grass roots, in vole runs, under stones, in leaf, pine and bracken litter, also obtained by sweeping the field layer and beating hawthorn, gorse, oak *etc*. Adults of both sexes recorded all months of the year. First county record Lowe, 1912.

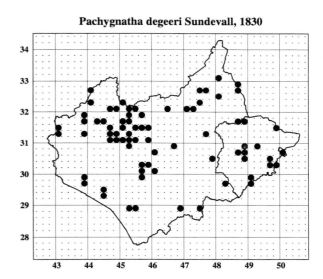

Pachygnatha degeeri Sundevall, 1830

Metellina segmentata (Clerck, 1757)

Metellina segmentata. Widespread and common in Leicestershire, under-recorded; abundant in orb webs, on shrubs, low vegetation, buildings and structures. A very successful spider which has exploited just about every suitable outdoor situation which will support its aerial web. Common in the field and shrub layers, in woodland, grassland, heathland, scrub and wetland habitats, gardens, wasteland, *etc*. In autumn, garden bushes and tall herbaceous plants are often attractively festooned with the webs of *M.segmentata*, *Zygiella x-notata* and *Linyphia triangularis*. Adults of both sexes all months May to November, highest numbers in September. First record Chalcraft (Rowley, 1897).

Atlas of Leicestershire and Rutland Spiders

METIDAE

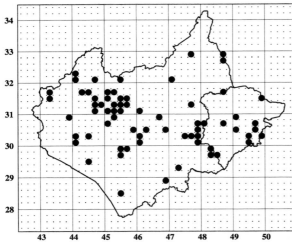
Metellina mengei (Blackwall, 1869)

Metellina mengei is widespread and common in the county, but is under-recorded. Females are indistinguishable in the field from the previous species, and the males only with the use of a hand lens. Care is needed when identifying these spiders as they often occur together in the same habitat. Whereas *M.segmentata* is adult in the late summer and autumn, *M.mengei* matures earlier, in spring and summer, but there is a considerable overlap. In Leicestershire, adult female *mengei* are recorded for all months February to November, males March to October, with highest numbers in May and June, as opposed to September for *segmentata*. First record D.G. Goddard, 1961, Leicester.

Metellina merianae. Widespread in the county and common, but never as numerous or conspicuous as the previous two species, preferring damp shaded sites, often close to water. Inside or at the entrance to culverts, caves, tunnels, drains, wells, hollow logs and trees; orb webs often over water with spider in retreat under stones, bark, or in crevices, some distance from the web. Amongst dense vegetation – gorse, box, holly – and wet marsh plants – flags, sedges, nettles – also in the chamber of a badger sett, in an old water tank, under corrugated iron sheet over water trough, and under turf overhanging ditches. Adults recorded February to November. First record Chalcraft (Rowley, 1897).

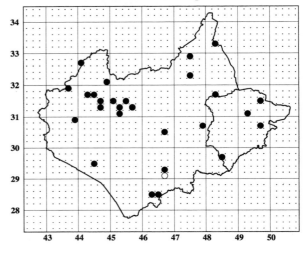
Metellina merianae (Scopoli, 1763)

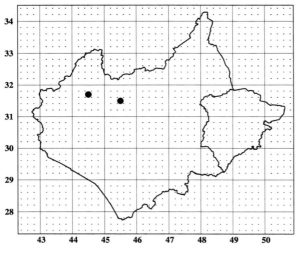
Meta menardi (Latreille, 1804)

Meta menardi. The 'cave spider'. Local, and encountered rarely in Leicestershire due to its subterranean habits. This large, glossy, chestnut-coloured orb-weaver is found in dark, damp situations such as caves, cellars, sewers *etc.*, well away from light. It is known from only two sites in the county, both from colonies with sub-adults and females in webs, and festooned egg-cocoons. Blackbrook Reservoir (May 1965), big orb-webs constructed across the long service tunnel under the dam in total darkness, whilst more recently (November 1993), a thriving colony was discovered in an underground, wartime Anderson shelter and adjoining dark culverts, at Swithland Reservoir. First record J. Crocker.

Atlas of Leicestershire and Rutland Spiders
METIDAE, ARANEIDAE

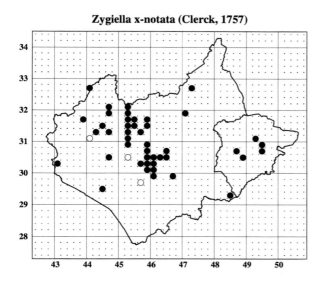

Zygiella x-notata (Clerck, 1757)

"The common geometrical spider of our window-panes", as the Rev. William Agar described ***Zygiella x-notata*** (Agar, 1890). This species is very common and widespread in the county but is under-recorded. Recorded in only 46 tetrads, this is a poor reflection of its true distribution. Found on buildings, inside and outside, on windows, walls, structures and artefacts, on bushes and herbaceous plants in gardens, churchyards, parkland *etc.*, also in rural situations, behind bark of trees, on fences and railings, on gorse (see *Z.atrica*), over water and in dry places. With urban expansion, *x-notata* may be replacing *atrica* in the countryside. Females throughout the year, males July, August and September.

Zygiella atrica. Widespread and common. Known from only 12 sites but probably under-recorded. From woodland rides, woodland edge, scrubby moorland, scrub grassland and heath, roadside verges, railway embankments, reservoir margins and rural gardens. Mostly on gorse where it is numerous and well concealed; also on oak, hawthorn, holly, beech, birch and larch, occurring in the shrub layer and on lower branches of trees. Specimens are often found in litter under bushes where their webs are situated. Occasionally in grass, bracken and heather. Adult females July to November, males July to October, peak activity September. First record Chalcraft (Rowley, 1897).

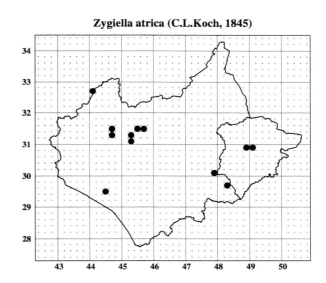

Zygiella atrica (C.L.Koch, 1845)

Gibbaranea gibbosa. Widespread but uncommon, this spider seems to have a very short season in mid-summer, both sexes being recorded only in July. It is much more common in the south than in Leicestershire, where only single individuals have been encountered, and none of these in webs! First recorded by J. Crocker, 1978, Sheet Hedges Wood, and subsequently from Buddon Wood, Croxton Park, Eyebrook Reservoir, Luffenham Heath and Wymondham N.R. Beaten from oak, hawthorn, hedgerow crab apple, pine trees and gorse, also swept from field layer grassland scrub, and hand collected from a wooden fence post, and on an oak trunk at night. In woodland and open country.

Gibbaranea gibbosa (Walckenaer, 1802)

Atlas of Leicestershire and Rutland Spiders
ARANEIDAE

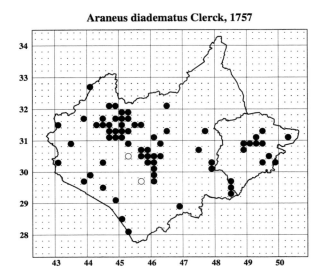

Araneus diadematus Clerck, 1757

Araneus diadematus. The 'garden spider'. One of our commonest spiders, widespread, but under-recorded. Pale orange and dark slatey grey colour extremes occur in Leicestershire. From a wide range of habitats: all kinds of woodland, scrub, hedgerows, roadside verges, parkland, churchyards, gardens, allotments, grassland, marsh, waterside vegetation, heathland, moorland, quarries, gravel pits, wasteland, all types of buildings (immatures often indoors), in fact on any structure that can support a web. Adult females from July to November, males July to October, sub-adults April to September and immatures every month of the year. First record (Agar, 1890).

Araneus quadratus. Widespread but uncommon. A very attractive spider reputed to be heavyweight champion of Britain; certainly well-fed gravid females can reach impressive proportions. In Leicestershire, the species is associated with damp, undisturbed, rough grassland with tall herbs, where it constructs a smallish orb-web in the field layer close to the ground (30–50cm) with a substantial tent-like retreat. Also in marshes on *Filipendula ulmaria*, fen, wet flushes on *juncus*, hay meadows, allotments and gardens. Recorded from only 16 sites, this local species is either under-recorded or declining. Adult females August to October, males August and September. First record Lowe, 1912.

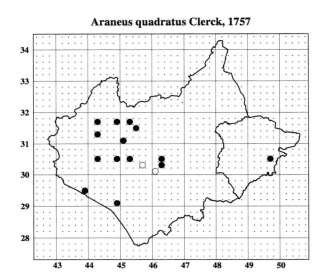

Araneus quadratus Clerck, 1757

Araneus marmoreus Clerck, 1757

Araneus marmoreus see page 48. Very rare in Leicestershire, established by a single old record (Rowley, 1897 and Bouskell, 1907) for Owston Wood. Quite possibly still present in this under-worked ancient woodland site, but not confirmed by any recent records. Elsewhere, outside the county, recorded from tall grasses, heather, bushes, brambles and trees, especially silver birch. The Leicestershire record, by G.B. Chalcraft, refers to the very distinctive form *Araneus marmoreus pyramidatus*, which has an unmistakable dark triangle on the anterior dorsal surface of the abdomen, and now considered the typical form of this species, which reaches maturity August–September.

Atlas of Leicestershire and Rutland Spiders
ARANEIDAE

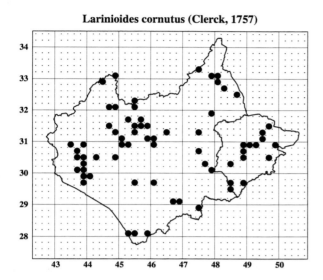

Larinioides cornutus (Clerck, 1757)

Larinioides cornutus. Widespread and common in Leicestershire, especially near water, in damp habitats, often very abundant. A field layer species of river, canal and open water marginal vegetation, in emergent vegetation in standing water, marshy vegetation in wet ditches, rough grassland with tall herbs, wet woodland, damp rides, boggy ground on cotton grass *etc.*, reed – swamp and marsh vegetation with good structure where orb-web and retreat are constructed in dead flower heads. Also in quarries and wet wasteland, on railings, fences, walls and other structures near water. Adult females March to October, males April to October. First record (Wild, 1952).

Larinioides sclopetarius. Widespread but local, only found over or near water. Unlike the previous species, this spider is never found far from open water, usually on structures over the water itself. Locally abundant. It is recorded in the county from bridges over the River Soar, canal structures, on railings, masonry and timber structures at sides of reservoirs, such as walls, parapets and inside a bird-watching hide. Also on waterside trees. First recorded I.M. Evans, 1962 from Eyebrook Reservoir, it is known from only 13 sites in 12 tetrads and is probably under-recorded. Adults of both sexes have been taken May to October and sub-adults April, May, June and October.

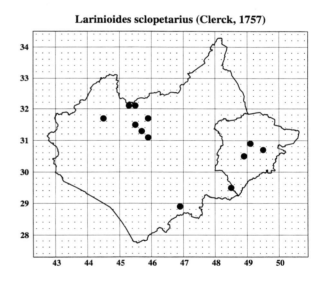

Larinioides sclopetarius (Clerck, 1757)

Larinioides patagiatus (Clerck, 1757)

Larinioides patagiatus. Rare and very local in the county. Specimens from Leicestershire have been very attractively marked, and all except one (a male) were beaten from gorse; this was taken by torchlight on a sycamore trunk at night. There are only 8 records for the county, from 3 sites: Swithland Reservoir from gorse, close to water, Buddon Wood, on gorse and sycamore along the woodland edge, and Luffenham Heath Golf Course from gorse in the rough at the side of the fairway. The first county record was established by Chalcraft (Rowley, 1897) and it is possible, indeed likely that this was from the Buddon site. Adult females June and September, males July, August and September.

Atlas of Leicestershire and Rutland Spiders
ARANEIDAE

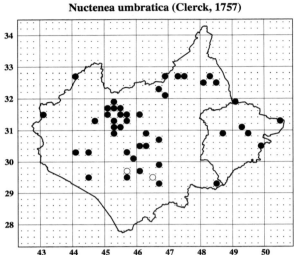

Nuctenea umbratica (Clerck, 1757)

Nuctenea umbratica. Widespread and common, but under-recorded due to its secretive habit. This black spider is nocturnal with a flattened body, enabling it to squeeze into confined spaces during the daytime: – behind bark, in crevices in masonry, under rocks and stones *etc*. From woodland, parkland and hedgerows, especially on over-mature and dead trees, also on fences and timber structures, gorse, churchyards, gardens, particularly on larch trelliswork and wooden structures, walls, buildings, inside houses, bedrooms. An active aeronaut, recognisable immatures can occur just about anywhere. Adult females all year, males June to September. First record Chalcraft (Rowley, 1897).

Agalenatea redii Rare in Leicestershire, this species is only known from a single site in the extreme east of the county. The national distribution of this local spider seems to be restricted to the southern half of Britain and Ireland, where it can be locally abundant on heather, gorse and in rough grassland. It is an early maturing species, adults being found between April and early June. First recorded for the county by H.N. Ball in October 1995, when three sub-adult females and a sub-adult male were swept from herb-rich limestone grassland at Ketton Quarry (SK977053). Several sub-adults of both sexes were collected from the same site in April 1996 and reared through to maturity.

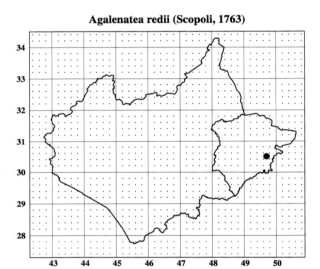

Agalenatea redii (Scopoli, 1763)

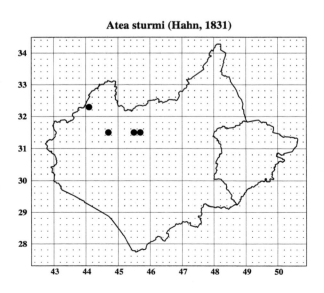

Atea sturmi (Hahn, 1831)

Atea sturmi appears to be rare in the county, with only 4 records from 3 sites in north-west Leicestershire, but is possibly overlooked elsewhere. This is a small orb-web spider, maturing in the early spring, and associated mainly with evergreen trees and bushes, where it is often abundant. First recorded for Leicestershire by P.H. Gamble in 1962 from Buddon Wood, this local spider has since been taken at Breedon Hill calcareous scrubland, Charnwood Lodge N.R. mixed woodland, and again in Buddon Wood, variously by beating young oaks, silver birch and spruce, also by sweeping low shrubs. Adult females are only recorded for May and June, with a male in July.

Atlas of Leicestershire and Rutland Spiders

ARANEIDAE

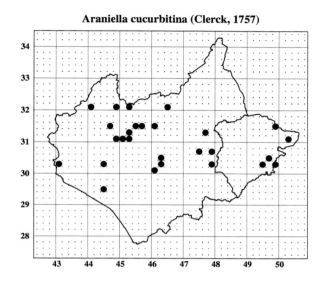

Araniella cucurbitina (Clerck, 1757)

Araniella cucurbitina. Widespread and common in the county, but probably confused in the past with *A.opistographa*, which are both indistinguishable at the immature and subadult stages. Though Chalcraft (Rowley, 1897) is credited with the first record, it is not certain to which species he was referring (see discussion p.27). *A.cucurbitina* has been confirmed from a wide variety of situations in Leicestershire, but mainly beaten from the foliage of oak trees and shrubs, also gorse, birch, yew, alder and heather. Occasionally also in marsh and scrub grassland field layers, where it has been swept from tall herbs and shrubs. Adults of both sexes May, June and July, with highest numbers in June.

Araniella opistographa. Widespread and common, in the same situations as the previous species, and often found together in the same habitat. This species has been obtained mostly by beating oak/birch/gorse scrub and young oaks, hawthorn, blackthorn, beech and briar. Also sweeping rough grassland with tall herbs, and occasionally grubbing under deciduous trees. First recorded in Leicestershire by H.A.B. Clements in 1962 from a disused brickpit at Shepshed (SK479181). Phenology as for the previous species but without the apparent June peak in numbers. All Leicestershire *Araniella* specimens have been checked and many early '*cucurbitina*' confirmed as *opistographa*.

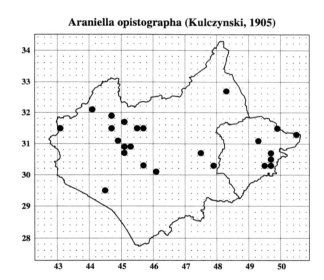

Araniella opistographa (Kulczynski, 1905)

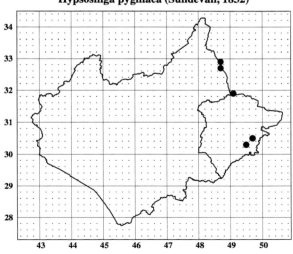

Hypsosinga pygmaea (Sundevall, 1832)

Hypsosinga pygmaea. Very local, restricted to eastern Leicestershire, on calcareous grassland, although not necessarily confined to this habitat. Widespread in Britain but uncommon. This distinctive spider is known in the county from only five sites; first recorded from King Lud's in 1964 by J. Crocker, and subsequently from Ketton Quarry, Luffenham Heath Golf Course, The Drift and Thistleton Gullet. These sites provide undisturbed herb-rich limestone scrubby grassland in which this species is established in the field layer, occasionally also on low shrubs. Adult females have been taken only in May and June, and subadults of both sexes are recorded for October.

Atlas of Leicestershire and Rutland Spiders
ARANEIDAE

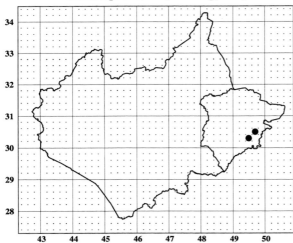

Cercidia prominens. Very local and rare in Leicestershire, only recorded from two sites in the extreme east of the county. Generally of southern distribution, this local and uncommon species has also been taken at Barnack Hills and Holes in Northamptonshire. There is also an unconfirmed report from Charnwood Forest. First record for Leicestershire J. Daws, 1995, Luffenham Heath; additionally from Ketton Quarry, where in both cases specimens were obtained sweeping the field layer of herb-rich and scrubby limestone grassland. A subadult male was taken in August and an adult female in October. At Barnack, which is not far from the Leicestershire sites, a female was taken in July.

Cyclosa conica. Widespread in the county but very local, preferring dark damp woodland with a good shrub layer. It is recorded from: Buddon Wood, Burley Wood, Burbage Wood, Barkby Holt Wood, Coppice Leys Wood, Knighton Spinney, Luffenham Heath, Little Dalby, Owston Wood, Pickworth Great Wood and Sheet Hedges Wood; mostly ancient woodland sites, but also a churchyard and scrub grassland. Beaten from oak, hawthorn, hazel, yew, and swept from rough grass. Adult females have been taken in April, May, June and July, and males June and July. First county record Chalcraft (Rowley, 1897). These spiders are cryptic and seldom discovered in their webs.

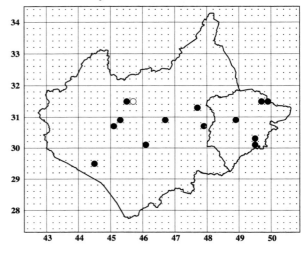

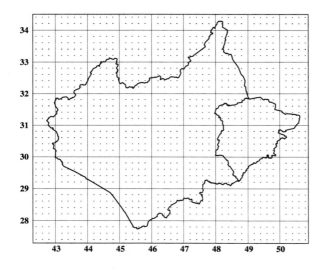

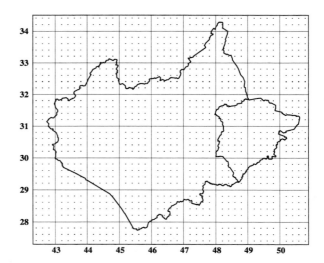

Atlas of Leicestershire and Rutland Spiders
LINYPHIIDAE

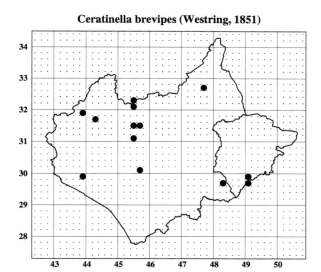

Ceratinella brevipes (Westring, 1851)

Ceratinella brevipes. Widespread but uncommon in the county. First record J. Crocker, 1975, Buddon Wood; since then from: Aylestone Meadows, Cademan Moor, Great Merrible Wood, Harby Hills, Loughborough Meadows, Lount Meadows (2 sites), Kendall's Meadow, Puddledyke, River Soar and Seaton Meadow. Generally in damp/wet situations in the litter zone. In Leicestershire, mostly from pitfall traps in wet woodland, *sphagnum*, leaf litter, and on bare ground under closed canopy. Also in wet marshy ground, reedswamp, flood meadows, damp hollows and on drier ground in hay meadows and rough grassland. Adult females May to September, males April to July.

Ceratinella brevis. Widespread and much more common than the previous species, and generally found in different locations, *brevis* has been taken in the same type of damp/wet habitats, but more records are from drier situations, and a lot more taken by hand collecting. In pockets of deep oak leaf litter, on bare ground, under stones and rocks, in *Nardus stricta* sward, heather detritus and *sphagnum*; also shaken from grass tussocks and occasionally by sweeping grassland field layer. From a wide variety of woodland, heathland, grassland and wetland sites. Adults of both sexes throughout the year with highest numbers April to June. First record J. Crocker, 1962, Bradgate Park.

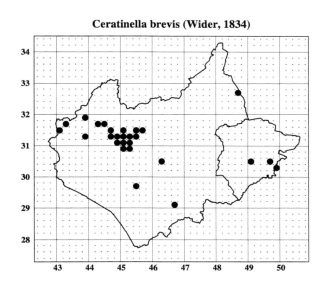

Ceratinella brevis (Wider, 1834)

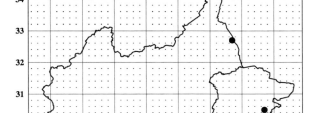

Ceratinella scabrosa (O.P.-Cambridge, 1871)

Ceratinella scabrosa. Widespread in the county, but rare, recorded from only four localities: King Lud's Entrenchments and The Drift, Geeston Quarry, Ketton Quarry and Narborough Bog. Specimens have been collected by pitfall trapping and hand collecting, in drier conditions than those for *C.brevipes*; in the litter zone of limestone grassland and from a dried-out reed bed. The maturity period is not clear from the data available but would appear to be similar to that of *C.brevis* Adult males are recorded for May, June and July, and females July, August and September. Highest numbers of both sexes were from Ketton in July. First county record J. Daws, 1992, King Lud's.

Atlas of Leicestershire and Rutland Spiders

LINYPHIIDAE

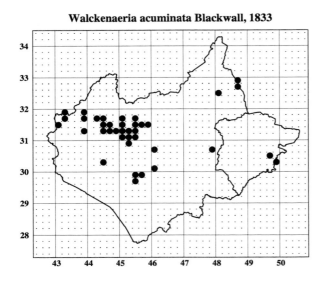
Walckenaeria acuminata Blackwall, 1833

Walckenaeria acuminata. Widespread throughout the county, and very common on Charnwood Forest. From woodland, heathland, grassland, marsh and reedswamp. A ground zone species, in leaf and pine needle litter, under stones and logs *etc.*, in moss, bracken litter, grass tussocks, vole runs, between rocks and in wet depressions – always in damp substrates. First recorded for Leicestershire by E. Duffey, 1953 from Beacon Hill, thereafter the species has been found at many sites on Charnwood Forest, also at The Drift, King Lud's, Ketton and Geeston Quarries, Narborough Bog and Owston Wood among others. Females every month of the year, males September to April.

Walckenaeria antica. Widespread and common in Leicestershire, frequent in damp grassland. All specimens have been examined carefully for *W.alticeps* but none found. *W.antica* is recorded from woodland, birch scrub, acid, neutral and calcareous grassland, rocky heathland and wasteland; mostly from pitfall traps in grass, moss, amongst sparse vegetation, under heather *etc.*, but also from under stones, cut grass, amongst grass roots, in a mole's nest, and sieved from straw around a new haystack. First record J. Crocker, 1963 from Bardon Hill. Adult females are recorded every month from February to November, males March – November, with peak activity April to July.

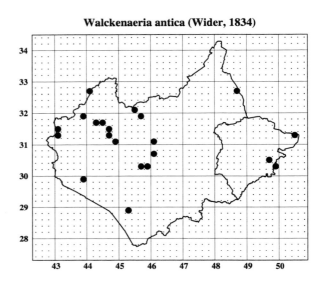
Walckenaeria antica (Wider, 1834)

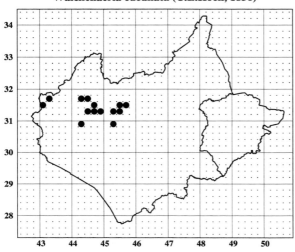
Walckenaeria cucullata (C.L.Koch, 1836)

Walckenaeria cucullata. Fairly common in the Charnwood Forest area, but few records from elsewhere. First record J. Crocker, 1964, Charnwood Lodge N.R., and subsequently from Bardon Hill, Buddon Wood, Bradgate Park, Cademan Moor, Drybrook Wood, Nailstone Wiggs, Newfield Heath, Norris Hill, Swithland Wood and Ulverscroft N.R. From deciduous and conifer woodland, heath scrub grassland and along linear features such as stone walls and hedgerows. Mostly in leaf (and pine needle) litter in deciduous (oak) woodland, but also in wet moss and damp grassland, under stones and in detritus. Adult females December to October, males January, February, April to June and September.

Atlas of Leicestershire and Rutland Spiders
LINYPHIIDAE

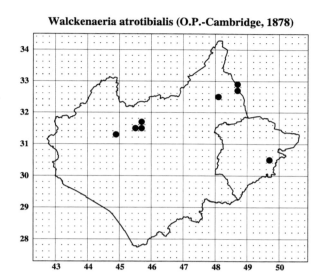

Walckenaeria atrotibialis (O.P.-Cambridge, 1878)

Walckenaeria atrotibialis is widespread but local in the county and is found amongst grass and heather. It has been collected also from amongst bluebells in woodland. Adult females are recorded from June to September, and males June to August, with highest numbers in July. First recorded for Leicestershire by J. Crocker, 1967 from Ulverscroft N.R., it has since been taken at Buddon Wood, Ketton Quarry, King Lud's Entrenchments, River Soar at Quorn, Stonesby Quarry and The Drift, often in pitfall traps but also by grubbing in the ground zone. This species has variously been listed as *Walckenaera melanocephala* and *Wideria melanocephala* in older records.

Walckenaeria capito is very rare in the county, only a single female being recorded. This was collected by J. Crocker in November 1964 from Charnwood Lodge N.R., and despite recent pitfall trapping across the site of this capture, no further specimens have been found. The spider was taken from the underside of a piece of rock, in a pile, on open *Vaccinium* moorland, on Timberwood Hill (250 m), one of the highest sites in Leicestershire. Nationally, *W.capito* is associated with mountains above 1000 m, in Wales, the Lake District and Scotland, where it is frequent, but has also been taken from widely separated sites in lowland counties in low vegetation.

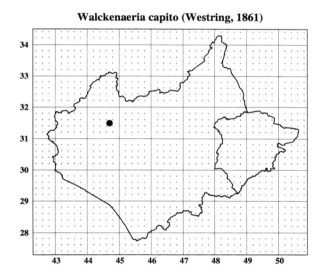

Walckenaeria capito (Westring, 1861)

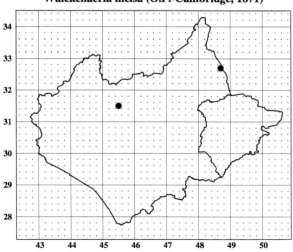

Walckenaeria incisa (O.P.-Cambridge, 1871)

Walckenaeria incisa. A nationally notable species, recorded from only three sites in Leicestershire; first collected by J. Daws in 1992 from rough calcareous grassland at King Lud's Entrenchments, and then in 1995 from a group of sites in or adjacent to Buddon Wood. Here a female was taken in a pitfall trap placed in acid grassland under open birch/oak canopy, and a male and female from a different site within the wood, from a pitfall trap set amongst bluebells in a bracken dominated clearing. At the third site, Swithland Reservoir, a female was collected from under a stone on bare mud at the waterside, which was at a very low level. All five specimens collected in July.

Atlas of Leicestershire and Rutland Spiders

LINYPHIIDAE

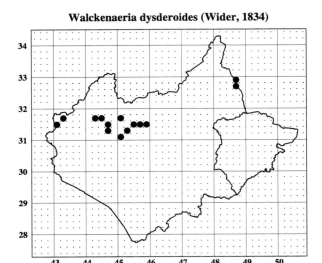

Walckenaeria dysderoides (Wider, 1834)

Walckenaeria dysderoides. Widespread but local in the county, this spider was previously known as *Wideria fugax*. First recorded by J. Crocker, 1962 from Charnwood Lodge N.R. Also on Charnwood Forest from Bardon Hill, Buddon Wood, Cademan Moor, High Sharpley, Outwoods, Stoneywell and Swithland Woods. Also from Newfield Heath and Norris Hill in the west, and King Lud's and The Drift in the east of the county. From woodland, acid and calcareous grassland, heathland and moorland. In pitfall traps amongst woodland vegetation and on bare ground; also hand collected from leaf and pine litter, grass and moss. Females February-September, males December-August.

Walckenaeria nudipalpis. Widespread and common in Leicestershire. A frequent aeronaut, first recorded for the county by P. C. Jerrard, 1962 from Cropston Reservoir margins. This litter zone species has been found in damp woodland, alder carr and grassland, often in very wet conditions; also marshes, reed-beds, waterside vegetation, flood meadows, boggy moorland and wet heathland. In pitfall traps and grubbed from ground zone vegetation, such as tussocky grass, moss, vole runs, leaf litter, detritus between rocks, in an old waterhen's nest *etc*. Adult females recorded for every month except December and January, males February to June, and September to November.

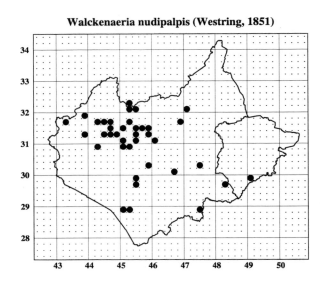

Walckenaeria nudipalpis (Westring, 1851)

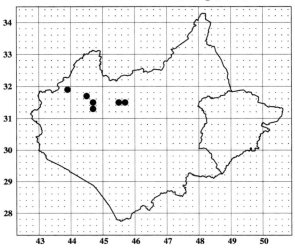

Walckenaeria furcillata (Menge, 1869)

Walckenaeria furcillata is very local, known from only five sites on Charnwood Forest and north-west Leicestershire. First recorded J. Crocker, 1963 from Charnwood Lodge N.R., thereafter it has been taken at Bardon Hill, two locations in Buddon Wood, at High Sharpley and at Lount (Coleorton SSSI), all heathy habitats. Habitat: siliceous woodland, birch/oak/gorse scrub, rocky heathland, wet heathland and heath grassland. From pitfall traps in bare ground in shaded woodland, amongst grass roots, under heather and in sphagnum; also hand collected from grass roots under oaks, from detritus between rocks and in heather litter. Females May-September, males June and July.

Atlas of Leicestershire and Rutland Spiders
LINYPHIIDAE

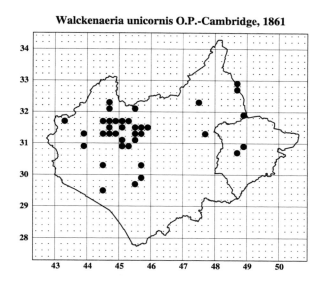

Walckenaeria unicornis O.P.-Cambridge, 1861

Walckenaeria unicornis. Widespread and common in Leicestershire, often in wet situations. First record J. Crocker, 1962, Shepshed. From alder carr, conifer plantation, rural hedgerow, rocky heathland, *Vaccinium* heath, rough grassland, damp/wet meadows, marsh, sedge-fen, reedswamp, waterside vegetation, arable field, disused allotments, farmyard and council refuse tip. Obtained by grubbing in ground zone, sorting leaf and pine needle litter; from under stones, old straw and rotted horse manure, from wet ground zone litter, also by sweeping marsh vegetation and beating hawthorn, gorse and spruce. Adults February to October, except no males taken July and August.

Walckenaeria cuspidata. Widespread but infrequent, from damp or wet situations. First record H. A. B. Clements, 1963, Shepshed. Habitat: damp woodland – in bracken litter, wet vegetation, leaf litter and on bare ground under trees; also in wet grassland, rough grass dominated by nettles, waterside nettle beds and marsh, generally from ground zone vegetation, but also from grass tussocks and by sweeping the field layer. Additionally from hedgerow detritus, amongst straw in a stable yard, on bare mud on a reedy riverside bank and from under bark. Females March to November, males February to May and August to November, highest numbers May and November.

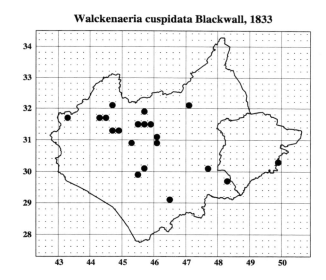

Walckenaeria cuspidata Blackwall, 1833

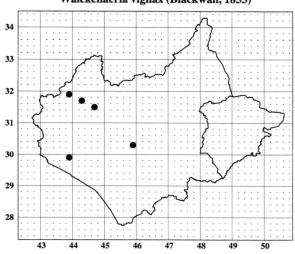

Walckenaeria vigilax (Blackwall, 1853)

Walckenaeria vigilax. Widespread but uncommon in west Leicestershire, in damp and wet situations. First county record from Cademan Moor, collected by M.G. Crocker in 1974; recorded subsequently from Charnwood Lodge N.R., Kendall's Meadow, Leicester Cattle Market, two re-seeded opencast sites at Lount, and wet heath grassland at Coleorton. Habitats: acid grassland with wet flushes, heath grassland with heather, herb-rich hay meadow, and waste ground within Leicester city; mostly from pitfall traps in damp hollows, but also in fairly dry grass and heather. Adult females have been taken in June, August and November, adult males in June and July only.

Atlas of Leicestershire and Rutland Spiders
LINYPHIIDAE

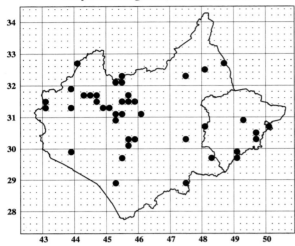

Dicymbium nigrum. Widespread and common in the county. Prefers a tussocky habitat. All Leicestershire male specimens checked for *D. brevisitosum* but none found. First record J. Crocker, 1964, Charnwood Lodge N.R. Habitat: undisturbed wet grassland, well vegetated marshes, reed-swamp, herb-rich grassland, waterside vegetation, *sphagnum* bog, bracken heath, heath grassland, damp woodland, disused allotments, set-aside rough grassland; from ground zone, mostly in long grass, in leaf litter and wet detritus, shaken from hay and cut bracken, sweeping field layer. Females February to November, males February to December, highest numbers in May/June.

Much less frequently encountered than the previous species, ***Dicymbium tibiale*** is widespread but very local. It was first recorded from Charnwood Lodge N.R. in 1963 by J. Crocker, and since then has been found at Buddon Wood, Cademan Moor, Great Merrible Wood, Coleorton wet heath grassland (Lount 2), Owston and Skeffington Woods. Habitat: undisturbed acid grassland with wet flushes, birch/alder wet woodland, deciduous and mixed woodland and unimproved hay meadow; in wet depressions, shaken from *sphagnum*, grubbed from damp grass, in pitfall traps, and swept from grass under birch trees. Females May/June and September/October, males April to October.

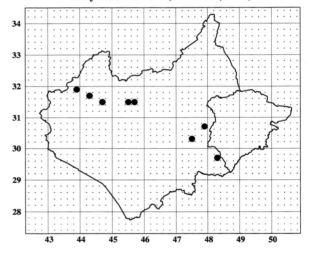

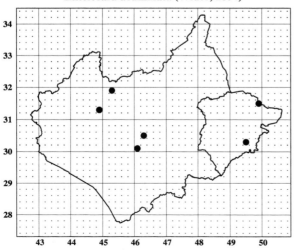

Entelecara acuminata. Widespread but uncommon, probably under-recorded; a frequent aeronaut. There are only ten records from six sites, the first from a garden in Loughborough, J. Crocker, 1962; thereafter from Luffenham Heath, Pickworth Great Wood, Ulverscroft N.R. and two gardens in Leicester. Those from gardens were collected from the outside walls of buildings, by beating shrubs, from amongst rockery plants and from an aerial trap. Otherwise specimens were from vole runs and sweeping rough grassland, also by beating shrubs and trees. Adult females are recorded for May, June and July, and males in May and June. Most of the records are of single specimens.

Atlas of Leicestershire and Rutland Spiders
LINYPHIIDAE

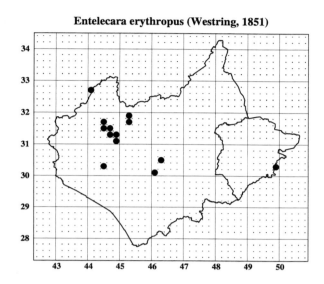

Entelecara erythropus (Westring, 1851)

Entelecara erythropus. Widespread but local in the county, from a variety of habitats. First record J. Crocker, 1962 from a Loughborough garden. This species also, would seem to be a frequent aeronaut. Habitats: grassland, heathland, a disused quarry, reservoir margins, farmyards, churchyards, town and city gardens, hedgerows, deciduous woodland and conifer plantation. Collected mostly from under rocks and stones, also in litter under pile of tiles, amongst rocks and under a corrugated sheet; also from grass cuttings, leaf litter and beaten from holly, hazel, hawthorn, yew and garden shrubs. Adult females March through to October, males in May, June and July only.

Moebelia penicillata. Although considered to be a common species throughout the British Isles, *Moebelia* is very rare in Leicestershire, recorded from only a single specimen. A male was collected from the trunk of an old pedunculate oak tree in the Loughborough Outwoods – mixed woodland on an ancient woodland site – in May 1963 by J. Crocker (SK514167). The typical habitat for this species is 'on the bark of trees, in crevices and amongst lichen on tree trunks' (L & M Vol. II, 1953). Though this is an admittedly under worked micro-habitat, sufficient collecting has been carried out on tree trunks around Charnwood Forest to suggest this is not a common species here.

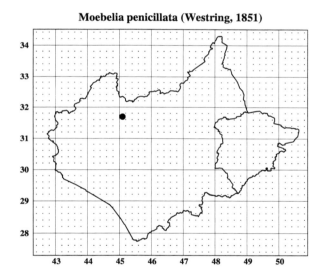

Moebelia penicillata (Westring, 1851)

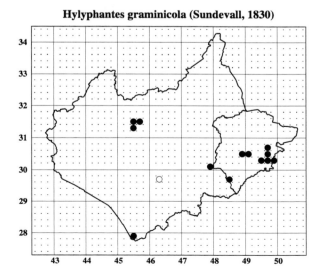

Hylyphantes graminicola (Sundevall, 1830)

Hylyphantes graminicola. Widespread but local, mostly on bushes and trees. First record for Leicestershire D.S. Fieldhouse, 1953, Wistow. Habitats range from well vegetated marsh, rough grassland with tall herbs and scrub, gorse scrub, hedgerows, wooded parkland, woodland, churchyards and reservoir margins. Most records are from beating shrubs and trees – oak, gorse, elm, blackthorn, hawthorn and evergreens including pine and yew; also by sweeping field layer vegetation, grubbing amongst heather and from pitfall traps in bare ground under brambles. Adult females recorded May through to September and males May, June and July, with highest numbers in June.

Atlas of Leicestershire and Rutland Spiders
LINYPHIIDAE

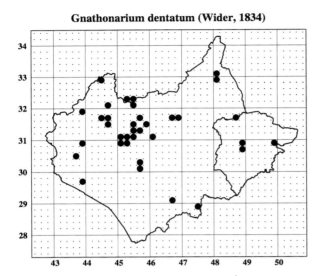

Gnathonarium dentatum (Wider, 1834)

Gnathonarium dentatum. Widespread in Leicestershire, but local where it is abundant in wet places. First recorded for the county I. M. Evans, 1962, Cropston Reservoir margins. Habitats: reedswamp, fen, boggy moorland, wet marshes, waterside vegetation, water-logged fields, alder carr, wet flushes in woodland, and occasionally from damp rough grassland; mostly in the wet ground zone, obtained by grubbing and shaking out wet litter, occasionally from pitfall traps in damp substrates and sweeping field layer vegetation in marshes and at the waterside. Both sexes have been recorded every month of the year except January, March and November. High numbers April/June and October.

Tmeticus affinis. Very localised. A riparian species, found also in lakeside vegetation. First Leicestershire record H.A.B. Clements, 1964 from the River Soar at Normanton; also from Cotes Bridge and Kegworth on the Soar, and at Barrow Gravel Pits, Foxton Locks on the Grand Union Canal, Groby Pool, Lockington Meadows and the River Trent, also at Swithland Reservoir; always in waterside vegetation, usually in standing or flowing water. On several occasions *Tmeticus* has been noted in *Tetragnatha striata* webs just above the water level, after dark, feeding on ensnared diptera. Adult females recorded April to October, males in February and May to October. High numbers April/May.

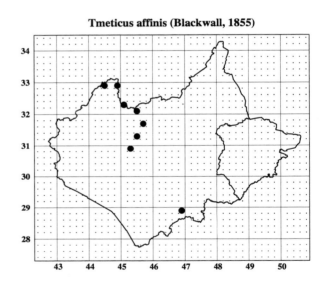

Tmeticus affinis (Blackwall, 1855)

Gongylidium rufipes (Linnaeus, 1758)

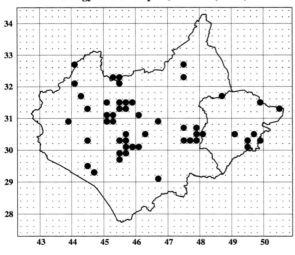

Gongylidium rufipes. Widespread and common, from damp and wet situations, but also from much drier sites. First record for Leicestershire J. Crocker, 1962, Loddington Reddish. Habitat: rough grassland with good structure, waterside vegetation, reedswamp, sedge beds, marshes and marshy meadows, damp woodland (birch/alder), deciduous and mixed woodland, mostly with good ground and shrub layers, also from suburban and urban gardens and disused quarries and allotments. From pitfall traps set in the ground zone, grubbing and sieving ground zone litter, sweeping field layer and beating shrubs and trees. Both sexes March to October, with high numbers May and June.

Atlas of Leicestershire and Rutland Spiders
LINYPHIIDAE

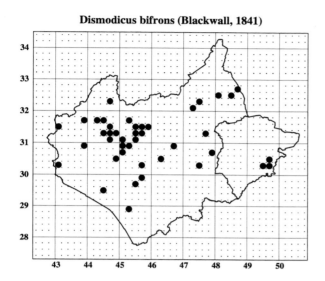

Dismodicus bifrons (Blackwall, 1841)

Dismodicus bifrons. Widespread and common in the county, first recorded by J. Crocker, 1962 from Cropston Waterworks grounds. Habitat: rough grassland, heather heath, marshes, marshy fields, reedswamp, sedge fen *etc.*, alder carr, deciduous and mixed woodland, hedgerows and gorse scrub, derelict quarries, disused allotments, suburban gardens, farmyards and churchyards. From wet litter, moss and amongst grass and heather, occasionally in pitfall traps, sweeping woodland ground flora and field layer vegetation, also beating shrubs and trees, particularly gorse. Adult females are recorded from April to September, males from April to July, with activity peak in May.

Hypomma bituberculatum. Widespread but local, in wet swampy areas, this species is able to withstand submergence by fresh water. First record for Leicestershire I.M. Evans, 1961 Groby Pool. Habitat: reedswamp, fen, sedge beds, *sphagnum* bog, brookside marshes, marshy fields and flood meadows, also rough grassland and nettle beds adjacent to water. From ground zone vegetation, sieving damp/wet litter, on bare ground and wet mud at waterside, and on emergent plant stems, also swept from field layer and occasionally beaten from marsh scrub. Adult females are recorded from April to November, and males from February to July. Highest numbers during May and June.

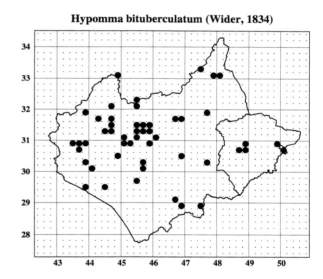

Hypomma bituberculatum (Wider, 1834)

Hypomma cornutum (Blackwall, 1833)

Hypomma cornutum. Widespread and local, in drier situations than the previous species and less frequent, on shrubs and trees. First recorded for the county by (Bristowe, 1939) unlocalised. There is a 1953 record from Wistow, of a male (aeronaut?) on a concrete post, and more recently of specimens (sex unknown) from a Malaise (aerial) trap in a suburban garden. Habitat: woodland, woodland edge, scrub – especially gorse and oak, churchyard trees, parkland and isolated trees. Collected almost entirely by beating: oak, birch, beech, blackthorn, cedars, pine, yew, box, gorse, *etc.*, also occasionally sweeping rough grass. Adult females May – August, males May – July; high numbers in May.

CHAPTER FIVE

Atlas of Leicestershire and Rutland Spiders
LINYPHIIDAE

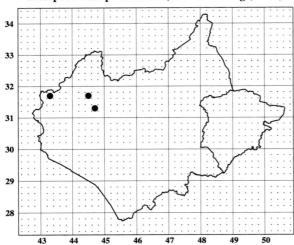

Metopobactrus prominulus (O.P.-Cambridge, 1872)

Metopobactrus prominulus. Widespread in Britain but local, often on calcareous grassland, this species is rare in Leicestershire, being recorded from only three sites. First county record J. Daws, 1992, Bardon Hill and subsequently also from High Sharpley on Charnwood Forest and Norris Hill in the extreme west of the county. All records are from pitfall trapping, set mostly amongst heather on rocky *Calluna* heathland, but also from an old rural hedgerow. Although adults are generally found in spring, summer and autumn, in Leicestershire the seven records for mature specimens show females from June to November and males in the months of June and September only.

Baryphyma pratense. Very localised, a riparian species restricted to the main rivers and their tributaries. First record Chalcraft (Rowley, 1897) from Buddon Wood, recorded as *Walckenaera pratensis* (see p. 35). This is doubtful but could have been taken at Buddon brook. Leicestershire records are from the River Soar, from Narborough Bog, through to Lockington Marsh at the confluence with the River Trent and also from the Trent on the Leicestershire border. Habitat: river fen, sedge beds subject to flooding, shingle bank, *Deschampsia cespitosa* damp grassland, riverside vegetation; ground zone and field layer. Adult females May to August, males May to July.

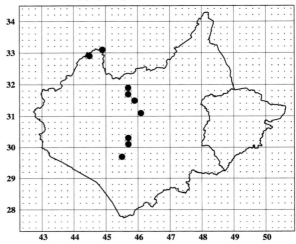

Baryphyma pratense (Blackwall, 1861)

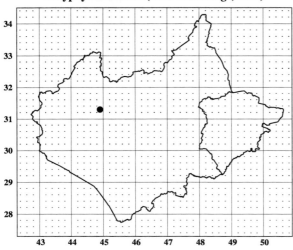

Baryphyma trifrons (O.P.-Cambridge, 1863)

Baryphyma trifrons is widespread in Britain, in wet marshy situations but uncommon; in Leicestershire this species is rare, recorded from only a single specimen, a male collected by J. Crocker in July 1967 at Ulverscroft N.R. (SK494134). The site is a wet marshy meadow, lying adjacent to the Ulverscroft brook near its source, and the specimen was in the 0-6 cm zone in a waterlogged depression. The meadow has a rich and varied flora including marsh horsetail, hard rush, ragged robin, water mint, meadowsweet and bog pimpernel. *Baryphyma trifrons* was taken with *Pirata latitans*, *Tallusia experta*, *Bathyphantes nigrinus* and *Oedothorax gibbosus*.

Atlas of Leicestershire and Rutland Spiders
LINYPHIIDAE

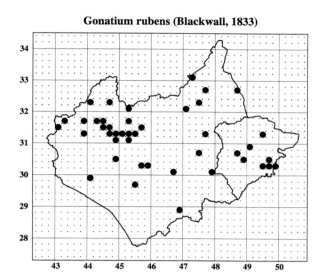
Gonatium rubens (Blackwall, 1833)

Gonatium rubens. Widespread and common, in acid neutral and calcareous grassland, heathland and moorland, occasionally from marshes, waterside vegetation and woodland. Mostly from the ground zone – grass roots, leaf and bracken litter, under heather, in moss *etc.*, in pitfall traps and by grubbing; also obtained by sweeping field layer vegetation – nettles, rough grass, heather – and beating shrubs and young trees. Generally in drier situations but occasionally in wet swampy conditions. First recorded for the county by J. Crocker, 1961 from Hose Grange. Adult females every month of the year, males in April and August to December. Highest numbers in November.

Gonatium rubellum. Less common than the previous species. Infrequent away from Charnwood Forest. All Leicestershire records are from deciduous and mixed woodland; in leaf litter and detritus, amongst short grass and moss, also swept from grass and dog's mercury, and beaten from young oaks. First county record J. Crocker, 1962 Charnwood Lodge N.R., thereafter also from Bradgate Park, Buddon Wood, Drybrook Wood, Great Merrible Wood, Owston Wood, Outwoods, Swithland Wood, Sheet Hedges Wood and woodland adjacent to the River Trent. Adult females are recorded from March to October, but are probably mature all through the year, males recorded only August to November.

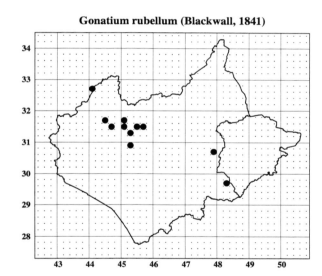
Gonatium rubellum (Blackwall, 1841)

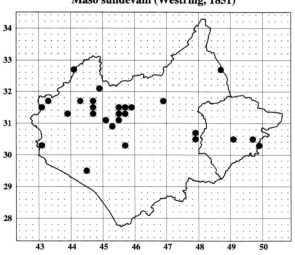
Maso sundevalli (Westring, 1851)

Maso sundevalli. Widespread and common in Leicestershire, in drier rather than damp situations. First record S.B. Scargill, 1960 from Leicester. Habitat: deciduous and mixed woodland, wood pasture, rural hedgerows, *Calluna/Vaccinium* heathland, limestone and acid grassland and dry fen. Amongst heather, bracken and leaf litter, grass roots, cut grass, moss and hedge-bottom detritus; also from a large *Carex paniculata* tussock in alder carr, under pieces of wood and stones, and swept from field layer vegetation. Adult females are recorded from February to November, and males in June/July and November. Highest numbers are recorded in July.

Atlas of Leicestershire and Rutland Spiders
LINYPHIIDAE

Maso gallicus Simon, 1894

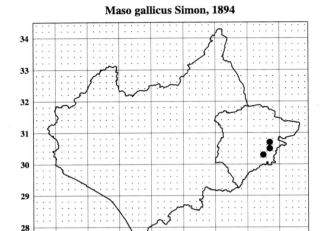

Maso gallicus is a relict wetland species of the once extensive Fen Basin. A nationally notable species recorded from only three sites in east Leicestershire. The first record was from Shacklewell Hollow, a small marsh on a tributary stream feeding the River Gwash, collected J. Crocker, 1962. Subsequently, also from Ketton Quarry – a working limestone quarry – and Luffenham Heath Golf Course, from calcareous grassland. The last two sites were much drier habitats than the first. With the exception of pitfall trap catches, most of the material was obtained by sweeping field layer grassland. Adult females are recorded in June, July and August, males in July only.

Peponocranium ludicrum. Not uncommon on Charnwood Forest, but infrequent elsewhere in the county. First recorded from Bardon Hill in 1963 by J. Crocker; also from Altar Stones, Billa Barra, Cademan Moor, Charnwood Lodge N.R., High Sharpley, Hill Hole and Rise Rocks, and at Geeston and Ketton Quarries. Habitat: acid grassland, rocky bilberry/heather heath, wet moorland, birch scrub, limestone grassland and waste ground; mostly in moss and grass roots, heather/bilberry litter and under rocks, also in ants' nest under an embedded stone, and beaten from gorse and low vegetation. Adult females April to September, males April to July, with higher numbers in June.

Peponocranium ludicrum (O.P.-Cambridge, 1861)

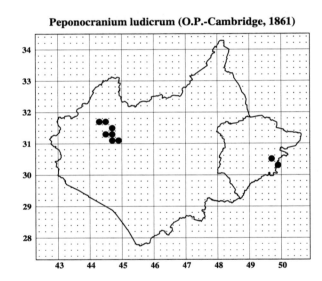

Pocadicnemis pumila (Blackwall, 1841)

Pocadicnemis pumila. Widespread in the county, but less common than *P. juncea* with which it is often confused. First record J. Crocker, 1961, Countesthorpe. Habitat: acid grassland, open heath grassland, rocky *Calluna* heath, bracken-dominated heathland, reservoir margins, birch scrub and deciduous woodland, also from a disused sandpit, abandoned allotments, roadside verges, a refuse tip and a suburban garden; amongst rocks, grassy tussocks, vole runs and leaf litter, under stones and rubbish, shaken from moss and grass, grubbed from ground zone and swept from heather and grass. Females April to October, males May to August, with high numbers in May and June.

Atlas of Leicestershire and Rutland Spiders
LINYPHIIDAE

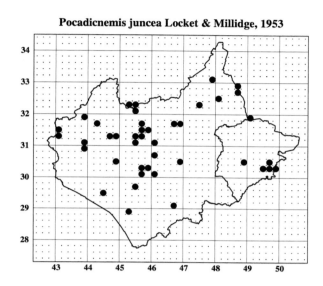

Pocadicnemis juncea Locket & Millidge, 1953

Pocadicnemis juncea. Widespread and common in the county. All Leicestershire *Pocadicnemis* specimens re-checked, and *P. pumila* (*sensu stricta*) found to be much less common than *juncea*, which seems to prefer wetter habitats. First record I. M. Evans, 1962, King Lud's Entrenchments. Habitats: damp grassland, flood meadows, marsh, fen, reed-swamp, sedge beds, peat bog and waterside vegetation; also from drier grassland, ley, calcareous turf and woodland. Mostly from pitfall traps in ground zone, amongst detritus and litter, wet flushes, also by sweeping field layer and beating gorse, oak and evergreens. Females May to November, males May to September, with high numbers in June.

Oedothorax gibbosus. Widespread but local, in wet places, usually in large numbers. First record J. Crocker, 1962, Tugby Wood. From swampy habitats, in woodland (alder carr), at waterside, waterlogged fields, flood meadows, grassland with wet flushes, reedswamp, fen, marsh and riverside nettle beds – mostly from the ground zone, but occasionally swept from the field layer. On bare mud and emergent plant stems, on plants growing out of the water, from wet litter and in vole runs. Sometimes in drier situations, usually subject to occasional flooding. Adult females are recorded for all months March to October, and males April to July, with activity peak May/June.

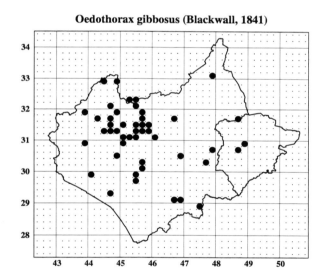

Oedothorax gibbosus (Blackwall, 1841)

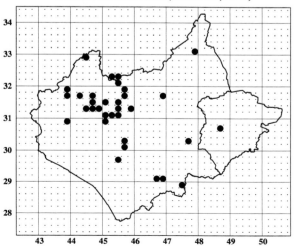

Oedothorax tuberosus (Blackwall, 1841)

Oedothorax gibbosus f. tuberosus. Since separate records have been kept for what are now known to be dimorphic males of the species *O.gibbosus*, these are presented here for interest rather than any scientific value. For further details of the conspecific relationship see: deKeer & Maelfait in *Newsl. Br.arachnol.Soc.* **53**, Nov. 1988, p.3. As with *O.gibbosus*, *tuberosus* is widespread but local in Leicestershire, in wet places and is found with *gibbosus*, but less frequently. In view of these findings, the first occurrence in Leicestershire is irrelevant. Adult *tuberosus* males have been taken between April and July, as with *gibbosus*, with recognisable subadults in April and October.

Atlas of Leicestershire and Rutland Spiders
LINYPHIIDAE

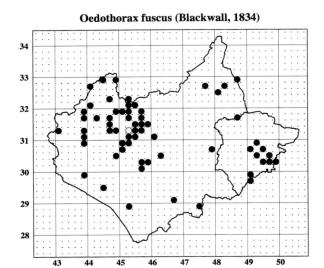

Oedothorax fuscus (Blackwall, 1834)

Oedothorax fuscus. Widespread and common in the county, in damp and occasionally wet situations. First record E. Duffey, 1952, The Brand. A frequent aeronaut. Habitat: damp grassland, flood meadows, arable and set-aside fields, ley, marshes, wet heathland, reservoir margins, river margins, waterside vegetation, deciduous and mixed woodland, rural hedgerows, disused quarries, allotments, urban derelict land and suburban gardens. Collected by sweeping field layer and beating shrubs and trees, also in pitfall traps and from ground zone detritus, leaf litter, on wet mud, under stones and dead wood. Adult females recorded February to November, males May to November.

Oedothorax agrestis. Widespread but local, in wet situations. First record H. A. Clements, 1964 Normanton-on-Soar; also from Barrow, Kegworth and Loughborough Meadows on the Soar, Ashby Canal, Bardon Hill, Buddon Wood, Cropston Reservoir, Grace Dieu Wood, River Chater – Leighfield, River Gwash – Braunstone, Stonesby Quarry and Swithland Wood; all waterside habitats. From wet marshy depressions in woodland, wet flushes, vegetation in standing water, reedswamp, waterside marginal vegetation, on riverside bare mud and shingle, under flood debris *etc*. Adult females are recorded for all months between April and September, males May to August and October.

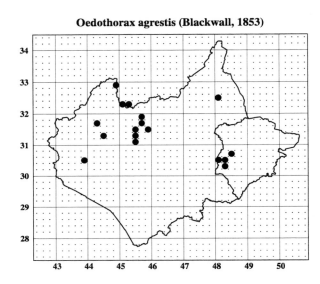

Oedothorax agrestis (Blackwall, 1853)

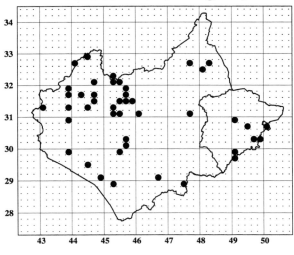

Oedothorax retusus (Westring, 1851)

Oedothorax retusus. A widespread and common spider, and a frequent aeronaut. First record for the county (Bristowe, 1939). In similar situations to *O. fuscus* but usually in the wetter habitats. From damp grassland and woodland with wet flushes, marshes, boggy moorland, wet heath bog, flood meadows, river banks, bare mud, shingle and waterside vegetation, also deciduous woodland, rural hedgerows, ley, rough grassland, disused quarries and allotments *etc*. In pitfall traps, from grass roots, under stones and detritus, on wet mud and amongst shingle, in leaf litter and occasionally swept from field layer vegetation. Both sexes recorded in all months of the year except January.

Atlas of Leicestershire and Rutland Spiders
LINYPHIIDAE

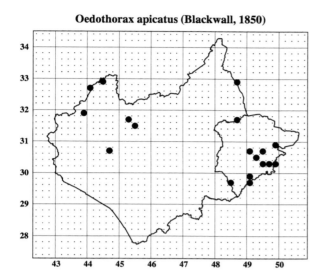

Oedothorax apicatus (Blackwall, 1850)

Oedothorax apicatus. Widespread in Leicestershire but very local; sometimes locally abundant. First county record J. Crocker, 1961, Pignut Spinney Marsh. Habitat: calcareous grassland, marsh, reseeded agricultural land, wheat field, set-aside field with tall herbs and rough grass, wet meadow, reservoir margins, river banks, disused quarries. From ground zone litter, often in pitfall traps, stubble, sparse grass, waterside vegetation, water mint *etc.*, on shingle and bare mud at waterside; also sweeping and beating field and shrub layer vegetation. This spider is a frequent aeronaut. Adult females recorded from March through to September, males April to November.

Pelecopsis parallela. A widespread but local species in Leicestershire. First record J. Daws, 1992 from the C.E.G.B. derelict site at Rawdykes, Leicester, then from the nearby St. Mary's Allotments, Acresford Sandpit, Dunton Bassett, Geeston Quarry, King Lud's Entrenchments and North Luffenham Quarry. Habitat: disused quarries, urban derelict land, abandoned allotments, rough calcareous grassland, and a vegetated sand bank. All 21 records are from pitfall traps, amongst short vegetation, on bare ground or sparsely vegetated substrate, under stones *etc.* Adult females recorded March to September, males February to November, with highest numbers July/August.

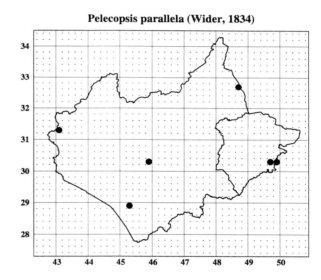

Pelecopsis parallela (Wider, 1834)

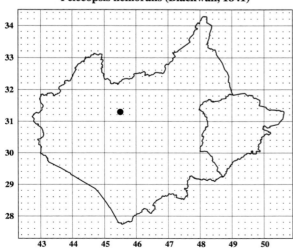

Pelecopsis nemoralis (Blackwall, 1841)

Pelecopsis nemoralis. Very rare in Leicestershire, with only two records from a single site in Swithland Wood. Nationally, this species appears to be widespread but with relatively few reliable records. At Swithland Wood, two females were collected by J. Crocker in May 1967 from the underside of pieces of granitic rock in a broken down dry-stone wall. The site (SK540127) is an unimproved grassland clearing, occasionally grazed, within ancient woodland. It is surrounded by a wall to keep stock out of the wood. During repairs to the wall the following year, in March, a further female was taken in the same micro-habitat, about 20 m distant from the original site.

Atlas of Leicestershire and Rutland Spiders
LINYPHIIDAE

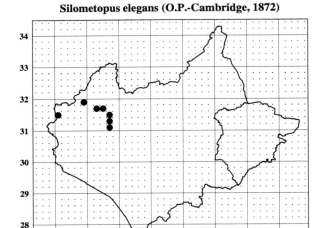

Silometopus elegans (O.P.-Cambridge, 1872)

Silometopus elegans. Local, from north-west Leicestershire only, generally in wet or marshy conditions. First county record M. G. Crocker, 1965, Charnwood Lodge N.R., thereafter also recorded from Bardon Hill, Billa Barra, Cademan Moor, High Sharpley, Lount (marshy meadow – Worthington SSSI) and Newfield Heath. Habitat: damp acid grassland with wet flushes, heath grassland, wet moorland, *Calluna/Vaccinium* heath, unimproved hay meadow. From *juncus*, grass, *sphagnum*, heather and bracken, also under a stone at pond side. This is a spring and summer maturing species. Females are recorded from May to August, males April to June with highest numbers May/June.

Silometopus reussi. Uncommon; breeding populations very localised, usually associated with straw. First record J. Crocker, 1962 from a Loughborough garden under litter. Also from an aerial trap in a garden in Leicester, Loughborough Meadows, the River Soar, Geeston Quarry and Newton Burgoland Marshes. Apart from the aerial capture all specimens were from the ground zone – disused ironstone quarry, amongst sparse vegetation, riverside bank on wet mud; and from two typical habitats, amongst straw in a pile of weathered cow manure in unimproved damp grassland, and from a pile of straw and horse manure at the roadside. Females April, May and October, males October.

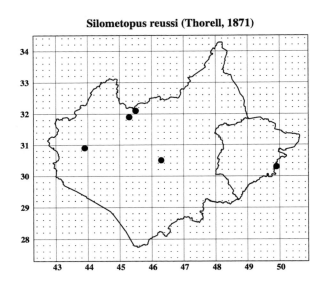

Silometopus reussi (Thorell, 1871)

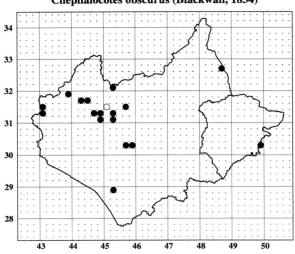

Cnephalocotes obscurus (Blackwall, 1834)

Cnephalocotes obscurus. Widespread but local, this is a litter zone species. First county record E. Duffey, 1953, Beacon Hill. Habitat: acid grassland with wet flushes, heath grassland, rocky moorland, *Calluna* heath, flood meadows, neutral grassland scrub, limestone grassland, marsh, reservoir margins, deciduous woodland and wood pasture, abandoned allotments, disused quarries and gravel pits, and a vegetated sand bank. In litter, grass roots, vole runs, pitfall traps, and also beaten from sallows at the edge of wet carr. Most records are from damp rather than wet micro-habitats. Adult females recorded all months except January, and males February to November.

Atlas of Leicestershire and Rutland Spiders
LINYPHIIDAE

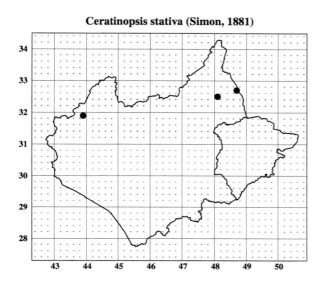

Ceratinopsis stativa (Simon, 1881)

Ceratinopsis stativa. Rare in Leicestershire, with only six records for the county from three sites. Nationally, a rather rare grassland species, usually on southern calcareous grassland. First record J. Daws, 1992, Lount (heath grassland Coleorton 'A' site), also from King Lud's Entrenchments and Stonesby Quarry. At the latter site, a disused limestone quarry with exposed bedrock, specimens were obtained from pitfall traps set on bare ground in shaded and unshaded positions; elsewhere from limestone grassland and an unimproved hay meadow – in pitfall traps in the ground zone amongst grass. Adult females were taken in June only, and adult males in June and July.

Evansia merens. This myrmecophilous spider is local in Leicestershire, recorded only from four heathy sites, mainly on Charnwood Forest. First recorded by J. Crocker, 1968 from Bardon Hill; subsequently from Charnwood Lodge N.R., High Sharpley and Newfield Heath – a derelict colliery site. Habitat: heath grassland, rocky heathland and *Calluna/Vaccinium* heath; frequently under rocks embedded in peaty soil in the nests of the black ant *Formica fusca*, occasionally with *Mastigusa macrophthalma*, also with the ant *Lassius fuliginosus*. Specimens have turned up in pitfall traps and occasionally on open ground. Females April to October, males April to August, and ova in May.

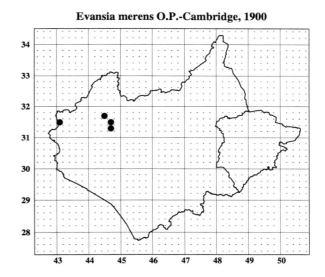

Evansia merens O.P.-Cambridge, 1900

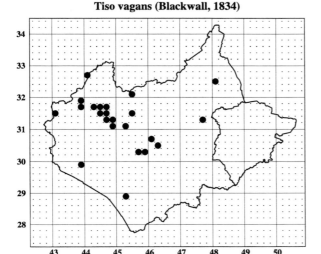

Tiso vagans (Blackwall, 1834)

Tiso vagans. Widespread but local in the county; a common aeronaut, showing a preference for short turf. First record J. Crocker, 1963, Charnwood Lodge N.R. Habitat: acid grassland with wet flushes, bracken-dominated furze heath, wet moorland, grassy heathland, *Calluna* heath, hay-meadow, marsh, birch/alder woodland, river bank on dry mud, suburban garden, also derelict allotments and disused quarries, in pitfall traps on open bare ground. Mostly from ground zone, amongst grass roots, vole runs and leaf litter; usually in dry or damp conditions, occasionally in damp hollows. Adult females March to October, males March to November with high numbers in June.

Atlas of Leicestershire and Rutland Spiders
LINYPHIIDAE

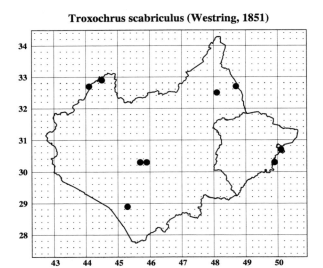

Troxochrus scabriculus (Westring, 1851)

Troxochrus scabriculus. Local but widespread in the county – a successful aeronaut. First record J. Crocker, 1964, King Lud's; also from Donington Park, Dunton Bassett, Geeston Quarry, Evesham Road Allotments, River Trent, St. Mary's Allotments, Stonesby Quarry and Tinwell Church. From calcareous grassland, abandoned allotments, cultivated grassland, disused limestone and gravel quarries, an open churchyard and a river shingle bank; swept from grass, grubbed from ground litter, sieved from straw around new haystack, also from pitfall traps in the ground zone of tall grass, open bare ground with arable weeds. Females May–September, males April–July.

Minyriolus pusillus. Nationally, said to be widespread and common in grass, moss, leaf litter, and on low vegetation and bushes, but this species appears to be very rare in Leicestershire, as it is recorded only from a single specimen. It seems unlikely that it has been overlooked. The record is of a female collected in a pitfall trap at Norris Hill (SK329161) on 18th June 1993 by J. Daws. The trap was set under an old rural hedgerow surrounded by farmland in the Ashby Woulds district, adjacent to a large area of industrial derelict land – including the Newfield Heath closed colliery site – from which it is separated by the Leicester to Burton-on-Trent railway line.

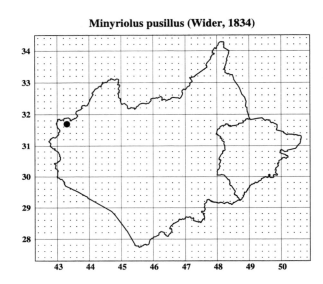

Minyriolus pusillus (Wider, 1834)

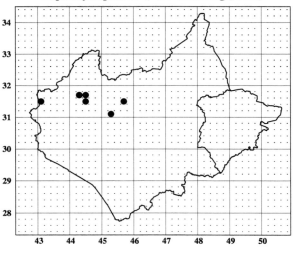

Tapinocyba praecox (O.P.-Cambridge, 1873)

Tapinocyba praecox. Local, north-west Leicestershire. Widespread throughout Britain but uncommon. First county record D.W. Mackie, 1964, Charnwood Lodge N.R., thereafter from Bradgate Park, Buddon Wood, Cademan Moor, High Sharpley, Moira Junction and Newfield Heath. From acid grassland with wet flushes, grass heath, rocky *Vaccinium* heathland, an old quarry in ancient woodland, birch coppice and wood pasture; collected from short grass and leaf litter, also from pitfall traps on open ground and in tussocky grassland, also sweeping bracken and grass in open areas. Females March–July and November, males April, June, September and November.

Atlas of Leicestershire and Rutland Spiders
LINYPHIIDAE

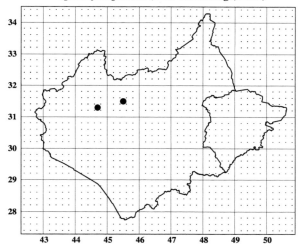

Tapinocyba pallens (O.P.-Cambridge, 1872)

Tapinocyba pallens. A northern woodland species, rare in the south; in woodland litter. Very rare in Leicestershire, known only from two ancient woodland sites on Charnwood Forest (also from Sherwood Forest in 1971 and 1977). First record for the county M.G. Crocker, 1975, Buddon Wood, where a female was shaken from *sphagnum* in wet flushes on the birch/alder woodland floor (SK557152) in June. A second female was taken at Bardon Hill (SK461132) around the summit, from a pitfall trap set in rocky *Calluna* heathland, July 1992. In the north of England *Tapinocyba pallens* is found among pine needles and dead leaves in woods, often abundant locally.

Tapinocyba insecta. Rare in Leicestershire, a species of ancient woodland, recorded from five sites in the county, mostly on Charnwood Forest, but also from Great Merrible Wood. First record J. Crocker, 1967, Swithland Wood. Since then, also from Brazil Wood, Buddon Wood and Ulverscroft N.R. Habitat: oak, birch/oak, birch/alder and alder/salix woodland. Mostly in leaf litter, often around the base of trees, shaken from *sphagnum* from wet flushes on the woodland floor and in pitfall traps amongst woodland ground flora and on bare ground in small-leafed lime coppice. Females April–July and October–January, males April–July and December/January.

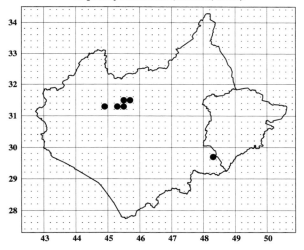

Tapinocyba insecta (L.Koch, 1869)

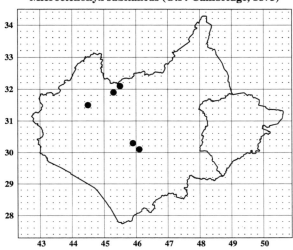

Microctenonyx subitaneus (O.P.-Cambridge, 1875)

Microctenonyx subitaneus. Infrequent, usually occurring amongst hay and straw. Rare but widespread in the county, probably more common than the few records suggest. First record J. Crocker, 1965, Charnwood Lodge N.R., subsequently from Leicester Cattle Market, Loughborough Meadows and two gardens – one in Loughborough, the other in Leicester. From a farmyard, under corrugated iron sheet near a straw rick, from a straw/manure heap at roadside, from a pitfall trap on waste ground on site of demolished cattle market, also under boxes, stones and amongst garden rubbish. Adult females March, May, September and October, males, May, June and October.

Atlas of Leicestershire and Rutland Spiders

LINYPHIIDAE

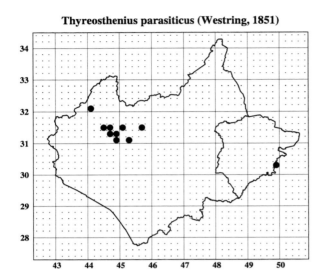

Thyreosthenius parasiticus (Westring, 1851)

Thyreosthenius parasiticus. Quite rare in the county, but widespread. Occurs in large numbers in birds' nests and inside hollow trees, but elsewhere only odd specimens are encountered. In Leicestershire, first record J. Crocker, 1965, Charnwood Lodge N.R., under stone on rubbish dump. Elsewhere from Bardon Hill, Bradgate Park, Buddon Wood, Cloud Wood, Coalbourne Wood, Geeston Quarry, Holly Hayes Wood, Loughborough Outwoods and Stoneywell Wood. Mostly from woodland, from detritus inside hollow trees, litter in twiggy growth at base of birch tree and from leaf litter under trees. Females December to May, males October, December, February and April.

Thyreosthenius biovatus is established for Leicestershire by one old record, but this spider is probably now extinct in the county. Recorded (Donisthorpe, 1927) from Buddon Wood where it was well established in the nests of the red wood ant *Formica rufa*, which was very common at this site (see comment pages 29 & 40) until the early 1940s, when the wood was clear felled and the ants disappeared. H.St.J. Donisthorpe records both sexes from Buddon Wood on 4th May 1909. This record also appears in (Bristowe, 1939). The ants lingered on at Buddon until the mid-1960s, and also in nearby Brazil Wood, but never in large numbers, and without any of the large nests which were a big feature.

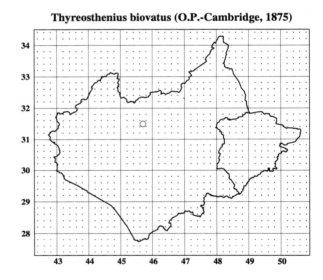

Thyreosthenius biovatus (O.P.-Cambridge, 1875)

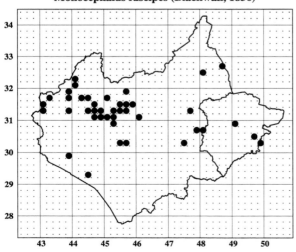

Monocephalus fuscipes (Blackwall, 1836)

Monocephalus fuscipes. Widespread and common in the county, in woodland. First record J. Crocker, 1962, Breedon Hill. The majority of records are from deciduous, mixed and conifer woodland, alder carr and wood pasture, also from grass heath, rocky *Calluna/Vaccinium* heathland, calcareous scrub grassland, open limestone grassland, rural hedgerows, abandoned allotments, disused quarries, farmyard and reservoir environs. Mostly from broadleaf litter, but also from pine needle, bracken and heather/bilberry litter, moss, under stones; also from pitfall traps on bare ground and sweeping grass under trees. Adults of both sexes taken every month of the year.

Atlas of Leicestershire and Rutland Spiders
LINYPHIIDAE

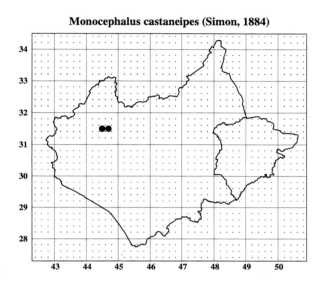

Monocephalus castaneipes. An uncommon spider nationally, with a patchy distribution, generally associated with higher open ground in more northern parts. In Leicestershire it is recorded only from a single locality, on two of the most elevated and exposed sites in the county. First record J. Crocker, September 1963, from bracken-dominated heathland at Timberwood Hill (SK470150) at Charnwood Lodge N.R., where a single female was collected from grass litter between rocks. In April the following year another three females were collected from bilberry litter at High Tor on Warren Hill (SK458152) – open, rocky, *Vaccinium* heathland with small amounts of *Calluna* and much less bracken.

Lophomma punctatum. Widespread in Leicestershire, but local, in wet places. First record Chalcraft (Rowley, 1897). From wet alder carr, boggy moorland, old acid grassland with wet flushes, waterlogged fields and flood meadows, wet fen, reedswamp, wet marsh and drier *Filipendula* marsh, canal, river, reservoir and pond waterside vegetation. In ground zone vegetation and amongst wet surface litter, often on plant stems emerging from the water. Where this species occurs away from open water, in wet flushes, the site is usually a long-established and undisturbed habitat. Adults have been recorded every month of the year with highest numbers in September and October.

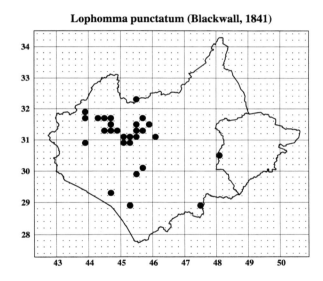

Saloca diceros. Rare but widespread, a nationally notable species of old damp woodland and fens. First record for the county J. Crocker, 1964, Owston Wood, and since then recorded from Barnsdale Wood, Burbage Wood, Gt. Merrible Wood, Harby Hills Wood and Narborough Bog. From deciduous woodland on ancient woodland sites, damp (not wet) reed-bed and sedge-fen (Narborough Bog), also from riverside nettle beds and unimproved grazed grassland subject to occasional flooding. Usually in damp ground zone litter and in pitfall traps. Adult females are recorded for April, May, July and November, and males in February, March, April and June.

CHAPTER FIVE

Atlas of Leicestershire and Rutland Spiders
LINYPHIIDAE

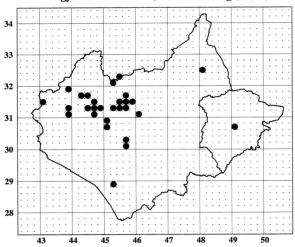

Gongylidiellum vivum (O.P.-Cambridge, 1875)

Gongylidiellum vivum. Widespread and common in Leicestershire, mostly in damp, often wet situations. A frequent aeronaut. First record J. Crocker, 1962, Charnwood Lodge N.R. Habitats: old dry grassland, wet acid grassland, flood meadow, ley, heath grassland, marsh, wet fen, rocky *Vaccinium* heathland, waterside vegetation, reedswamp, sedge-bed, birch/alder woodland with wet flushes, deciduous and mixed woodland, also from disused allotments, quarries, gravel pits, and from under stones. In wet depressions, in ground zone litter, amongst grass, in vole runs and under cut bracken, on bare ground and swept from low vegetation. Females January to November, males April to November.

Gongylidiellum latebricola is much less common than the previous species, and is found in similar situations, preferring damp rather than wet habitats, but is often found in drier places. This spider is very local in Leicestershire, known only from a single site. First recorded from Ketton Quarry by J. Daws in June 1994, when a male was taken in a pitfall trap placed in short grazed turf in herb-rich limestone grassland (SK977053). In the following month, July, another male was taken in similar circumstances (SK976053) a short distance from the original site. *G.latebricola* is fairly widespread in England but appears to be quite rare in the Midland counties and Wales.

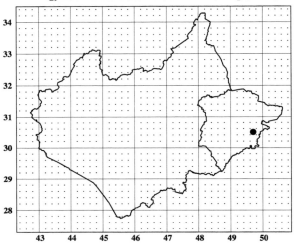

Gongylidiellum latebricola (O.P.-Cambridge, 1871)

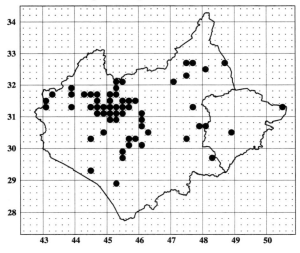

Micrargus herbigradus (Blackwall, 1854)

Micrargus herbigradus, a leaf litter spider frequenting the ground zone, is widespread and common in Leicestershire. First record J. Crocker, 1962, Cropston Waterworks grounds. From a wide variety of habitats, mostly in damp, tending towards drier situations, but occasionally in wet substrates. Woodland, wood pasture, hedgerows and scrubland, grassland, heathland, marshes, fen, peat bog, reed and sedge beds, waterside vegetation, also derelict industrial sites, quarries and allotments, churchyards, gardens *etc*. Common in pitfall traps, and hand collecting in a wide range of ground zone micro-habitats, grass roots, leaf litter *etc*. Adults every month of the year.

Atlas of Leicestershire and Rutland Spiders
LINYPHIIDAE

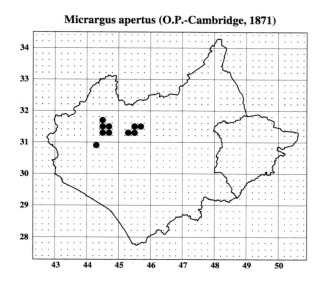

Micrargus apertus (O.P.-Cambridge, 1871)

Micrargus apertus. Uncommon and local in the county; well established on Charnwood Forest, occurring occasionally with *M.herbigradus*. All Leicestershire specimens of *herbigradus* examined and a number of the older records found to be *apertus*. First record J. Crocker, 1964, Charnwood Lodge N.R., also from Bardon Hill, Buddon Wood, Drybrook Wood, Nailstone Wiggs and Swithland Wood. From birch/alder woodland, with wet flushes, oak/birch/beech, conifer and mixed woodland, grassy heathland, open moorland and a well vegetated marsh; on bare ground under trees, in leaf and pine litter, *sphagnum*, ground zone vegetation and sedges in water. Phenology as for *herbigradus*

Micrargus subaequalis. Much more widespread in the county than the previous species, but local, in drier situations; tall grass, low plants *etc*. First record J. Crocker, 1962 from a Loughborough garden. Habitat: rough grassland, unmanaged and grazed, hay meadows, well-structured marsh, mown churchyard, disused allotments and quarries, also urban derelict land, a vegetated sand bank and riverside mud banks. From the ground zone, amongst grass roots, pitfall traps, and swept from field layer vegetation. Adult females recorded May to August, and males June to September, with highest numbers of each in July. In both pitfall and hand collecting males have predominated.

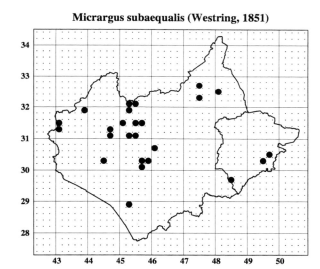

Micrargus subaequalis (Westring, 1851)

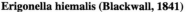

Erigonella hiemalis (Blackwall, 1841)

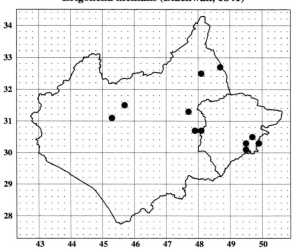

Erigonella hiemalis. Widespread in the county, but infrequent. First record J. Crocker, 1964, Owston Wood. Also from Bradgate Park, Buddon Wood, Burrough Hill Wood, Coppice Leys Wood, Geeston Quarry, Ketton Quarry, Luffenham Heath, King Lud's Entrenchments, Stonesby Quarry and The Drift. Deciduous woodland, calcareous grassland, disused limestone/ironstone quarries; in ground zone litter, grass roots *etc*., shaken from *sphagnum*, in pitfall traps on bare ground, also swept from field layer, amongst woodland grass tussocks, and in damp litter in a culvert. Adult females March, April, June and September, males March to August. High numbers March/April.

Atlas of Leicestershire and Rutland Spiders
LINYPHIIDAE

Erigonella ignobilis (O.P.-Cambridge, 1871)

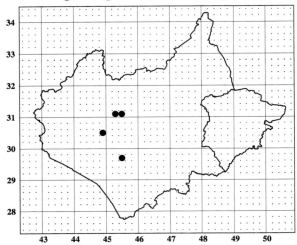

Erigonella ignobilis. Very local; there are only six records from four sites, generally in damp, marshy situations. First recorded for the county by I.M. Evans, 1962, Cropston Reservoir margins; also from Botcheston Bog, Bradgate Park and Narborough Bog. From damp litter at the base of an old brick wall and from the bottom of a 1 m deep culvert filled with accumulated litter – both sites at Bradgate Park, also from ground zone vegetation at waterside sites, including a marsh and a wet hay meadow. These six records span a period from 1962 to 1993. Adult females are recorded for March, May and July, whilst males have been taken only in December and March.

Savignia frontata. Widespread and common in the county, in a variety of situations. A frequent aeronaut. First record J. Crocker, 1962, Bradgate Park. Habitat: woodland, damp meadows, rough neglected grassland, cultivated land, marsh, waterside vegetation, reedswamp, sedge beds, wet flushes in drier grassland, moorland, rocky heath, disused quarries, farmyards, gardens, churchyards and buildings. From leaf and pine needle litter, on tree trunks, beating trees and shrubs, sweeping field layer and amongst grass, moss and debris, sieved from straw and on walls of buildings *etc.*, also inside houses. Adults of both sexes have been taken every month of the year.

Savignia frontata (Blackwall, 1833)

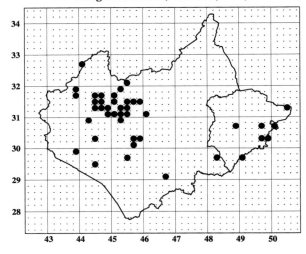

Diplocephalus cristatus (Blackwall, 1833)

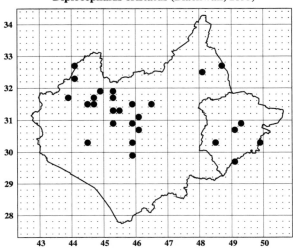

Diplocephalus cristatus. Widespread and common in Leicestershire; first recorded for the county by G.& A. Smith, 1960, Birstall. Habitat: Reservoir margins, river and canalside vegetation, calcareous and neutral grassland, flood meadows and marshy ground, disused quarries and a flooded brick pit, mixed and deciduous woodland, urban derelict land, farmyards, churchyards and gardens. From amongst ground zone vegetation, under stones, pieces of wood and rubbish, on bare ground, on the underside of a manhole cover; also swept from grass and low vegetation. Adults of both sexes have been recorded for each month of the year except November and January.

Atlas of Leicestershire and Rutland Spiders
LINYPHIIDAE

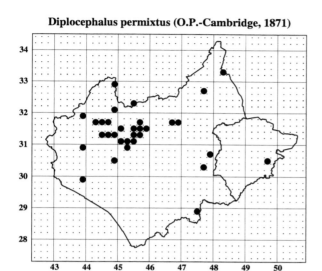

Diplocephalus permixtus (O.P.-Cambridge, 1871)

Diplocephalus permixtus. Widespread and common in Leicestershire, in damp and wet situations. First recorded for the county by P.C. Jerrard, 1962, Cropston Reservoir margins. Habitat: Lakeside and riverside margins, brookside marsh and marshy meadows, wet flushes in grassland, reedswamp, peat bog, limestone and acid grassland, and deciduous woodland – particularly alder carr. From pitfall traps and hand sorting ground zone litter at waterside, shaken from *sphagnum* and amongst *juncus*, in leaf litter and sifted from detritus at the bottom of a culvert, also shaken from litter in a swan's nest. Adults of both sexes are recorded from February to October.

Diplocephalus latifrons. Widespread and common in the county. First Leicestershire record E.Duffey, 1955, Loddington Reddish. Habitat: woodland, wet alder carr, grassland, marsh, damp sedge bed, disused quarries, abandoned allotments, urban derelict land, farmyards and waterside vegetation. From the ground zone amongst grass tussocks and woodland flora, and in leaf and pine needle litter; from pitfall traps in ground zone vegetation and on bare ground, also under pieces of wood and other bits of rubbish, on the underside of a manhole cover, and on several occasions swept from field and shrub layers. Both sexes every month of the year, with high numbers in April and July.

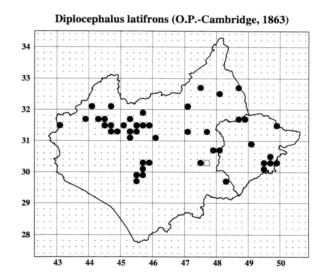

Diplocephalus latifrons (O.P.-Cambridge, 1863)

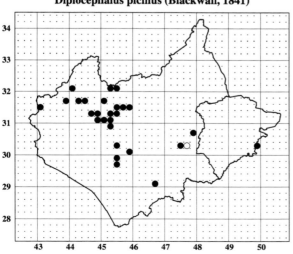

Diplocephalus picinus (Blackwall, 1841)

Diplocephalus picinus. Widespread and common in Leicestershire. First county record E. Duffey, 1955, Loddington Reddish. Habitat: deciduous woodland, wet alder carr, damp grassland, marsh, waterside vegetation, reedswamp, *Calluna* heath and a suburban garden. In leaf litter, under stones and pieces of wood, shaken from moss, in litter between rocks, from pitfall traps amongst ground zone vegetation and on bare ground, also swept from woodland rides and beaten from oak, birch, rhododendron and sycamore. Adult females are recorded from February to October, and adult males from April to July, with highest numbers of both sexes recorded in May and June.

Atlas of Leicestershire and Rutland Spiders
LINYPHIIDAE

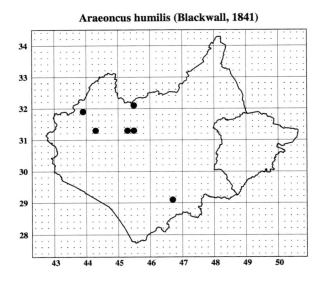

Araeoncus humilis (Blackwall, 1841)

Araeoncus humilis. A frequent aeronaut which, although widespread, is uncommon, with only six records for the county. First Leicestershire record D.B. Forgham, 1961, Hugglescote, and subsequently from Lount grassland, the River Soar at Stanford, Saddington Reservoir, Swithland Reservoir and Swithland Wood. Habitat: deciduous woodland, under dead birch bark; waterside vegetation, under drift litter, among reeds and other wet vegetation, and in pitfall traps; also from farmland, amongst thrashed corn. Adult females have been recorded for the months of June, August and September, and adult males in July and August only, based on these six records.

Panamomops sulcifrons. Widespread but local in Leicestershire. First record J. Daws, 1992, Loughborough Meadows; since then also from Aylestone Meadows, Carlton Curlieu churchyard, Geeston Quarry, Harby Hills, Kendall's Meadow, Ketton Quarry, Seaton Meadow and Watermead Country Park. Habitat: flood meadows and drier hay meadows, limestone grassland, open sedge bed adjacent to the River Soar, and disused quarries. From pitfall traps in wet vegetation, damp hollows, fairly dry ridge and furrow grassland, amongst grass roots, also in tall dry grassland. Adult females recorded from February to September, and males from April to October, with peak numbers in June.

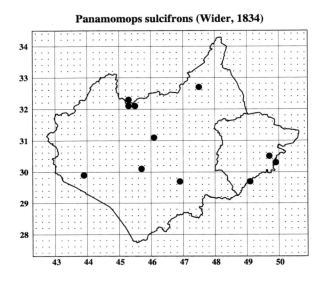

Panamomops sulcifrons (Wider, 1834)

Lessertia dentichelis (Simon, 1884)

Lessertia dentichelis. A cave spider, often found in sewers and stone-filled sewage filter beds. There are only three records from two tetrads in Leicestershire, but this species is certainly much more widespread and common than is apparent from these few records. First county record J. Crocker, 1966, from the underside of a sewer manhole cover at Coleorton Hall in March. Two years later also in March, further specimens were obtained from the underside of a different sewer cover some distance from the original site. The third record is from three-year-old foul drains at Loughborough, on underside of a manhole cover. Both sexes taken together March and May, with ova in March.

Atlas of Leicestershire and Rutland Spiders

LINYPHIIDAE

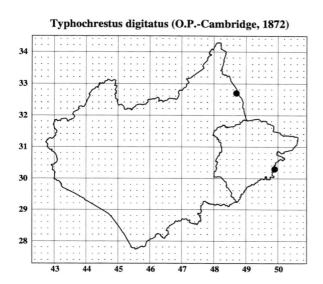

Typhochrestus digitatus (O.P.-Cambridge, 1872)

Typhochrestus digitatus. Local, known from only two sites in east Leicestershire, one of which has been sampled systematically by pitfall trapping over a twelve month period. First record (Bristowe, 1939) from an unknown locality. Recorded from calcareous grassland at King Lud's Entrenchments, where two females were swept from grass amongst thorn scrub in May 1964. In 1995 regular pitfall trapping at Geeston Quarry, a disused ironstone quarry abandoned in 1950, was carried out with several lines of traps set on bare ground and amongst sparse vegetation. Adults of both sexes were taken in every month from November 1994 to May 1995, with peak activity January to March.

Milleriana inerrans. Widespread but local in the county. This spider's natural habitats are open stony ground and sand dunes. It is a successful aeronaut and colonises new open habitats such as agricultural land. First county record J. Crocker, 1972, Donington Park, subsequently also from Ravenstone (Coalfield West), Leicester Cattle Market, Evesham Road Allotments, Lount Grassland and Luffenham Heath Golf Course. From disused allotments, reseeded grassland, ley and calcareous grassland; on open bare ground with arable weeds, and swept from field layer; also from open wood pasture in litter under oak tree. Both sexes June to September and females also in November.

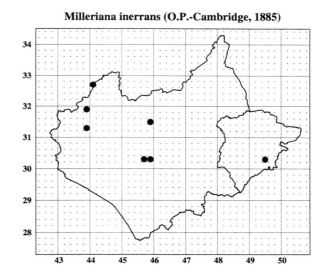

Milleriana inerrans (O.P.-Cambridge, 1885)

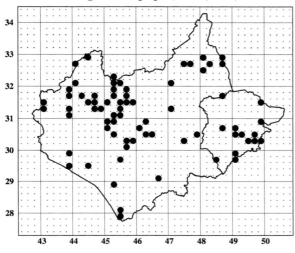

Erigone dentipalpis (Wider, 1834)

Erigone dentipalpis. Widespread and very common, but less so in the county than *Erigone atra*; in a wide variety of habitats, and a common aeronaut. First Leicestershire record R. Lee, 1960, Scraptoft. Habitat: woodland rides, parkland, hedgerows, heath, permanent grassland, ley, marsh, waterside vegetation, farmland, disused quarries, urban derelict land, churchyards, gardens, buildings *etc*. In dry, damp and occasionally wet places in ground zone amongst grass, often in small sheet webs in depressions, under stones and pieces of wood, and frequently swept from field layer and beaten from shrubs and trees. Both sexes recorded every month of the year, with the activity peak from May to August.

Atlas of Leicestershire and Rutland Spiders
LINYPHIIDAE

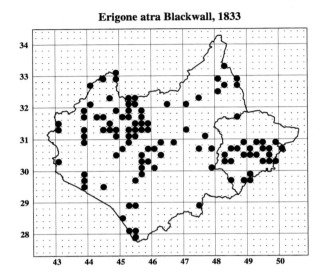

Erigone atra Blackwall, 1833

Erigone atra. Widespread, and one of the county's commonest spiders; a frequent aeronaut, found in a wide variety of habitats, in similar situations to *Erigone dentipalpis*; also on tree trunks, on wet mud – often in small sheet webs – amongst straw and garden rockery plants, and in leaf litter; also inside houses, swept from heather, grass, nettles, reedswamp *etc.*, and beaten from birch, oak, hawthorn, blackthorn, briar, ash, crab apple, pine, beech and gorse. First Leicestershire record I.M. Evans, 1959, Puddledyke. Phenology probably the same as *dentipalpis*, with adults all through the year, but high numbers have been recorded for a longer period, from May through to October.

Erigone arctica. A coastal species, rare in Leicestershire, but spreading along inland waterways. First county record H. Ikin, 1983, Lockington Meadows, since then from Geeston Quarry (adjacent to the River Welland), Rutland Water, River Trent and River Soar. From river flood meadows, river and reservoir margins, a disused ironstone quarry; in pitfall traps on bare ground, also amongst waterside vegetation in wet ground zone litter, on bare mud and amongst stones. Adult females May and July, males February, March, May and July; highest numbers July. Found on mountains in Sweden and around our coasts, this species disperses over long distances but survives in only a few places.

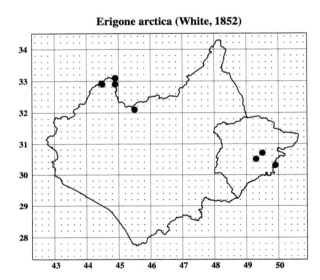

Erigone arctica (White, 1852)

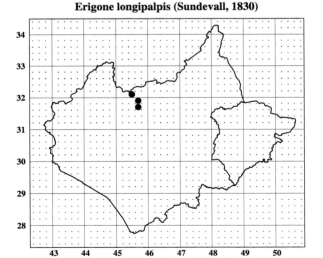

Erigone longipalpis (Sundevall, 1830)

Erigone longipalpis. Very rare in Leicestershire, known only from four recent records on the River Soar, all in 1994. First record D.A. Lott, July – a male in pitfall trap set in an island of sedimentary mud in the River Soar at Cotes Bridge (SK553206); and since then from three similar situations on the River Soar, between Cotes and Quorn (April– September) – 1 male (SK570175), 15 males & 4 females (SK566178) and 1 male (SK562189). This species is an aeronaut, and is often found with the previous species, it disperses over long distances but, like *Erigone arctica*, survives only in wet habitats which are subject to inundation. It appears to be extending its range in Britain.

Atlas of Leicestershire and Rutland Spiders
LINYPHIIDAE

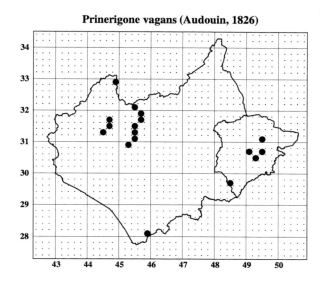
Prinerigone vagans (Audouin, 1826)

Prinerigone vagans. Widespread in the county but local, in wet places; it has been recorded elsewhere in very large numbers in stone-filled sewage filter beds. First record I.M. Evans, 1959, Puddledyke. Habitat: waterside vegetation – rivers, ponds, lakes, reservoirs *etc.*, on bare mud and shingle, under drift litter and stones, in waterside vegetation – ground zone and field layer, on plant stems growing out of the water, and occasionally in pitfall traps in the drier situations. This species is an aeronaut with rather specialist habitat requirements. Females April to October, also December and February; males May to October, with high numbers in July and October.

Leptorhoptrum robustum. Widespread in Leicestershire, but local, in wet or very damp situations; a frequent aeronaut whose natural habitat is freshwater marshes and wet meadows. It prefers stable temperatures and high humidity, and like the previous species is often found in filter beds. First record I.M. Evans, 1964, Swithland Wood. County habitats: wet depressions in flood meadows, alder carr, reedswamp, wet marsh, waterside vegetation, stream and riverside wet mud; in the ground zone, also single specimens are beaten from the shrub layer occasionally. Adult females recorded February to December, males April to December; highest numbers recorded in July and November.

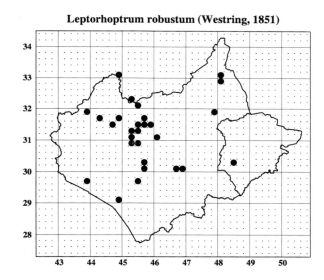
Leptorhoptrum robustum (Westring, 1851)

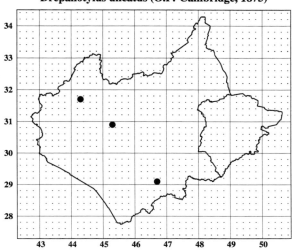
Drepanotylus uncatus (O.P.-Cambridge, 1873)

Drepanotylus uncatus. Rare in the county, four records only, each of single specimens, from three sites; very local, in wet situations. First record J. Crocker, 1967, Cademan Moor; also from Groby Pool and Saddington Reservoir. The Cademan site was a dried-up pond in unmanaged acid grassland with wet flushes (SK433173), where females were collected in September and October 1967 from cut *juncus* litter in the damp depression. This site has now been ploughed. At Groby Pool (SK522083), a female was collected in April 1970 from waterside ground zone vegetation, and at Saddington Reservoir (SP664914) a male was taken in March 1993 in ground zone litter from a marginal marsh.

CHAPTER FIVE

Atlas of Leicestershire and Rutland Spiders

LINYPHIIDAE

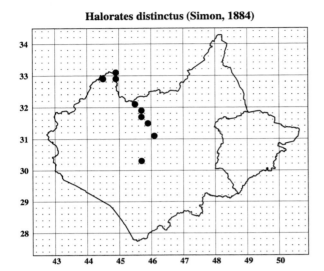

Halorates distinctus (Simon, 1884)

Halorates distinctus. Rare in the county and very local, apparently restricted to riparian habitats. First Leicestershire record A.E. Squires, 1973, Barrow Gravel Pits; also from the River Trent, Lockington Meadows, and various sites along the River Soar from Lockington to Leicester. Habitat: river flood meadows, river cut-off, riverside mud and shingle banks and waterside vegetation, including a nettle bed subject to occasional flooding; from wet riverside grass, reeds and water mint, under stones and on mud, also from a wet marsh vegetation; mostly from ground zone pitfall traps. Adult females are recorded from May to October, and males from May to July, October and December.

Asthenargus paganus. A rare species of ancient woodland, recorded only from the Charnwood Forest area. Very local. First Leicestershire record J. Crocker, 1967, Nailstone Wiggs; and subsequently also from Bardon Hill, Drybrook Wood, Stoneywell Wood and Ulverscroft N.R. Usually only single specimens encountered. Habitat in Leicestershire: old broadleaf woodland sites with pockets of deep leaf litter, which seems to be the favoured micro-habitat; specimens have been collected from leaf litter around stands of oak, hazel, oak/beech and beech/rowan. Adult females are recorded for the months of February, May and October, and an adult male for April only.

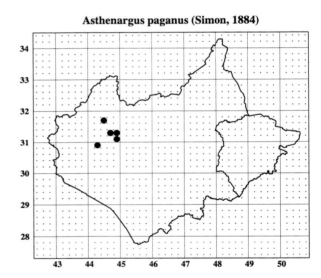

Asthenargus paganus (Simon, 1884)

Ostearius melanopygius (O.P.-Cambridge, 1879)

Ostearius melanopygius. Widespread but infrequent; a naturalised species from New Zealand, probably under-recorded in the county. Cosmopolitan, on rubbish heaps, in gardens, on buildings – often indoors, and a frequent aeronaut. First European records – 1906, Kent and Essex. First Leicestershire record – J. Crocker, 1962, from a garden in Loughborough, and subsequently from a number of other disturbed habitats: haystacks, various dry manure heaps, garden rubbish, under stones, pieces of wood, plaster board *etc.*, on bare ground, in a mole's nest, in leaf litter, and beaten from low vegetation. Adult females throughout the year, males July to November. Most records September.

Atlas of Leicestershire and Rutland Spiders
LINYPHIIDAE

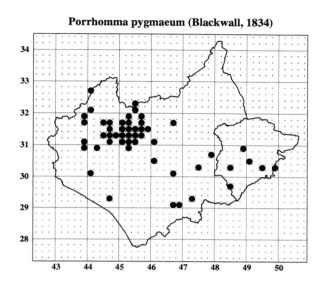

Porrhomma pygmaeum (Blackwall, 1834)

Porrhomma pygmaeum. A common aeronaut, often beaten from shrubs and trees, but the main habitat of this species is in damp marshy places. It is widespread and common in the county. First record P.C. Jerrard, 1962, Bradgate Park. From woodland, especially birch/alder/willow, waterlogged grassland, marsh, reedswamp, boggy moorland, waterside vegetation, also heathland, rough grassland, churchyards, gardens and wasteland. Usually in ground zone vegetation, under stones and on bare ground, but also in field, shrub and canopy layers. Both sexes adult throughout the year, with activity peaks recorded between February and May.

Porrhomma convexum. Very rare in Leicestershire, but possibly under-recorded. Often confused with *Porrhomma pygmaeum* which is a smaller spider, but can be separated by its size and distinctive eye pattern. *Porrhomma convexum* is a subterranean woodland species, often in the chambers of badger setts and in culverts, but occasionally occurs on agricultural land during dispersal activities. There are only two records for the county, from the same site. Recorded by J. Crocker, June 1967, (female) and May 1968, (male), from under slates and stones in a dark culvert carrying a woodland stream under a road at Swithland Wood (SK539119). Presumably adult throughout the year.

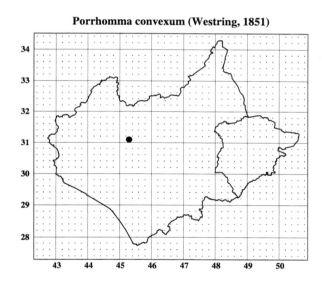

Porrhomma convexum (Westring, 1851)

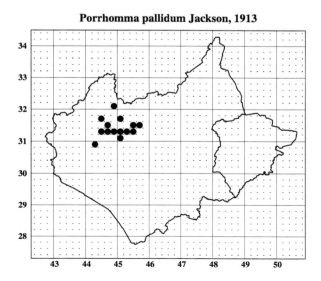

Porrhomma pallidum Jackson, 1913

Porrhomma pallidum. A local, woodland spider, apparently restricted to the Charnwood Forest area. First record J. Crocker, 1963, Charnwood Lodge N.R.; also from Bardon Hill, Benscliffe Wood, Buddon Wood, Drybrook Wood, Nailstone Wiggs, Oakley Wood, Outwoods, Stoneywell Wood, Swithland Wood and Ulverscroft N.R. Habitat: broadleaf and conifer woodland, also one record from under a stone on rocky heathland adjacent to a conifer plantation. Amongst leaf and pine needle litter, under stones and shaken from *sphagnum* from wet flushes on woodland floor. Webs noted amongst leaf litter. Adult females September to June; males February to June and October.

Atlas of Leicestershire and Rutland Spiders

LINYPHIIDAE

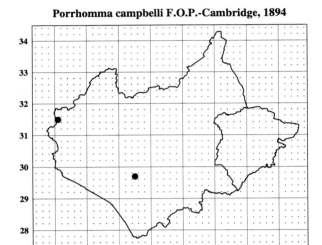

Porrhomma campbelli F.O.P.-Cambridge, 1894

Porrhomma campbelli. Very rare in Leicestershire, recorded from only two widely separated and different habitats. It is also uncommon and widely distributed throughout the British Isles. First record for the county J.Mathias, 1977, Narborough Bog, where two females were collected from damp ground zone litter in a rather dry reed bed adjacent to the River Soar (SP549979) in June. The only other record is of a female caught in a pitfall trap placed between clumps of heather on coal measures heathland which has developed, over the past 100 years, on an abandoned colliery site at Newfield Heath (SK381154). This second record is from collections made in July 1992.

Porrhomma microphthalmum. Though local, this spider is widespread throughout the county and not uncommon. It is a fairly frequent aeronaut. First Leicestershire record J. Crocker, 1962, Bradgate Park. From mixed and deciduous woodland, calcareous and neutral grassland, marsh and reservoir margins, also heather/bilberry heath. Swept from woodland rides and field layer vegetation, and beaten from shrubs and trees; also in damp ground zone litter, on bark of an old oak tree, under stones and in pitfall traps amongst ground flora and on bare ground. Adult females have been recorded from January to August, and adult males from April to July.

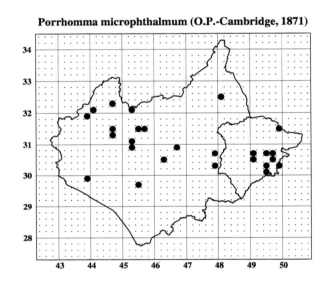

Porrhomma microphthalmum (O.P.-Cambridge, 1871)

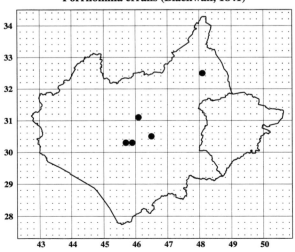

Porrhomma errans (Blackwall, 1841)

Porrhomma errans. This nationally notable species is extending its range and has been found at a number of different sites within the city of Leicester, on disturbed ground. First record I.M. Evans, 1983, Scraptoft; also from Evesham Road Allotments, St. Mary's Allotments, and Watermead Country Park; additionally from Stonesby Quarry. From grazed pasture under a piece of cardboard, hawthorn/elder woodland, various disused allotments abandoned between 1990 and 1992, and a disused limestone quarry. In pitfall traps in open ground with arable weeds, amongst nettles, and in unmanaged grassland. Females June to November, males January, May, June and November.

Atlas of Leicestershire and Rutland Spiders
LINYPHIIDAE

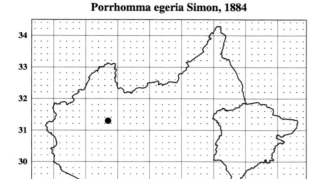

Porrhomma egeria Simon, 1884

Porrhomma egeria. Very rare, known in the county from only a single female, collected on 15th February 1970 by J. Crocker, from under a 9 cm frozen 'crust' of leaf litter around the base of a mature oak tree at Bardon Hill. This site is a small remnant of ancient oak wood (SK461129) which once covered Bardon Hill, the highest point in Leicestershire. *Porrhomma egeria* is a predominantly caverniculous species with an extensive European distribution. Very little is known about its biology and ecology. Adults have been collected in the Bange Caves, France, in January 1977, with the greatest number of males (peak activity?) recorded at this time of the year.

Agyneta subtilis. Frequent and widespread throughout the British Isles, but rare in Leicestershire, with only four records from three sites in the north-west. First county record J. Crocker, 1973, Donington Park, where females were sieved from straw around an old haystack on cultivated ground taken out of the deer park. Subsequently recorded from deciduous woodland (Stoneywell Wood – 2 records) where a female was grubbed from leaf litter around oak trees, and another female beaten from oak; also recorded from Charnwood Lodge N.R., a female from damp grass under mixed woodland open canopy on former moorland. Females in May and September, no males taken.

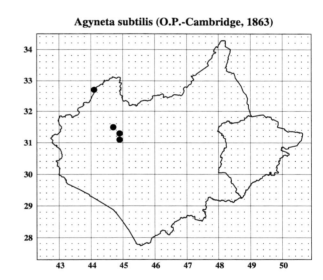

Agyneta subtilis (O.P.-Cambridge, 1863)

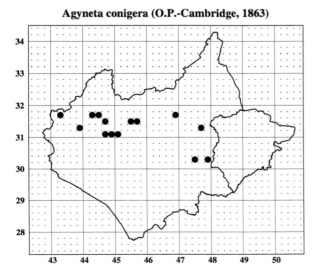

Agyneta conigera (O.P.-Cambridge, 1863)

Agyneta conigera. Widespread but infrequent in the county. First Leicestershire record J. Crocker, 1962, Skeffington Wood. Habitat: broadleaf woodland, rural hedgerows with and without trees, heather and bilberry heathland, heath grassland, and a marshy field with tussock sedge. These records are mainly from pitfall traps amongst woodland flora and on bare ground, under heather and amongst grass; *Agyneta conigera* has also been collected from leaf litter, swept from field layer herbs and low vegetation along woodland rides, and beaten from bushes and trees. Adult females have been recorded in May, June and August, and adult males in May, June, July and September.

CHAPTER FIVE

Atlas of Leicestershire and Rutland Spiders
LINYPHIIDAE

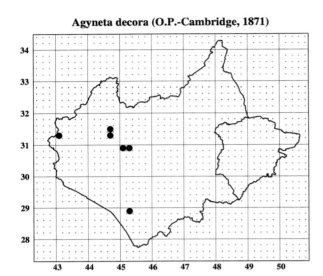

Agyneta decora (O.P.-Cambridge, 1871)

Agyneta decora. Widespread but rare in the county, only from west Leicestershire. First record J. Crocker, 1965, Bradgate Park, and subsequently from Acresford Sandpit, Bardon Hill, Charnwood Lodge N.R., Dunton Bassett and Groby Grassland. Habitat: disused gravel pit (marsh), heath birch scrub, heather and bilberry heath, bracken-dominated acid grassland, neutral grassland and a sparsely vegetated sand bank. From pitfall traps in rough marsh vegetation, on sand bank and between *Calluna* and *Vaccinium*, also grubbed from *Nardus* grass and amongst wet *juncus*, and swept from the field layer. Adult females recorded in June and July, males in May, June and August.

Agyneta cauta. A local species, very rare in the county, recorded only from a single site. It occurs more frequently in northern, exposed situations, in moss and litter.
J. Daws collected both sexes of *Agyneta cauta* in pitfall traps at Bardon Hill in July 1992. At site 'A' (SK460131) the trap was set in rank bracken/heather heathland on the southern flank (260 m) adjacent to conifer woodland; at site 'B' (SK461132) the trap was in open, rocky, exposed *Nardus* grass heath on the summit ridge (270 m). A male and a female (confirmed J. Crocker) were obtained from site 'A', and a female (confirmed S.Dobson) from site 'B'. Bardon Hill (278 m) is the highest point in Leicestershire.

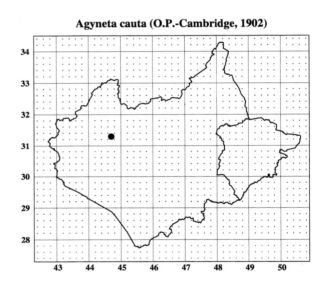

Agyneta cauta (O.P.-Cambridge, 1902)

Agyneta ramosa Jackson, 1912

Agyneta ramosa. Very local and rare in Leicestershire, in wet situations. First record for the county, M.G. Crocker, 1975, Buddon Wood, where the species appeared to be well established. *Agyneta ramosa* is also recorded from Moira Junction and Swithland Reservoir. Habitat: damp broadleaf woodland with wet flushes, particularly birch/alder, also wet marsh and reservoir margins. From pitfall traps in bare ground under fallen trees, shaken from *sphagnum* in wet flushes, from ground zone marsh litter, and on well-drained *Nardus* grass heath. Adult females are recorded for the months of May and June, and males for May, June and July, with highest numbers recorded in June.

Atlas of Leicestershire and Rutland Spiders
LINYPHIIDAE

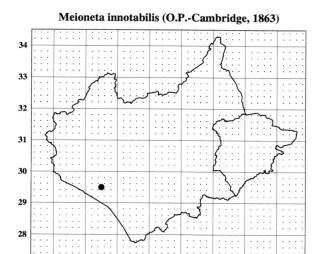

Meioneta innotabilis (O.P.-Cambridge, 1863)

Meioneta innotabilis. Widespread throughout Britain but with a rather patchy distribution; very rare in the county, known only from a single record – a male collected in June 1994 by J. Daws at Burbage Wood (SP451941), from ground zone litter amongst tussocky grass in three-year-old coppice. This is an ancient woodland site, partly on ridge and furrow, which has been traditionally coppiced in the past, and the practice has recently been reintroduced. Locket and Millidge (1953) state that *Meioneta innotabilis* occurs "on tree trunks, in crevices in the bark and among the dead leaves near the base of trees"; they also state that this species is "adult in summer".

Meioneta rurestris. Widespread and common, from a wide variety of habitats; a successful aeronaut. First Leicestershire record J. Crocker, 1961, Coleorton Hall. Habitat: natural and cultivated grassland, heathland, farmland, parkland, hedgerows, gardens, allotments, churchyards, marsh, waterside vegetation, gorse scrub, woodland, disused quarries, urban and rural derelict land, buildings *etc*. On bare ground, in ground zone vegetation, in leaf litter, under stones, also swept from the field layer and beaten from bushes and trees, on walls of buildings and inside houses, offices *etc*. Adults of both sexes have been recorded for every month of the year.

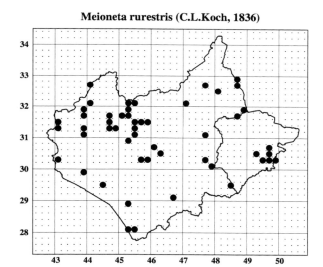

Meioneta rurestris (C.L.Koch, 1836)

Meioneta saxatilis (Blackwall, 1844)

Meioneta saxatilis. A litter zone species, frequent in damp grassland. Widespread and common in Leicestershire, but less so than *M.rurestris*. First county record J. Crocker, 1962, Breedon Hill. Habitat: rough scrubby grassland, herb-rich meadows, marsh and marshy fields, grass heath, heather heathland, *sphagnum* bog, reed and sedge beds, mixed and deciduous woodland, wood pasture, disused quarries, also gardens and allotments. From pitfall traps in bare ground and in short vegetation, sifted from leaf litter, grass, *sphagnum, juncus etc.*, and has been swept from heather. Adult females are recorded from May to September, and males from May to August, high numbers June and July.

CHAPTER FIVE 169

Atlas of Leicestershire and Rutland Spiders
LINYPHIIDAE

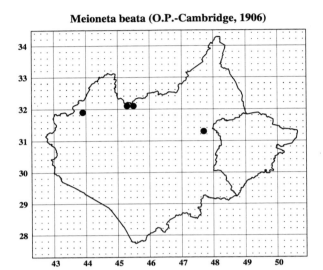

Meioneta beata (O.P.-Cambridge, 1906)

Meioneta beata. An uncommon but widespread species in Britain, amongst moss and grass, but rare in Leicestershire, with only seven records. First recorded by D.A. Lott, 1991 from Lount Grassland – reseeded opencast site (Worthington 'A'). Since then the species has also been recorded from Burrough Hill and Loughborough Meadows. Habitats: new grassland (three-year-old ley), herb-rich hay meadow, drier part of riverside flood meadow, and gorse scrub. From pitfall traps set in the grassland ground zone, and also beaten from gorse. Adult females have been recorded for the months of May, June, August and September, and males for June, August and September.

Microneta viaria. Widespread and common in the county, especially in woodland. First Leicestershire record H.A.B. Clements, 1962, from Charnwood Lodge N.R. Habitat: woodland, scrub and hedgerows, *Calluna* heathland, grassy heathland, acid grassland, disused limestone and ironstone quarries, limestone grassland, urban wasteland, churchyards and gardens. In leaf, bracken and grass litter, pitfall traps in bare ground and ground zone vegetation, also under stones and pieces of wood. Most records are from hand-sorting woodland litter. Adult females recorded from January to November, males January to June and September to November, high numbers in April and May.

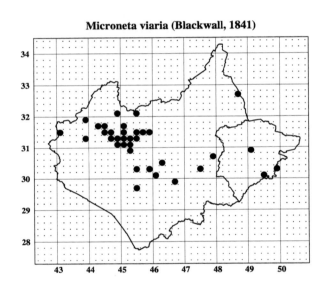

Microneta viaria (Blackwall, 1841)

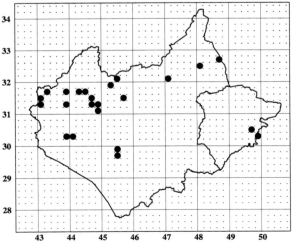

Centromerus sylvaticus (Blackwall, 1841)

Centromerus sylvaticus. Widespread and common, particularly in woodland. First Leicestershire record H.I. James, 1961, Bosworth Park. Habitat: deciduous, mixed and conifer woodland; acid, neutral and calcareous grassland, heath scrub grassland, *Calluna* heath, marsh, peat-bog, riverside; hedgerows, disused quarries, gardens and a vegetated sand bank. In leaf and pine needle litter, under cut bracken, grass and pieces of wood, amongst waterside detritus, garden rubbish and on underside of small webs in vole runs; frequent in pitfall traps. Adult females recorded for every month of the year, males October to January and April. High numbers in November and December.

Atlas of Leicestershire and Rutland Spiders
LINYPHIIDAE

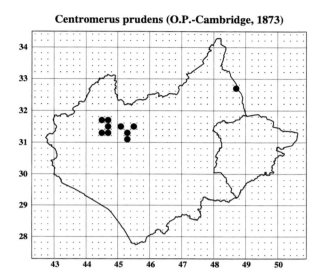

Centromerus prudens (O.P.-Cambridge, 1873)

Centromerus prudens. More common in the north of Britain than in the south; common in the Charnwood Forest area but few records from elsewhere. First county record J. Crocker, 1962, Charnwood Lodge N.R., also from Buddon Wood, Bradgate Park, Bardon Hill, Beacon Hill, High Sharpley, Ives Head, Swithland Wood and King Lud's Entrenchments. Habitat: rocky *Vaccinium/Calluna* heathland, heath grassland, woodland, and an isolated record from calcareous grassland. Under stones, amongst *Nardus* grass, bracken litter, and in detritus between rocks. Adult females are recorded from October to June, males October/November and April/May.

Centromerus dilutus. Widespread and common; a web spinner, requiring small open spaces in relatively loose ground zone litter. From a wide variety of habitats, mostly in the litter layer. First Leicestershire record J. Crocker, 1963, Charnwood Lodge N.R. Habitat: grassland, heathland, moorland, farmland, marsh; deciduous, mixed and conifer woodland, scrub and hedgerows, also from a farmyard, disused quarry and a garden. In leaf, pine needle, bilberry, heather, bracken and grass litter, under stones, amongst moss, from litter between rocks; also swept from field layer and occasionally in pitfall traps. Females recorded all the year, males September to April.

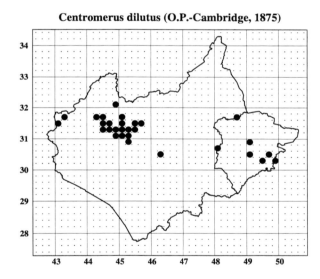

Centromerus dilutus (O.P.-Cambridge, 1875)

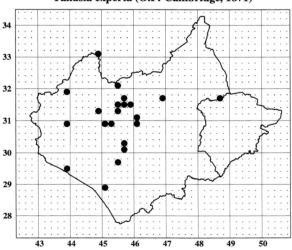

Tallusia experta (O.P.-Cambridge, 1871)

Tallusia experta. Widespread but local in the county, in damp and wet situations; a frequent aeronaut. First record I.M. Evans, 1963, Wymondham. Habitat: waterlogged marsh, marshy fields, flood meadows, rough grassland, ungrazed pasture, waterside vegetation, raised fen, reedswamp, osier beds, alder carr, deciduous woodland with wet flushes, disused riverside allotments. In wet moss, grass and ground zone litter, in wet depressions, on stems of water plants, on wet mud at waterside, in pitfall traps in sedge beds *etc.*, and occasionally swept from marsh and waterside vegetation. Females recorded February to December, males February to May and September to December.

Atlas of Leicestershire and Rutland Spiders
LINYPHIIDAE

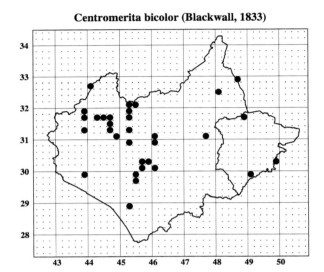

Centromerita bicolor (Blackwall, 1833)

Centromerita bicolor. Widespread and common in the county. First record Chalcraft (Rowley, 1897). Habitat: damp grassland, riverside wet hay meadows, marsh, waterside vegetation, damp woodland, heathland, disused allotments and quarries, gravel pits, urban derelict land, manure heaps, limestone grassland, churchyards, farmyards and gardens. In short grass and ground zone vegetation, vole runs, under stones, pieces of wood, cut grass *etc.*; common in pitfall traps, in bare ground, amongst sparse vegetation and leaf litter. Occasionally swept from field layer and beaten from bushes. Adult females September to June, males September to March, activity peak October to December.

Centromerita concinna. Widespread and common in Leicestershire, but less so than *C.bicolor*. A litter zone species, preferring tall, dense grass. First record J. Crocker, 1963, Charnwood Lodge N.R. Habitat: calcareous and acid grassland, heath grassland, bracken heath, rocky heather and bilberry heathland, disused ironstone and limestone quarries, urban and rural derelict land, birch scrub and oak/birch woodland. Mostly from pitfall traps in drier, rough grassland, and on bare ground amongst sparse vegetation; also in grass tussocks and unmanaged *Nardus* grassland. Adult females recorded from September to June, males from October to April, with high numbers recorded November to January.

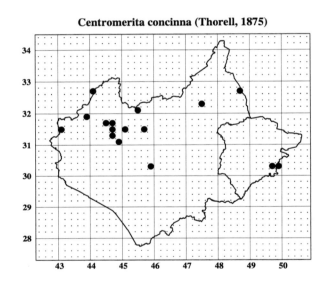

Centromerita concinna (Thorell, 1875)

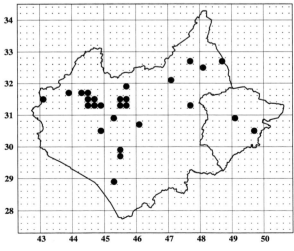

Saaristoa abnormis (Blackwall, 1841)

Saaristoa abnormis. Widespread and common, in damp situations, often in woods. First record J. Crocker, 1962, Coleorton Hall Woods. Habitat: deciduous, mixed and conifer woodland, scrub grassland, wet heathland, peat bog, marsh and dry fen, moorland, waterside vegetation, rough grassland, disused limestone, gravel and clay quarries *etc*. In wet flushes, shaken from moss and grass, amongst grass roots and other ground zone vegetation, in vole runs and leaf litter, frequently in pitfall traps in these situations, also under stones and swept from woodland floor vegetation. Adult females recorded February to October, males April to October, activity peak June and July.

Atlas of Leicestershire and Rutland Spiders
LINYPHIIDAE

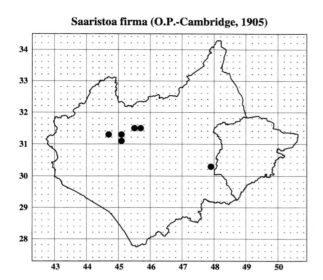

Saaristoa firma. A rare spider in the county, restricted to ancient woodland sites. First record J. Crocker, 1970, Bardon Hill; since then this species has also been collected from Benscliffe Wood, Buddon Wood, Launde Big Wood and Stoneywell Wood. From birch/alder, birch/oak, oak and mixed woodland. Shaken from *sphagnum* from wet flushes on the woodland floor, from damp leaf litter under oaks, and amongst grass under open canopy, also sifted from bracken litter. Although, elsewhere, adult females are known to occur throughout the year, in Leicestershire they are recorded only for February, April, June and September, with adult males recorded in June only.

Macrargus rufus. Common in the Charnwood Forest area, mostly in woodland, but few records from elsewhere. First county record J. Crocker, 1962, Charnwood Lodge N.R. Habitat: deciduous, mixed and conifer woodland, wood pasture, bracken-dominated heathland, and *Calluna* and grass heathland. In leaf and pine needle litter, under stones and cut bracken, amongst grass tussocks, in wet moss under alders, occasionally swept from grass under deciduous trees, and in pitfall traps in heathland ground zone amongst heather. Adult females are recorded from September to June, and males September to November, February and April. Highest numbers are recorded for October.

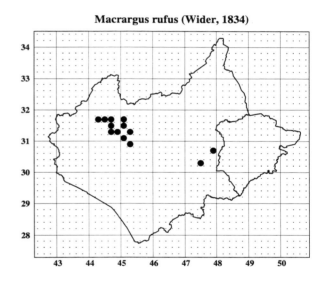

Bathyphantes approximatus. Widespread and common in wet marshy places. First record I.M. Evans, 1962, Cropston Reservoir. Habitat: reedswamp, waterside vegetation, riverside fen, osier beds, river shoreline, alder carr, birch/alder wet woodland, waterlogged fields, flood meadows, marsh, unimproved damp grassland, and occasionally in drier rough grassland. On wet mud, amongst plant stems emerging from water, amongst wet ground zone litter, under pebbles on shoreline, in moss, sedges, reeds *etc.*, and damp grass, also in old waterhen's nest, and swept from marsh field layer. Adults every month of the year. High numbers recorded February, May and September/October.

CHAPTER FIVE

Atlas of Leicestershire and Rutland Spiders
LINYPHIIDAE

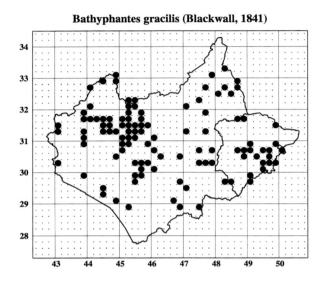

Bathyphantes gracilis (Blackwall, 1841)

Bathyphantes gracilis. Widespread; ubiquitous; a common aeronaut, and one of the county's most numerous spiders; in a wide variety of damp grassland habitats. First record (Bristowe, 1939). Habitat: deciduous and mixed woodland, hedgerows, heathland, calcareous, acid and neutral grassland, marsh, fen, bog, waterside vegetation, farmland, parkland, urban and rural derelict sites, gardens, churchyards, farmyards, disused quarries, buildings *etc*. Under stones, on bare ground, in ground zone vegetation often in very wet conditions, in leaf and pine needle litter, moss, grass and straw, also swept from field layer, beaten from shrubs and trees, and inside buildings. Adults throughout the year.

Bathyphantes parvulus. Widespread and common, usually in drier habitats than *B.gracilis*. First record J. Crocker, 1962, Breedon Hill. Habitats: heathland, heath grassland, neutral and limestone grassland, marsh, sedge beds, reservoir margins, urban derelict land and disused quarries, roadside verges, birch/oak woodland and hedgerows. Amongst ground zone litter, grass roots, moss, vole runs; under cut grass and bracken litter, pieces of corrugated iron and bits of fibre board, in ground zone pitfall traps, and occasionally swept from woodland rides and marsh field layer. Adult females recorded May to September, males February and May to October. High numbers July.

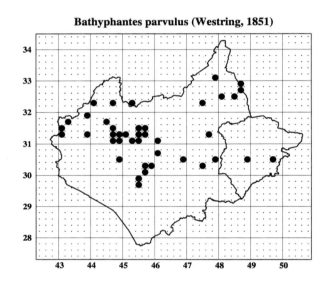

Bathyphantes parvulus (Westring, 1851)

Bathyphantes nigrinus (Westring, 1851)

Bathyphantes nigrinus. Widespread and common; mostly in woodland and damp or wet situations. First record H.A.B. Clements, 1962, Owston Wood. Habitat: wet woodland, alder carr, deciduous and mixed woodland, wood pasture, marsh, reedswamp, sedge fen, reservoir and river margins, and wet grassland. From wet litter at base of trees, under wet logs, in litter in small webs at ground level, in short grass, moss, straw, and detritus in wet depressions, amongst *juncus*, sedges, nettles *etc*., also occasionally swept from waterside vegetation and woodland rides. Adult females recorded from February to November, and males from February to October. High numbers recorded in April.

174   SPIDERS OF LEICESTERSHIRE AND RUTLAND

Atlas of Leicestershire and Rutland Spiders
LINYPHIIDAE

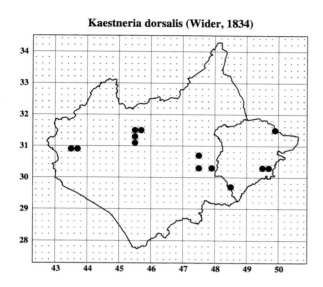

Kaestneria dorsalis (Wider, 1834)

Kaestneria dorsalis. Infrequent in Leicestershire; on bushes and trees. First record J. Crocker, 1962, Skeffington Wood. Habitat: deciduous and mixed woodland, hedgerows and thorn scrub, rough calcareous grassland with scrub gorse and oak, open churchyards with yew trees and ornamental shrubs, marsh and waterside trees, also bushes on railway embankments. Beaten from trees and shrubs, including oak, ash, birch, gorse, pine, willow and hawthorn, also swept from the field layer in woodland rides, along woodland edge and on rough grassland. This species seems to have a short season; adults of both sexes are recorded for the months of May, June and July only.

Kaestneria pullata. Widespread and common in the county, amongst grass, in wet situations. First record J. Crocker, 1962, Cropston Reservoir margins. Habitat: wet flushes in acid, calcareous and neutral grassland, peat bog, marsh, reedswamp, waterside vegetation, river and reservoir banks, deciduous woodland – especially alder carr. From wet ground zone vegetation, on mud and amongst plant stems emerging from the water, under cut *juncus*, cut grass and drift litter; also swept from field layer grass and herbs, and occasionally beaten from hedgerow and marsh bushes. Adults of both sexes recorded from February to November, with highest numbers recorded in July.

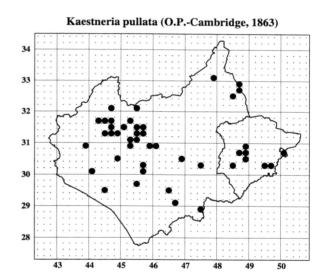

Kaestneria pullata (O.P.-Cambridge, 1863)

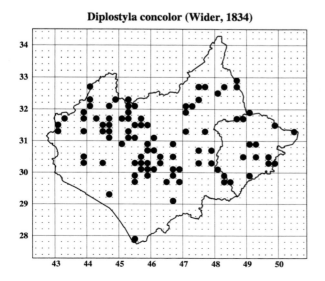

Diplostyla concolor (Wider, 1834)

Diplostyla concolor. Widespread, and one of the county's commonest spiders; in a wide variety of habitats, a frequent aeronaut. First record (Bristowe, 1939). Habitat: grassland, woodland, hedgerows, marsh, waterside vegetation, parkland, churchyards, gardens, farmyards, cultivated farmland, abandoned allotments, disused quarries, railway sidings, sand and brick pits, waste ground and refuse tips. On bare ground, under stones, logs and rubbish; in leaf and pine needle litter, and in ground zone vegetation and detritus; in pitfall traps and occasionally beaten from shrubs and low vegetation. Adults of both sexes have been recorded throughout the year, with highest numbers from August to October.

Atlas of Leicestershire and Rutland Spiders
LINYPHIIDAE

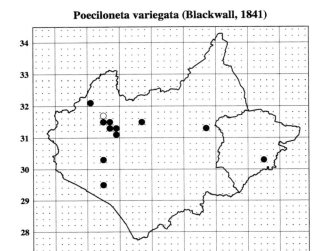

Poeciloneta variegata (Blackwall, 1841)

Poeciloneta variegata. Infrequent in the county, except on Charnwood Forest where it is very local, but not uncommon on open ground. First record E. Duffey, 1953, High Sharpley. Habitat: rocky *Calluna/Vaccinium* heathland, heath grassland, open moorland, bracken-dominated heath, gorse/oak scrub, broadleaf woodland, disused quarry, farmyard and rural garden. Under stones and rocks, in grass turf, under grass cuttings and in dry woodland litter, also in pitfall traps on bare ground; and occasionally swept from the field layer and beaten from gorse. Adult females are recorded from March to July, October and December, males May to July. Highest numbers have been recorded during May.

Drapetisca socialis. Widespread and common; on tree trunks, but under-recorded. First record J. Crocker, 1962, Charnwood Lodge N.R. Habitat: deciduous, mixed and conifer woodland; also on isolated trees in wood pasture, heathland, rural gardens, parkland and at the roadside. Mostly on tree trunks of silver birch, but also on sycamore, oak, alder, holly, beech and pine trees. Frequently in leaf litter under trees and occasionally beaten from low canopy and young trees. Sub-adult females are recorded June, July and August, sub-adult males June and July. Adult females August to November, adult males August to October. Highest numbers October. See also (Crocker, 1963).

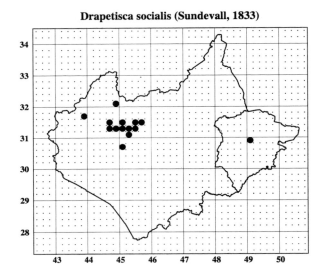

Drapetisca socialis (Sundevall, 1833)

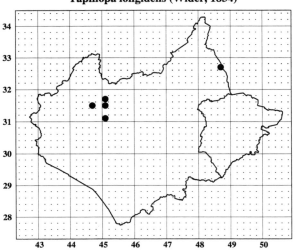

Tapinopa longidens (Wider, 1834)

Tapinopa longidens. A field layer spider, uncommon in Leicestershire. First record A.E. Squires, 1961, Heyday Hays (SK505102), and since then from Beacon Hill Plantation, Charnwood Lodge N.R., King Lud's and Outwoods. Habitat: mature pinewood, mixed woodland and limestone grassland. Under a granite rock embedded in peaty soil, in woodland litter, on the trunk of a pine tree, on a web inside a bleached badger skull amongst woodland debris, and from a pitfall trap in calcareous grassland. This web spinner requires small open spaces in relatively loose vegetation. Females recorded in January, August and September, males in September and November.

Atlas of Leicestershire and Rutland Spiders
LINYPHIIDAE

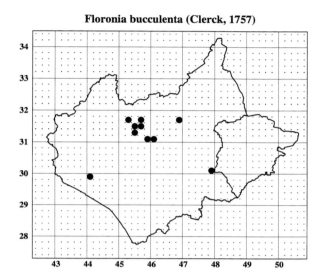

Floronia bucculenta (Clerck, 1757)

Floronia bucculenta. Infrequent, a local species with few records in the county. First record (Wild, 1952). From the Ashby Canal at Sutton Cheney, Barrow Gravel Pits, Buddon Brook Marsh, Frisby-on-the-Wreake, Loddington, Pignut Spinney Marsh, Swithland Reservoir, and Wanlip Gravel Pits. Habitat: marshes, marshy grassland, sedge beds, waterside vegetation, rough unmanaged grassland with scrub and tall herbs, birch/alder and siliceous woodland. Amongst wet vegetation, on wet mud at waterside, also swept from heather, scrub oaks and field layer herbs. Adult females have been recorded for August, September and October, with adult males in September only.

Labulla thoracica. Infrequent in Leicestershire, probably under-recorded. First record (Wild, 1952). Habitat: deciduous and mixed woodland, wood pasture, hedgerows, unmanaged grassland, rural, suburban and urban gardens, farmyards and buildings. Amongst woodland litter, in grass beneath trees, under loose dead bark, on tree trunks and dry-stone walls, in hollow trees, under logs, cut bracken, garden seedboxes, pieces of masonry, garden rubbish *etc.*, inside houses and outhouses; also occasionally swept from woodland field layer. Adult females are recorded from August through to February, males in March and August to October. Activity peak September.

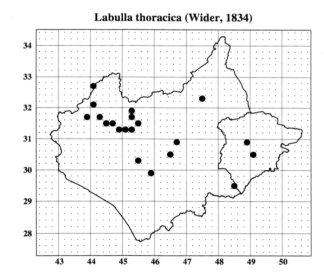

Labulla thoracica (Wider, 1834)

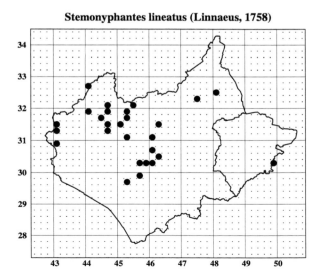

Stemonyphantes lineatus (Linnaeus, 1758)

Stemonyphantes lineatus. Widespread and common in the county, frequent in gardens. First record (Wild, 1952). Habitat: gardens, churchyards, urban derelict land, abandoned allotments, disused quarries, sand pits, brick pits, rubbish tips, domestic and industrial buildings, farmyards and buildings, farmland, grassland and heathland, and occasionally woodland. In pitfall traps in grassland and heathland, on bare ground and amongst litter, but recorded mostly from ground zone vegetation, under stones and pieces of rubbish; also inside houses and other buildings, and sometimes swept from woodland rides. Adults of both sexes are recorded for every month of the year.

Atlas of Leicestershire and Rutland Spiders
LINYPHIIDAE

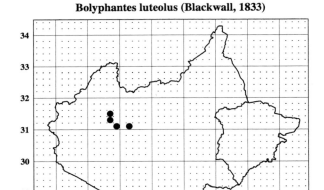

Bolyphantes luteolus (Blackwall, 1833)

Bolyphantes luteolus. An uncommon spider in Leicestershire, with few records from four sites in Charnwood Forest. This is a field layer species which spins webs in small open spaces in loose vegetation and litter. First record J. Crocker, 1961, Charnwood Lodge N.R., also recorded from Bradgate Park, Bardon Hill and Hill Hole, Markfield. Habitat: rocky heathland and grass heath on exposed sites, under stones and pieces of rock in acid grassland; in the ground zone beneath heather and amongst short grass and loose litter, occasionally in pitfall traps. Adult females have been recorded from April to September, but males have been taken in September and October only.

Bolyphantes alticeps. An uncommon spider in the county with a similar distribution to the previous species, with which it can easily be confused. There does, however, seem to be a preference by *B.alticeps* for damper habitats. First record I.M.Evans, 1962, Charnwood Lodge N.R., also from Bardon Hill, Cademan Moor and Swithland Wood. Habitat: bracken-dominated wet heathland, undisturbed acid grassland with wet flushes, marsh, birch and thorn scrub, and open woodland. In bracken litter, in damp/wet ground zone vegetation, under logs, swept from grass and beaten from lower branches of conifer. Adult females August to November, males September and October.

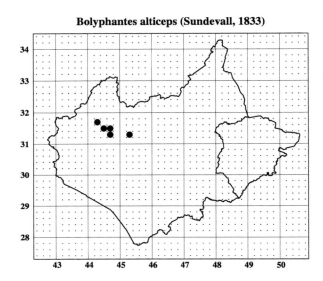

Bolyphantes alticeps (Sundevall, 1833)

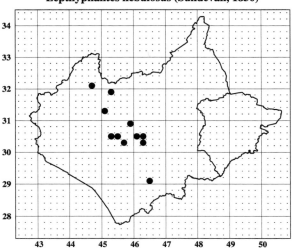

Lepthyphantes nebulosus (Sundevall, 1830)

Lepthyphantes nebulosus. Widespread but few records for Leicestershire, in and around houses, probably under-recorded. First record G.& A. Smith, 1960, Birstall. Habitat: buildings, gardens and farmyards, also from amongst rubbish on a council refuse tip. There is one record from woodland – under loose dead bark. Typically *L.nebulosus* is recorded from inside houses, outbuildings, garages and greenhouses, usually in webs, under rubbish, wet sacking, stones, bricks, and also in webs in an upturned tin bath and amongst garden seedboxes and plant pots. Adult females have been taken in February, March, May, June and August, and males in March, May and October only.

Atlas of Leicestershire and Rutland Spiders
LINYPHIIDAE

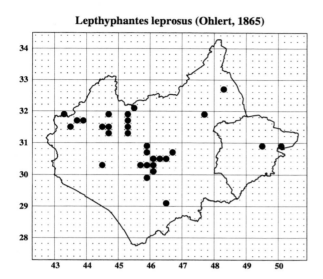

Lepthyphantes leprosus (Ohlert, 1865)

Lepthyphantes leprosus. Widespread and fairly common in the county, in similar situations to the previous species. First record G.& A. Smith, 1960, Birstall. Habitat: as for *L.nebulosus*, also churchyards, urban derelict land, industrial sites, ruins of old buildings, rural artefacts, disused quarries, and old oak trees in parkland and deciduous woodland. In small triangular webs in corners of buildings, cupboards *etc.*, in clefts of rock faces, in cracks in bark of trees, under logs, on underside of manhole covers, under stones, rubbish and garden litter, also amongst grass tussocks and rockery plants. Adult females recorded throughout the year, males recorded October to March and July.

Lepthyphantes minutus. Widespread and common in Leicestershire. First record J. Crocker, 1961, Pignut Spinney Marsh. Habitat: woodland, woodland edge, wood pasture and hedgerow trees, dry-stone walls and rural artefacts, scrub grassland and rural and suburban gardens with trees. Mainly inside hollow trees, in crevices in bark on outside of old trees, under dead wood and fallen bark around bases of trees, in leaf litter inside and under oak trees, under loose bark of dead trees, under stones, in crevices on walls and buildings, and in an aerial trap. Adult females July to November and February to May, males July to October. High numbers recorded August/September.

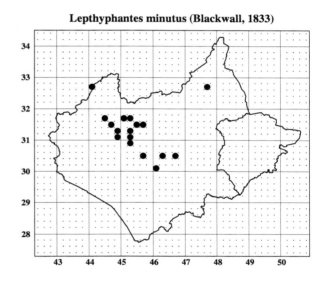

Lepthyphantes minutus (Blackwall, 1833)

Lepthyphantes alacris (Blackwall, 1853)

Lepthyphantes alacris. Widespread and common in the county, especially in woodland. First record H.A.B. Clements, 1962, Owston Wood. Habitat: deciduous, mixed and conifer woodland, alder carr, heath birch scrub, rocky heath grassland, marsh, and rough grassland adjacent to woodland. From woodland ground zone vegetation, damp litter, grass tussocks and moss under or adjacent to trees, shaken from *sphagnum* from wet flushes on the woodland floor, in vole runs, under stones and logs, and occasionally swept from the field layer along woodland rides. Adults have been recorded for all months except August and December, with high numbers recorded for April.

Atlas of Leicestershire and Rutland Spiders
LINYPHIIDAE

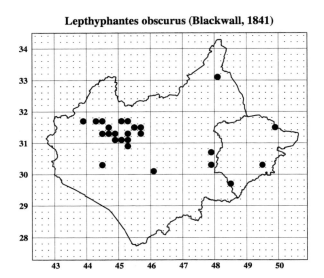

Lepthyphantes obscurus (Blackwall, 1841)

Lepthyphantes obscurus. Widespread but local, not uncommon in the Charnwood area; primarily a woodland species. First record J. Crocker, 1962, Charnwood Lodge N.R. Habitat: deciduous, mixed and conifer woodland, lakeside pine trees, hedgerows, scrubby heath grassland, rocky bilberry heath, marsh, rough calcareous grassland with scrub oak, and a rural garden. Beaten from broadleaf and evergreen trees and shrubs, including oak, holly, hawthorn, pine and spruce, and swept from grass, heather, *juncus*, nettles, *etc.*, in open glades and rides; also under stones and cut grass and in *sphagnum* and ground zone litter. Females April to September, males May to August.

Lepthyphantes tenuis. Widespread and ubiquitous; a frequent aeronaut, with far more records for Leicestershire than any other species. A field layer spider, preferring open grassland conditions; found more commonly on the drier substrates, but also found in damper marshy conditions. *L.tenuis* has been recorded from a very wide range of habitats: under stones, on bare ground, in ground zone and field layer vegetation, in leaf litter, on shrubs and trees, also on outside walls and inside buildings, and is frequently taken in pitfall traps. First county record (Bristowe, 1939). Both sexes mature all year, with a build up of activity April onwards; peak numbers August/September.

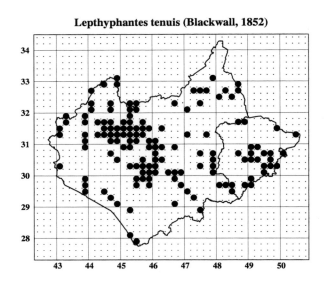

Lepthyphantes tenuis (Blackwall, 1852)

Lepthyphantes zimmermanni Bertkau, 1890

Lepthyphantes zimmermanni. Widespread and common, but less so than the previous species. A frequent aeronaut and one of the county's commonest spiders. It is most often found in the upper zone of the field layer of a wide variety of herbaceous and woody plants, and is essentially a woodland species. However, it is recorded from a wide variety of habitats in much the same situations as *L.tenuis*, ranging from bare ground to the woodland canopy and inside buildings to reedswamp. First recorded for Leicestershire by E.Duffey, 1953, from High Sharpley. Adult females are recorded throughout the year, males every month except January and February. High numbers June to October.

Atlas of Leicestershire and Rutland Spiders
LINYPHIIDAE

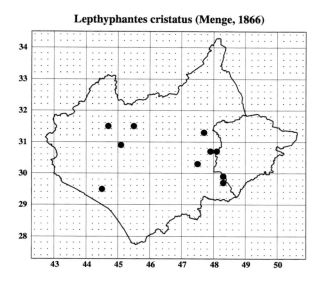

Lepthyphantes cristatus (Menge, 1866)

Lepthyphantes cristatus. Widespread but uncommon in the county, this local species is associated with old woodland. First recorded for Leicestershire by J. Crocker in 1961 from Charnwood Lodge N.R., since then it has also been recorded from Allexton Wood, Buddon Wood, Burbage Wood, Burrough Hill Wood, Great Merrible Wood, Groby Pool, Owston Wood and Skeffington Wood. Habitat: ancient woodland sites with rich herb and shrub layers; deciduous woodland with open canopy and grassy clearings. Amongst grass and leaf litter. Adult females have been recorded from March through to November, but males have been taken only in the months of March and November.

Lepthyphantes mengei. Widespread and common, especially in the Charnwood Forest area. First record J. Crocker, 1962, Bradgate Park. More frequent than the previous species; records from gardens and allotments suggest efficient dispersal. Found in the same places as *L.cristatus* but more usually in damp open grassland and heathland, particularly in the middle zone of the field layer, but also often in the ground zone. Recorded from 21 sites (23 tetrads), including Eyebrook Reservoir, King Lud's Entrenchments, Aylestone Holt, Scraptoft, St.Mary's Allotments and Loughborough. Adult females are recorded from February to November, and adult males from April to November.

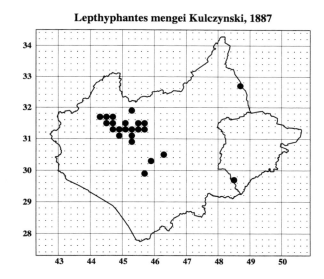

Lepthyphantes mengei Kulczynski, 1887

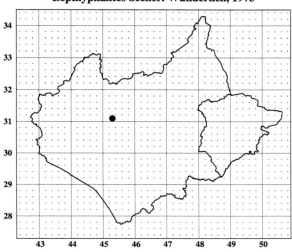

Lepthyphantes beckeri Wunderlich, 1973

Lepthyphantes beckeri. This spider is new to Britain, and is known here only from a single female collected in Bradgate Park on 22nd March 1970 by J.& M. Crocker from damp leaf litter at the base of a low brick wall surrounding a sunken lawn, within the Bradgate House ruins (SK535102), near Leicester. The species is discussed more fully and described on pages 22 and 74-75. An effort was made to locate additional specimens by sampling the site at different seasons over several years, but none were found. It would be useful to set pitfall traps at both the original site and in adjacent unmanaged grassland, in the hope of resolving some of the questions surrounding this species.

Atlas of Leicestershire and Rutland Spiders

LINYPHIIDAE

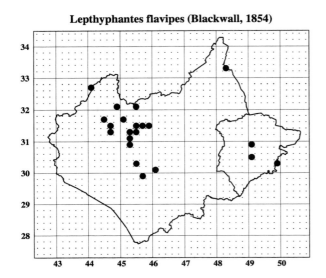

Lepthyphantes flavipes (Blackwall, 1854)

Lepthyphantes flavipes. Infrequent in Leicestershire; a spider of woodland and scrub. First record J. Crocker, 1962, Farnham's Wood, Quorn. Habitat: broadleaf and mixed woodland, alder carr, birch scrub and wooded parkland, also from heath grassland with heather, a flood meadow and a river bank (on the wet shoreline). On bare ground, under stones and amongst pebbles, but predominantly from leaf litter, amongst grassy ground zone vegetation, and occasionally swept from the field layer beneath trees and along rides. Adult females are recorded from February to September, and males from April to September, with highest numbers recorded in June.

Lepthyphantes tenebricola. Very local in the county, widespread but rare; associated with ancient woodland. First record J. Crocker, 1966, Briery Wood, Belvoir; also from Bardon Hill, Barnsdale Wood, Buddon Wood, Burbage Wood, Owston Wood, Pickworth Great Wood, Stoneywell Wood and Swithland Wood. From oak leaf litter, in ground zone amongst bluebells, from pitfall traps on bare ground in small leafed lime coppice, shaken from sedge litter in a dried-up woodland pond and from *sphagnum* in wet depressions; also on an oak tree trunk, swept from woodrush and beaten from the shrub layer. Adults of both sexes recorded for the months April, May, June and July.

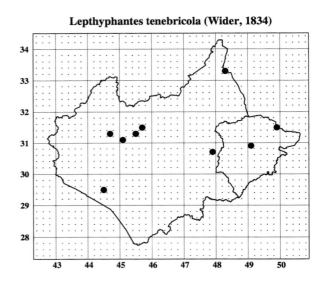

Lepthyphantes tenebricola (Wider, 1834)

Lepthyphantes ericaeus (Blackwall, 1853)

Lepthyphantes ericaeus. Widespread, and one of Leicestershire's commonest species; a frequent aeronaut. Usually found close to the ground in damp grassland. First record E. Duffey, 1953, High Sharpley. Habitat: rough grassland, moorland, grassy heathland, rocky *Calluna/Vaccinium* heath, marsh, fen, water meadows, agricultural land, waterside vegetation; woodland, parkland, hedgerows, scrub, churchyards, gardens, allotments and disused quarries. In ground zone vegetation, leaf and pine needle litter, amongst grass tussocks, in vole runs, detritus, and under stones. Frequently in pitfall traps, also swept from the field layer. Both sexes adult all year, highest numbers recorded July to October.

Atlas of Leicestershire and Rutland Spiders
LINYPHIIDAE

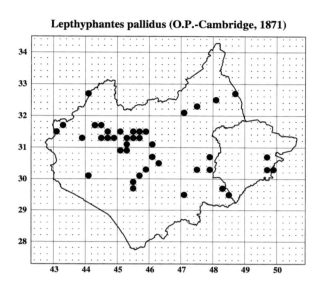

Lepthyphantes pallidus (O.P.-Cambridge, 1871)

Lepthyphantes pallidus. Widespread but local; prefers short open grassland and scrubby areas, often in dry situations. First county record J. Crocker, 1963, Bardon Hill. Habitat: deciduous, mixed and conifer woodland, scrub grassland, limestone, neutral and acid grassland, rocky scrub heathland, marsh, peat bog, sedge fen, urban derelict land, churchyards, gardens and disused quarries. In pitfall traps and amongst ground zone vegetation, under stones, in moss and litter; also inside hollow oak trees in root cavities and amongst dry litter. Adult females have been recorded throughout the year, and males from April to December, with highest numbers in June and July.

Lepthyphantes insignis. A nationally notable (Nb) species, very rare in the county, with only two records. There are less than 40 records nationally, from grass and heathland, often in disturbed habitats, in the south of England, East Anglia, Yorkshire and Scotland. Although widespread, it is never taken in numbers. First county record D.A. Lott, 1991, a female in a pitfall trap in ley, on disturbed ground (26.6.91) Coalfield West, Ravenstone (SK393125). The second record is of a male in a pitfall trap placed between clumps of heather on exposed heath grassland (21.8.95) Charnwood Lodge N.R. (SK468150). Further trapping at this site has failed to produce any more specimens.

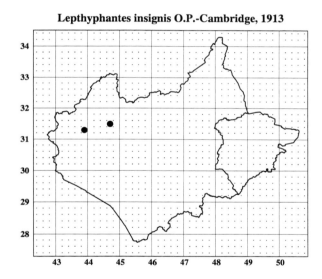

Lepthyphantes insignis O.P.-Cambridge, 1913

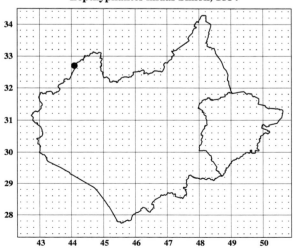

Lepthyphantes midas Simon, 1884

Lepthyphantes midas. A Red Data Book species (RDB2) known only from a single site in Leicestershire, at Donington Park (SK414268). First record J. Crocker, 1971. This species is discussed in more detail on page 72. The site is open wood pasture grazed by deer, with isolated over-mature oak trees. Adult females are recorded from July to September, and males July and August. On pieces of dead wood inside a hollow oak tree, in litter under dead bark on the ground, under pieces of dead wood lying on the ground against the tree, from loose litter inside the hollow trunk, and in a sheet web inside the rotten root limb at the base of the ancient decaying oak tree.

Atlas of Leicestershire and Rutland Spiders
LINYPHIIDAE

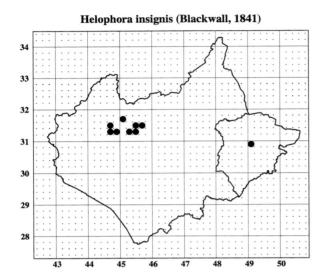

Helophora insignis (Blackwall, 1841)

Helophora insignis. Infrequent in Leicestershire, probably under-recorded. First record J. Crocker, 1962, Charnwood Lodge N.R., also from Barnsdale Wood, Bardon Hill, Farnham's Wood Quorn, Outwoods, Swithland Wood, Swithland Reservoir and Ulverscroft N.R. Habitat: deciduous, mixed and conifer woodland, wooded parkland, and marsh with alder and elder. Numerous in leaf litter; also from pine and larch litter, shaken from *sphagnum*, swept from grassy and bracken field layers in open areas of woodland, and occasionally beaten from bushes and trees. Adult females are recorded from August to November, and males September to November.

Pityohyphantes phrygianus. Rare in the county with only one record, a female beaten from conifers in Pickworth Great Wood (SK985148) by J. Daws, 26.6.1994. This is a mixed plantation with recently cleared rides, on the Rutland/Lincolnshire border. *Pityohyphantes phrygianus* is a conspicuous and distinctive spider, first recorded new to Britain in 1978 from Peebles, Roxburgh, Northumberland and Yorkshire where it had become established in conifer plantations. It has a marked preference for species of spruce, notably Norway spruce and Sitka spruce, and is often found with *Linyphia peltata* and/or *Lepthyphantes expunctus*. It appears to be extending its range in Britain.

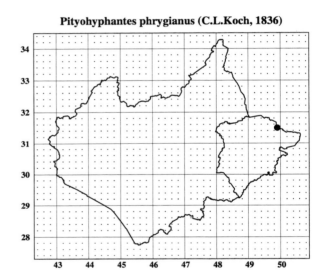

Pityohyphantes phrygianus (C.L.Koch, 1836)

Linyphia triangularis (Clerck, 1757)

Linyphia triangularis. Widespread and common in Leicestershire, and certainly under-recorded. On low vegetation, bushes and trees, late summer and autumn. First record Lowe, 1912. Habitat: woodland with a good shrub layer, scrub grassland, furze heath, bracken-dominated heathland, rough grassland with tall herbs, marsh, waterside vegetation, shrubs and trees, roadside verges, parkland, gardens *etc*. Primarily a shrub layer species, but will take advantage of any low vegetation which will support its sheet web and aerial trip lines. Often swept from field layer and beaten from trees. Occasionally in dead leaves and damp litter. Adult females July to November, males July to October.

Atlas of Leicestershire and Rutland Spiders
LINYPHIIDAE

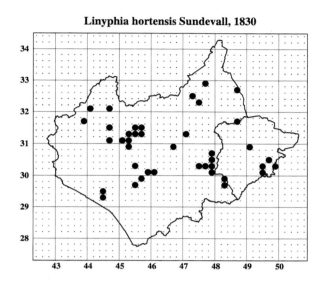

Linyphia hortensis Sundevall, 1830

Linyphia hortensis. Widespread and common, on low vegetation in woods and hedgerows, especially on dog's mercury. First record for the county (Bristowe, 1939). Habitat: damp broadleaf woodland, woodland rides, open glades and woodland edge, hedgerows, calcareous grassland, marsh, streamside vegetation, roadside verges, disused limestone and ironstone quarries, and urban and suburban gardens. Mostly swept from low vegetation – woodrush, yellow rattle, dog's mercury, ground ivy *etc.*, also in grass, moss and leaf litter. Adult females have been recorded from April to July and September, and males from April to July, with highest numbers recorded for May.

Linyphia montana. Widespread and common. First county record (Wild, 1952). Much less obvious than *L.triangularis*, usually in shady places. From broadleaf and mixed woodland, parkland trees, hedgerow trees, scrub grassland, furze heath, gorse scrub, churchyards, gardens, farmyards, rubbish tips, structures – wooden fences, timber buildings. On low vegetation, shrubs and trees especially in litter caught up in twiggy outgrowths on tree trunks, under logs and loose bark, amongst litter in hollow trees, and in old birds' nests, amongst rubbish, piles of bricks *etc*. Beaten from gorse, holly, ivy and rhododendron. Adults recorded April to September, sub-adults throughout the year.

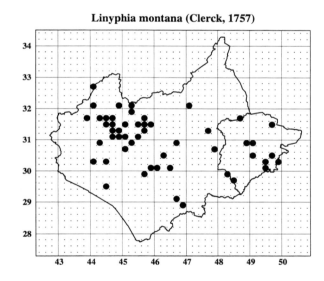

Linyphia montana (Clerck, 1757)

Linyphia clathrata Sundevall, 1830

Linyphia clathrata. Widespread, and one of the county's commonest spiders, in the field layer and on low vegetation. First record Lowe, 1912. Habitat: broadleaf woodland, spinneys, hedgerows, parkland, rough grassland, marsh, reedswamp, sedge fen, waterside vegetation, wet heathland, boggy moorland, wasteland, farmyards, churchyards, stables, disused allotments, gardens, disused quarries, refuse tips, wet ditches, *etc*. Under cut grass, bracken, straw, and bits of rubbish, amongst ground zone vegetation, tussocky grass, in vole runs, on bare ground, and under stones, shaken from *sphagnum*, swept from field layer and occasionally beaten from birch and bramble. Adults at all seasons.

CHAPTER FIVE

Atlas of Leicestershire and Rutland Spiders
LINYPHIIDAE

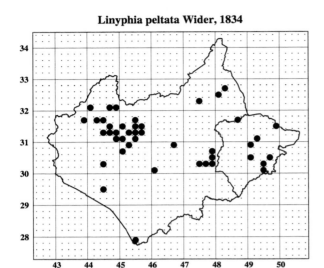

Linyphia peltata Wider, 1834

Linyphia peltata. Widespread and common in Leicestershire, in woods, on low vegetation and on the lower branches of trees. First record J. Crocker, 1961, Swithland Wood. Habitat: deciduous, mixed and conifer woodland, wooded parkland; gorse, bramble and hawthorn scrub, churchyards and rural gardens with trees, also scrubby or wooded areas of disused quarries. Beaten from trees and bushes, swept from low vegetation including nettles, dog's mercury, heather, meadowsweet, woodrush, yellow rattle *etc.*, and occasionally on bare ground and amongst ground zone plants under trees. Adults of both sexes have been recorded from April to August, with peak activity in May.

Microlinyphia pusilla. Widespread throughout the county but local, on low vegetation, grass *etc.* First record Lowe, 1912. From calcareous, neutral and acid grassland, *Calluna* heathland, rocky *Vaccinium* heath, *sphagnum* bog, marsh, wet fen, riverside reedswamp, gorse/hawthorn scrub, broadleaf woodland rides and glades, disused quarries, roadside verges and gardens. In ground zone vegetation, wet litter, on underside of small sheet webs in reeds just above water level, swept from grass and marsh vegetation, also beaten from ivy, hawthorn, heather and young trees. Adult females recorded from May to September, males from May to August and October.

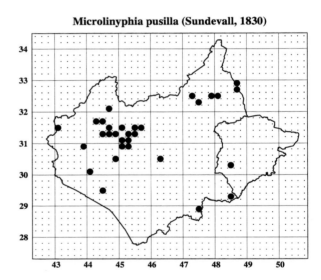

Microlinyphia pusilla (Sundevall, 1830)

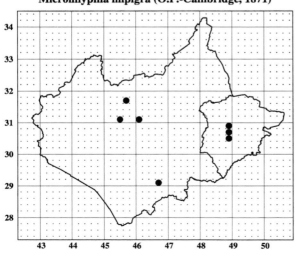

Microlinyphia impigra (O.P.-Cambridge, 1871)

Microlinyphia impigra. Widespread but rare in Leicestershire; a wetland spider with a local distribution. First recorded for the county by M.G. Crocker, 1973, from Barrow Gravel Pits; and subsequently recorded from Lyndon N.R., Hambleton, Puddledyke, Rutland Water, Saddington Reservoir and Watermead Country Park. From damp *Deschampsia cespitosa* grassland, wet marsh, *glyceria* swamp, reedmace bed, sedge fen, *Phragmites* swamp and rough grassland dominated by *Elymus repens* and *Urtica dioica*. In wet ground zone vegetation and amongst litter. Adult females have been recorded for May, June and August only, and adult males for May and July.

Atlas of Leicestershire and Rutland Spiders
LINYPHIIDAE

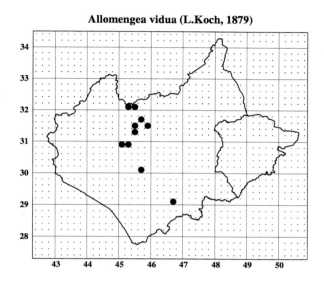

Allomengea vidua (L.Koch, 1879)

Allomengea vidua. Fairly widespread in Leicestershire, but very local, only in permanent wetland habitats, especially riverside sites; in swampy situations. First record J. Crocker, 1964, Saddington Reservoir. Habitat: birch/alder carr with wet flushes, flood meadows with wet depressions, lakeside *Phragmites* fen with wet substrate, reservoir margins, reedmace swamp, ox-bow and brookside vegetation. From wet ground zone litter, on plant stems growing out of the water, under drift litter, and from pitfall traps in marshy ground. Adult females have been recorded for the months of August, September, October and November, and adult males for August and September.

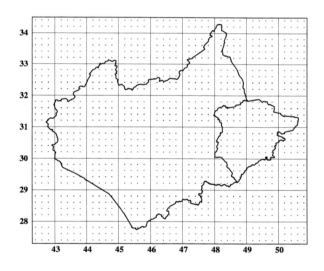
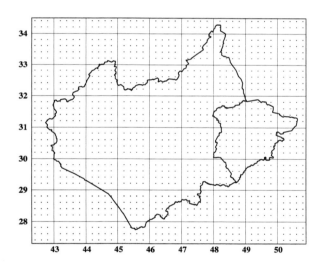
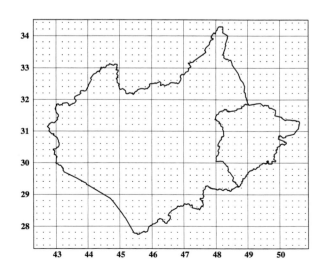
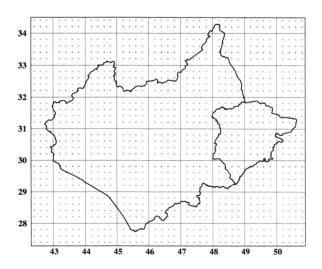

CHAPTER SIX

Habitat Evaluation

Introduction

The purpose of this book has been primarily to gather together all known occurrences of spider species in the county and to present this information as a graphical statement, but it would be remiss not to attempt an evaluation of the wealth of detail so far accumulated. As has been mentioned earlier, there has been very little experimental work on ecological relationships carried out in the county, so there is no baseline from which to work. However, the species lists resulting from such work are incorporated into this present overview of the spiders of Leicestershire. Collecting on Charnwood Forest has been more intensive, and has been carried out over a longer period than in any other natural area of Leicestershire, therefore, this would seem to influence any value judgements when comparing sites. On the other hand, at any time there is a finite fauna present in any one place and the law of diminishing returns, in terms of new species, must apply to continued collecting at that site. Climax woodland would less readily yield up its secrets than, say, open grassland; and there is also the question, already discussed in chapter four, of the efficacy of different collecting techniques. It is sufficient to state at this point that the limitations in any value analysis are readily acknowledged. However, one cannot work closely with a group of invertebrates without forming a subjective opinion of their relevance to current conservation issues, and this has guided our approach to qualifying the results subjectively. To this end, the basis for consideration is the species list for each site, irrespective of how well, or otherwise, the site has been worked. For instance, there are many woodland sites which are known to be faunistically rich but which have not been worked for spiders, therefore no comment can be made on their araneological value. Conversely, the outstanding list of species from Buddon Wood should not be under-rated because most of the wood has been quarried away since the original work was carried out, for a recent survey has shown that the remnant woodland still carries a remarkably high percentage of the original species list drawn up in 1974. Then again, one of the surprises has been the richness of some derelict industrial sites where new habitats have been created either by naturalisation (*e.g.* Newfield Heath, C.E.G.B. Rawdykes, Geeston Quarry *etc.*) or by planned restoration (*e.g.* Coalfield West, Lount Grassland, Watermead Country Park *etc.*). The recolonisation of these sites has relied on the ability of the flora and fauna of surrounding habitats to overflow into the new ones. Herein lies the importance of such wildlife reservoirs as SSSIs and nature reserves. However, these cannot survive as isolated entities, but must form a part of a wider conservation concept. English Nature has now adopted this approach to conservation and is looking at the countryside of the whole of England as an interlocking mosaic of areas, each defined by natural physical characteristics. In Leicestershire there are three such distinctive natural areas. The acidic uplands of Charnwood Forest stand out clearly from the surrounding countryside (Crocker, 1981), whilst the less obvious boundary of the eastern limestone country is defined by the geology of the Lincolnshire Limestone. The rest of the county, comprising mainly neutral soils, with isolated acidic and limestone outcrops and local patches of heath and calcareous grassland, is further divided into three separate areas: Trent Valley and Levels, the Welland basin and the Avon catchment area.

The Natural Areas Concept

A larger range of plant and animal species survives in an integrated mosaic of habitats than in a series of isolated fragments of pristine sites, a case of 'the whole is greater than the sum of its parts'. Each important site needs a buffer zone around it and corridors connecting it to similar sites if it is to retain a healthy balance and be able to withstand natural disasters. Many sites have become devalued due to activities just outside their boundaries. Modern river maintenance can cause the drying out of valuable wetlands, chemical drift from agricultural practices and other forms of aerial pollution are modifying the character of sensitive habitats. Field drains in arable land adjacent to marshland deprive these sites of essential ground water, and nearby quarrying and opencast mining will also affect natural drainage. Residential development near ancient woodland will have an inevitable downgrading effect on the very essence of its rich and varied habitat systems.

English Nature has developed a new concept for looking after the countryside; as well as recognising the importance of the best sites and habitats (SSSIs), it is also now taking an holistic view and considering the landscape in its entirety. To achieve this, England has been divided into ninety-two terrestrial natural areas, using physical boundaries such as geology and river catchment areas, with most of these natural areas crossing man-made county and regional boundaries. Additionally, the off-shore coastline has been divided into twenty-four maritime natural areas. The objective is to define a landscape in terms of its geology, land-form and plant and animal communities, and to look at nature conservation across the whole ecosystem. In this way, SSSIs and nature reserves will become stable reservoirs of wildlife, rather than vulnerable isolated islands in decline. In order to maintain the full range of indigenous species the physical constraints that underline traditional ranges for many plants and animals, and limit their occurrence in Britain, must be taken into account. This should enable English Nature to maintain and enhance the characteristic biological diversity and features of each natural area. Examples of the ninety-two natural areas are: Breckland, Broadland, Charnwood Forest, Coal Measures, Dartmoor, Dorset Heaths, Greater Cotswolds, Hampshire Chalk, Lincolnshire Limestone, London Basin, Middle England, New Forest, Sherwood Forest, Thames Marshes, Trent Valley & Levels and Yorkshire Dales. The natural areas will be described using their key features, species and habitats, to set out their characteristics and to place them in a national overview. Then areas can be compared with one another as a basis for policy formulation and for setting management priorities. Although several natural areas may have similar constituents, such as limestone grassland, their management may differ. There will be a bias toward local need, with an area's history of traditional management also taken into account. The aims of English Nature, working with others, are a) to enable endangered or threatened species to recover viable populations, b) to set measurable nature conservation targets and c) to set objectives in a national context. The success of this approach will depend upon the measurement, through recording, of changes in the status of key species. It is hoped that a useful contribution to this end has been made by the presentation of material in this publication.

Leicestershire Natural Areas

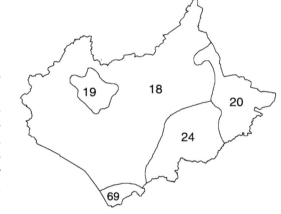

There are five natural areas that fall within Leicestershire's administrative boundary, but only one – Charnwood Forest – lies wholly in the county.
NA18 Trent Valley & Levels (excluding Charnwood Forest) incorporates two-thirds of Leicestershire which is drained into the River Trent, mostly via the River Soar and its tributaries, but also via the Mease and Anker in the west and the Devon and Smite in the north-east. The wider boundary of this area takes in parts of Nottinghamshire, Lincolnshire, Warwickshire, Staffordshire and Derbyshire.
NA19 Charnwood Forest is situated entirely within north-west Leicestershire and is undoubtedly one of the best areas of Central England for wildlife. The boundary was established by a joint working party set up by the Nature Conservancy Council in 1973 (Nicholson, 1975).
NA20 Lincolnshire Limestone. This area lies in the far east of the county comprising land where the Great and Inferior Oolites are close to the surface; its boundary extends into Lincolnshire, South Humberside, Northamptonshire and Cambridgeshire.
NA24 Middle England. In Leicestershire this embraces the watershed of the River Welland, except where it crosses on to the Lincolnshire Limestone. The natural area covers most of the central Midland Plain and incorporates parts of Lincolnshire, Northamptonshire, Cambridgeshire, Bedfordshire and Buckinghamshire.
NA69 Greater Cotswolds. Touching the southern tip of the county, this natural area is defined by the watershed of the River Avon. It extends in a south-westerly direction, cutting through the counties of Warwickshire, Buckinghamshire, Oxfordshire, Gloucestershire, Avon and Wiltshire.

On the ground the boundaries of these natural areas would be hard to detect as they merge, one into another, rather than there being a definite line between them. Differentiation of the natural areas in Leicestershire is more likely to be seen in the vegetation and plant-eating insects than in the spiders, of which few species seem to be clearly associated with this natural areas concept.

With these species, the structure of the habitat and the history of land use appear to be more convincing determinants than the geology, altitude or temperature. It will be necessary to consider data from a much wider area than Leicestershire to draw any firm conclusions about spiders of natural areas. However, sufficient is known about certain species to make some comment on the few that have affinities with Charnwood Forest and the Lincolnshire Limestone.

The greater part of Leicestershire is that natural area forming the catchment of the River Trent, draining nearly three quarters of the county via its many tributaries, including the Soar, Wreake, Mease and Anker (see Map 4, p.18). Large expanses of neutral grassland and intensively managed farmland fall within this natural area, as do the county's main conurbations. Wetland and woodland are the most important semi-natural habitats represented, with some residual lowland heath in the west. From an arachnological point of view, there is nothing to differentiate the Greater Cotswolds area (NA69) from the Trent Valley and levels (NA18). The same might be said about the Middle England area (NA24) on the basis of the small number of spider records available from this part of the county; but this is the least-worked part of Leicestershire and is likely to hold some surprises when more work is done here. Of particular interest is the series of woodland sites in the east, including Skeffington Wood, Owston Wood, Launde Big Wood, Tugby Wood, Loddington Reddish, Allexton Wood, Wardley Wood and Great Merrible Wood. These, if worked thoroughly, should turn up good species lists which are likely to compare favourably with those of some of the Charnwood Forest woodlands.

So, at present, we are left with three main zones in the county which contain distinctive spider faunas: the Lincolnshire Limestone (NA20) accounting for around 11.5% of the surface of Leicestershire, Charnwood Forest (NA19) 4.8%, and the rest (NA18, NA24 and NA69) 83.7%, which contains the typical wetland fauna of the Soar Valley.

The Lincolnshire Limestone (NA20)

Characteristic sites on the eastern limestone include King Lud's Entrenchments, The Drift, Thistleton Gullet, Bloody Oaks Quarry, Pickworth Great Wood, Shacklewell Hollow, Ketton Quarry, Geeston Quarry, Luffenham Heath, North Luffenham Quarry, Tixover Quarry and Seaton Meadow. The main feature of this area is the calcicolous vegetation (combined in places with patches of lowland heath), particularly herb-rich limestone grassland, marsh, open stony habitats and short rabbit-grazed turf. Seaton Meadow lies close to the River Welland, partly on calcareous clay, but is overlain mostly with river alluvium, characterised by marsh-marigold *Caltha palustris* and great burnet *Sanguisorba officinalis*.

Table 9: Spiders taken only on Lincolnshire Limestone (NA20)

Species	Sites	Habitat
Drassyllus pusillus	(3 sites +Harby Hills)	Calcareous and neutral grassland also short sparse grass.
Clubiona subtilis	(1 site)	Seaton Meadow, damp herb-rich grassland, with tall herb.
Diaea dorsata	(3 sites +Wymondham Rough)	Woodland, rough calcareous and neutral grassland.
Ozyptila scabricula	(1 site)	Bare stony ground with sparse vegetation.
Ozyptila sanctuaria	(5 sites)	Calcareous scrub grassland with tall herb.
Philodromus collinus	(4 sites)	Rough calcareous scrub grassland, woodland.
Philodromus albidus	(1 site)	Shrubby, wooded Motte & Bailey site.
Pardosa agrestis	(2 sites)	Bare ground, damp marsh, rough grass.
Alopecosa barbipes	(1 site)	Open ground with sparse vegetation.
Trochosa robusta	(1 site)	Open stony ground with patches of rough grass.
Agelena labyrinthica	(5 sites)	Herb-rich limestone grassland.
Agalenatea redii	(1 site)	Herb-rich limestone grassland.
Hypsosinga pygmaea	(5 sites)	Calcareous scrub grassland, herb-rich limestone pasture.
Maso gallicus	(3 sites)	Relict Fen Basin species.
Gongylidiellum latebricola	(1 site)	Damp calcareous grassland.
Typhochrestus digitatus	(2 sites)	Herb-rich calcareous grassland.

Charnwood Forest (NA19)

Unlike the limestone country, where there is no obvious physical boundary, Charnwood Forest stands high above the surrounding lowlands of the Coal Measures and the Soar Valley. The geology is complex, and in many places exposures of weathered Pre-Cambrian and volcanic rocks thrust

Lincolnshire Limestone (NA20) and Trent Valley and Levels (NA18)

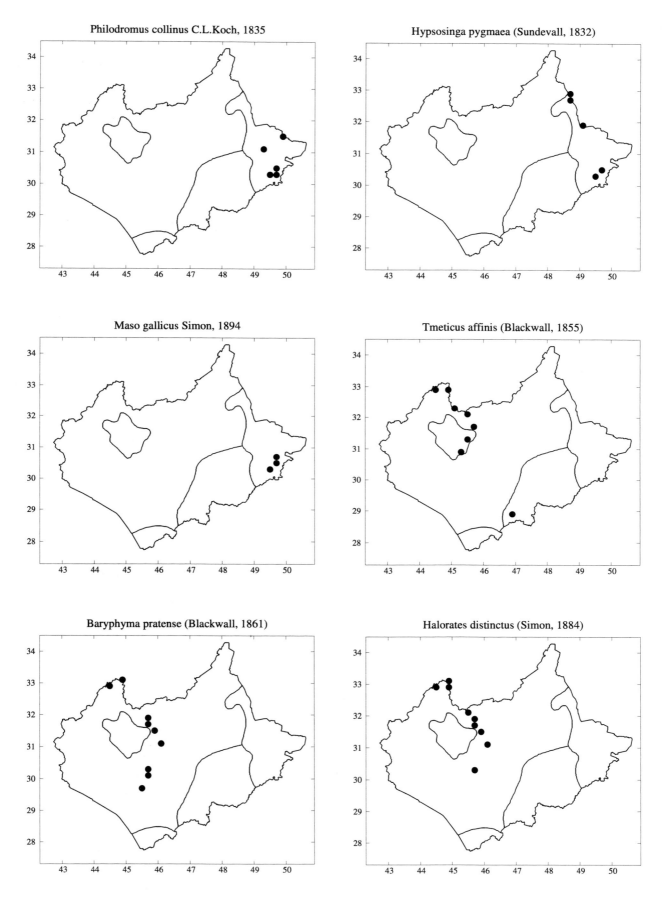

Charnwood Forest (NA19)

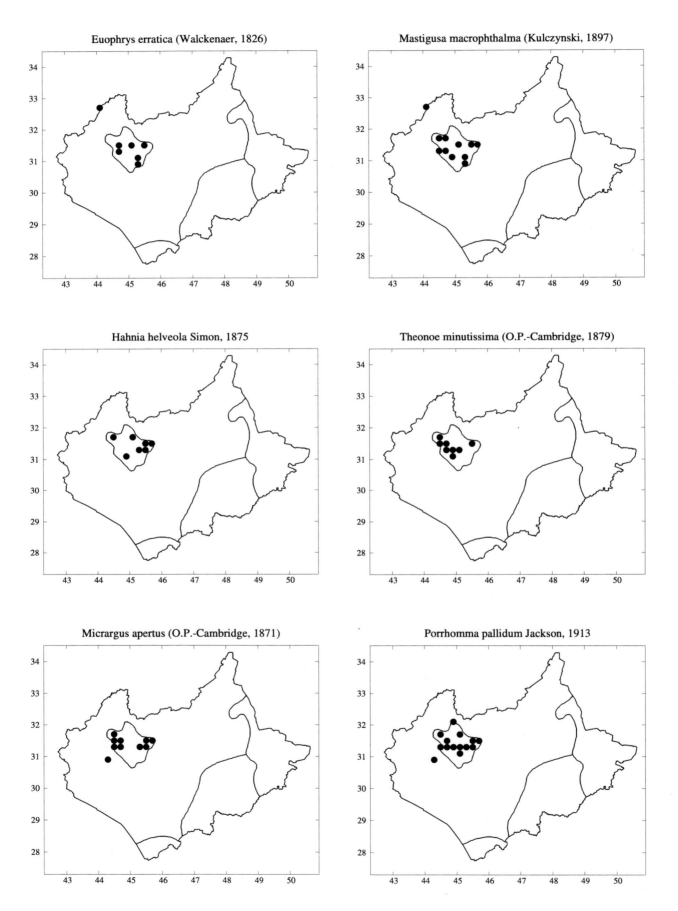

through the overlying Triassic deposits, to form the characteristic rugged landscape of some of the prime sites in this area such as Bardon Hill, High Sharpley, High Cademan, Ives Head, Timberwood Hill, Beacon Hill, Benscliffe and Bradgate Park. Although there is an outcrop of carboniferous limestone in the extreme north-west corner of Charnwood Forest at Grace Dieu, and some isolated patches of chalky boulder clay deposits around The Brand, the essential character of the Forest is derived from its acidic rocks, and gives rise to *Calluna* and *Vaccinium* heathland, moorland, and where the soil is shallow to siliceous woodland, much of these habitats now dominated by bracken. In places, however, the ancient rocks of Charnwood Forest are overlain with varying thicknesses of clay – the so called Mercian mudstone or Keuper marl – and this accounts for some marked changes in the vegetation. Many of the interesting spiders on the Forest are associated with habitats that have a long history of unchanged land use. Thus relict species of the ancient forest managed to survive in this isolated manorial waste. Charnwood Forest was not physically enclosed until 1829, though the Act of Inclosure had been passed in 1808. Thereafter, agricultural reclamation of the uplands, residential development, industrial exploitation, roads and services have largely changed the face of Charnwood. Nevertheless, many good sites remain as public open spaces, nature reserves, SSSIs and sensitively managed private estates. The current spider fauna bears witness to the importance of these sites.

Table 10: Spiders found only on Charnwood Forest (NA19)

Species	Sites	Habitat
Haplodrassus silvestris	(1 site)	Buddon Wood.
Clubiona trivialis	(2 sites)	*Calluna* heath.
Agroeca brunnea	(2 sites)	Ancient woodland.
Ballus chalybeius	(1 site)	Buddon, ancient birch/oak woodland.
Mastigusa macrophthalma	(9 sites +Donington Park)	Ancient oakwoods, rocky heath grassland, pasture woodland, acid grassland.
Hahnia helveola	(5 sites)	Ancient woodland (most records from Buddon Wood).
Theonoe minutissima	(7 sites)	Ancient woodland, moorland, *Calluna* heath.
Walckenaeria capito	(1 site)	C.L.N.R. moorland.
Moebelia penicillata	(1 site)	Outwoods, ancient woodland.
Baryphyma trifrons	(1 site)	Ulverscroft N.R., ancient marsh.
Pelecopsis nemoralis	(1 site)	Swithland Wood, ancient woodland.
Evansia merens	(3 sites +Newfield Heath)	Heath grassland (often with ants).
Tapinocyba pallens	(2 sites)	Bardon Hill, Buddon Wood, ancient woodland.
Monocephalus castaneipes	(1 site)	C.L.N.R. moorland.
Micrargus apertus	(5 sites +Nailstone Wiggs)	Ancient woodland, heathland.
Asthenargus paganus	(4 sites +Nailstone Wiggs)	Ancient woodland.
Porrhomma egeria	(1 site)	Bardon, ancient woodland.
Agyneta cauta	(1 site)	Bardon Hill.
Bolyphantes alticeps	(4 sites)	Rocky heathland, damp acid grassland, open woodland.
Lepthyphantes beckeri	(1 site)	Bradgate, ancient deer park.
Lepthyphantes insignis	(1 site + Coalfield West)	C.L.N.R. moorland.

Trent Valley and Levels (NA18)

A dramatic feature of the largest 'natural area' in the county is the range of artificial habitats created by the city, towns and villages with their associated industries and demands for communications and services. These dominate central and western Leicestershire. Characteristic spiders of this zone are the synanthropic species shown in chapter two, and those listed in appendix IIg which include many of the larger common opportunistic species.

Table 11: Ubiquitous Urban Spiders

Amaurobius similis	*Tegenaria domestica*	*Nuctenea umbratica*
Amaurobius ferox	*Steatoda bipunctata*	*Erigone atra*
Xysticus cristatus	*Enoplognatha ovata*	*Bathyphantes gracilis*
Salticus scenicus	*Metellina segmentata*	*Diplostyla concolor*
Pardosa pullata	*Zygiella x-notata*	*Lepthyphantes tenuis*
Pardosa amentata	*Araneus diadematus*	*Linyphia triangularis*

Much more interesting, however, are those uncommon and rare species which have become established on urban and rural derelict industrial sites (Acresford sandpit; C.E.G.B., Rawdykes; Leicester Cattle Market; Lount Colliery; Newfield Colliery and Lount Opencast sites; Moira Junction and St Mary's Allotments, Leicester).

Table 12: Uncommon spiders from disturbed sites

Species	Sites	Habitat
Zelotes latreillei	(23 sites)	Wide range of habitats.
Cheiracanthium virescens	(2 sites)	Dry and sandy places.
Euophrys aequipes	(8 sites)	Open areas, amongst short vegetation.
Sitticus pubescens	(10 sites)	On buildings, trees, fences.
Arctosa perita	(7 sites)	Dry heaths and sandy places.
Tegenaria agrestis	(10 sites)	Stony sparsely vegetated areas.
Tegenaria silvestris	(4 sites)	Damp woodland, grassland & stony ground.
Cicurina cicur	(5 sites)	Damp & dark places, stony sparse vegetation.
Hahnia nava	(14 sites)	Grassland and under stones.
Euryopis flavomaculata	(2 sites)	Damp heathland.
Achaearanea simulans	(5 sites)	Woodland, scrub heath.
Enoplognatha thoracica	(14 sites)	In grass and under stones.
Walckenaeria vigilax	(7 sites)	Damp/wet grassland.
Pelecopsis parallela	(7 sites)	Amongst short vegetation, under stones.
Cnephalocotes obscurus	(21 sites)	Damp grassland, moss and detritus.
Troxochrus scabriculus	(9 sites)	Sandhills, open dry places.
Microctenonyx subitaneus	(5 sites)	Amongst straw, detritus *etc.*
Micrargus subaequalis	(26 sites)	Tall grass.
Milleriana inerrans	(6 sites)	Sandhills, disturbed ground.
Porrhomma errans	(5 sites)	On disturbed ground.
Lepthyphantes pallidus	(42 sites)	Short open grassland.
Lepthyphantes insignis	(2 sites)	Taller grassland.

Whereas built-up areas have the greatest impact on the environment of the natural area, a characteristic feature of the Trent Valley and Levels is its complex network of watercourses, open water and wetlands. Though reservoirs, lakes and ponds are important habitats (see appendix IIe), the series of flood meadows, marshes, fens and swamps linked by main rivers has produced a creditable list of specialist spiders which indicates a healthy and thriving fauna. Species lists are given in appendix IId for the river sites on the Trent and Soar and the main specialist spiders are pinpointed here.

Table 13: Wetland Species – Trent Valley & Levels (NA18)

Tetragnatha striata (1 site = Groby Pool). Included here, as this species is worthy of special comment. It has been found on only one site in Leicestershire, at the edge of Charnwood Forest, on a small tributary of the River Soar. The nearest known station for this species is adjacent to the River Trent in Derbyshire just over the county boundary with Leicestershire. This spider is probably under-recorded due to its specialist habitat – *Phragmites* stands in water away from the shore.

Baryphyma pratense. Many sites along the River Soar and River Trent. Riverside fen conditions, wet and marshy ground, sedge-beds *etc.*

Halorates distinctus. As for *Baryphyma pratense*.

Tmeticus affinis (9 sites). Riverside reedswamp, also similar situations at Swithland Reservoir and Groby Pool.

Allomengea vidua. Riverside fens, wet and marshy ground, sedge beds, reedswamp also reservoir and lake margins in similar habitats.

Prinerigone vagans (17 sites). Widespread in waterside habitats, rivers, lakes, reservoirs, ponds.

Erigone arctica. 3 sites on River Trent/River Soar (also at Geeston Quarry – River Welland, and Rutland Water – 2 sites).

Erigone longipalpis (4 sites). River Soar, Quorn to Cotes Bridge, on wet muddy substrate.

Factors Affecting the Distribution of Spiders

The flora and fauna is in a constant state of change, responding to all the varying influences in the environment. Landscape modification by man, and changing land use have been the greatest factors influencing the distribution of spiders in Britain. The gradual conversion of ancient forest and undrained wetlands to the present, largely man-made, intensively managed agrarian landscape is reflected in the spider fauna. Despite many losses there are some obvious gains, with species such as *Pholcus phalangioides*, *Psilochorus simoni*, *Euophrys lanigera* and many of the other synanthropic species (see chapter two) being brought into the county by human agency. *Tegenaria agrestis*, *Ostearius melanopygius* and *Pityohyphantes phrygianus* are foreign spiders which have recently established themselves in Leicestershire, in habitats created by man, whilst species of sand dunes and open stony ground – such as *Troxochrus scabriculus* and *Milleriana inerrans* – have colonised agricultural land. The type of habitat preference may depend to some extent on where the spiders occur in their geographic range, and Duffey (1993) classifies the life strategies of spiders in Britain according to their adaptability to environmental diversity as follows:

Table 14: Classification of life strategies of British spiders (Duffey, 1993)

Pioneer Species: Active aeronauts which disperse freely, exploiting newly created open ground where competition is low. Widely distributed in temporary or changing habitats such as agricultural cropland, gardens, urban areas, leys and other types of disturbed ground or vegetation.

Generalists: Common species with a capacity to adapt to a wide range of semi-natural habitats and permanent artifacts in the man-made environment. May be difficult to assign to a particular habitat.

Broad Specialists:

A: Widespread, euryoecious, or 'characteristic' species associated with major habitats such as deciduous woodland, marshes, heaths or ancient grassland, but which may be found in many different variants of the chosen major formations.

B: Diplostenoecious species, mostly widespread and associated with two different habitats but usually more common in one than the other. Occasionally much more abundant in man-made habitats than in the natural habitat. This grouping grades into species successful in three or more different environments.

Narrow Specialists: Stenoecious species which seem confined to clearly defined habitat. Includes rare species in low numbers and others which may be locally abundant, but confined to a restricted area because the habitat is scarce.

Habitat Diversity

The suitability of a habitat for a particular spider species is closely related to the history of the site, and ultimately to micro-structure, abiotic factors and biotic competition. Structure will be influenced by vegetation, site continuity, lack of disturbance and geology, whilst the micro-climate will be determined by the effect of altitude, humidity, temperature and wind. Niche domination by an established fauna in a stable habitat will present little opportunity for pioneer species, whereas recently cleared woodland, drained fen or ploughed grassland offer excellent potential for colonisation by pioneers, as indeed do most established habitats which have been subjected to radical change.

Geological influences on spider distribution are more subtle than in the case of plant feeding insects but, as has been discussed, certain species have marked preferences for alkaline or acidic conditions. The complexities of solid and drift geology, from which are derived a habitat's substrate and soils, are often compounded by the re-distribution of large (and small) volumes of minerals as part of industrial and agricultural processes. It is useful therefore, to observe the composition of the flora as an indication of geological changes on a site.

Calcareous grassland (*e.g.* King Lud's Entrenchments) in Leicestershire contains calcicoles such as common rock-rose *Helianthemum nummularium*, field scabious *Knautia arvensis*, greater knap-weed *Centaurea scabiosa*, common restharrow *Ononis repens*, quaking-grass *Briza media* and tor-grass *Brachypodium pinnatum*.

Heath grassland occurs at Luffenham Heath, where acidic conditions have developed in a limestone area on a thin bed of upper estuarine clays and sandy soil. This rests directly on Lincolnshire Limestone and is typified by calcifugous plants, such as heather *Calluna vulgaris*, on the band of sandy soil. Lowland heath is represented in west Leicestershire by acidic dry and wet heath, containing *Calluna* and formerly bell-heather *Erica cinerea*, on soils derived from the surface coal measures.

Moorland upland habitats on Charnwood Forest (acid heath, bog and acid grassland), are best represented at Charnwood Lodge N.R. and a smaller area at High Sharpley, with examples of degenerate moorland at Bradgate Park, Beacon Hill and Bardon Hill (almost gone). Typical moorland calcifugous plants are: purple moor-grass *Molinia caerulea* (wet moorland), cross-leaved heath *Erica tetralix* (relict of wet heath – e.g. C.L.N.R.), heather *Calluna vulgaris*, bilberry *Vaccinium myrtillus*, mat-grass *Nardus stricta* (higher ground), wavy hair-grass *Deschampsia flexuosa*, sheep's-fescue *Festuca ovina*, common bent *Agrostis capillaris*, tormentil *Potentilla erecta* and heath bedstraw *Galium saxatile*.

Alkaline conditions can be found in areas of otherwise acidic or neutral soils, on small isolated deposits of chalky boulder clay, and calcicoles occur on such local patches within Charnwood Forest (e.g. The Brand) including traveller's-joy *Clematis vitalba*, hairy St John's-wort *Hypericum hirsutum* and wild marjoram *Origanum vulgare*, but are too restricted to have any influence on the spider fauna. The very acid siliceous deciduous woodland at Buddon, and particularly at Swithland, contains a complex mixture of tree species ranging from sessile oak *Quercus petraea* (acid loving) to small-leaved lime *Tilia cordata* (typical of Northamptonshire and Lincolnshire Limestone), and reflects something of the subtle mysteries of these ancient ecosystems. Neutral grassland – the traditional and most widespread grassland habitat in Leicestershire – occurs throughout the Trent Valley and Levels (NA18) on Triassic, Liassic and alluvial soils.

Charnwood Forest is noted for its ancient forest and upland (northern) fauna, typified by *Haplodrassus silvestris, Mastigusa macrophthalma, Theonoe minutissima, Walckenaeria capito, Monocephalus castaneipes, Bolyphantes alticeps* and *Lepthyphantes midas* (Donington Park – the link between Charnwood and Sherwood). Apart from topographical considerations, one of the reasons Charnwood Forest has such a good spider fauna is that it has more ancient sites – which are relatively undisturbed and have a long history of unchanged land use – than other areas in the county of comparable size. It has also been very well worked. Many of the interesting sites elsewhere, with similar composition and history, will produce equally impressive lists of species, each with their own specialist spiders.

The Lincolnshire Limestone area shows an eastern influence and is generally warmer (see chapter one). It is characterised by such species as *Diaea dorsata, Philodromus albidus* and *Agelena labyrinthica* (southern species); *Ozyptila scabricula* and *Trochosa robusta* (Breckland species); *Pardosa agrestis* (stony calcareous places), *Hypsosinga pygmaea* (which favours calcareous grassland), *Clubiona subtilis* (a wetland spider, found also on sandhills) and *Maso gallicus* (a relict Fen Basin spider).

The river systems of inland Leicestershire are typified not only by established specialist riparian spiders, including *Baryphyma pratense* and *Halorates distinctus*, but also by coastal species such as *Erigone arctica* and *Erigone longipalpis* both of which are extending their range inland. These aeronaut species disperse over long distances but survive in only a few places (Duffey, 1993).

Site Evaluation

Limited space has restricted the number of species lists for different sites which it has been practical to present in this book. The sites selected are those with highest qualitative scores, in other words those with the most interesting spider faunas, assessed on currently available data, and are tabulated in appendices I and II, grouped on the basis of habitat similarity. There is no overall consistency in the way this information has been obtained, since a number of people have been working independently, and their work cannot necessarily be related to other collecting across the county. For example, Buddon Wood, Geeston Quarry and the River Soar are probably the three most thoroughly worked sites in Leicestershire, but the only comparisons that may be made between these essentially dissimilar habitats are the total number of records for each site, the number of species and the ranking of these species.

It is surprising how much valuable information can be obtained about a site from the results of only a few hours intensive collecting, and the proportion of indicator species in a small sample can be a fair statement of ecological diversity and 'quality'. Site scores have therefore been derived by ignoring the common and synanthropic species, and evaluating only the specialist spiders, as defined in table 14. The score values are subjective, but this concept could be developed, based on national statistics, and expanded to include an element related to the total number of species, the complexity of the site, the ratio of species to records and ratio of scorable species to total species.

However, the proposition presented here is simple and appears to be effective. All local species are scored as unity; nationally notable species Nb= 4, Na= 6; and Red Data Book spiders RDBK= 8, RDB3= 8, RDB2= 12, RDB1= 20.

Appendix I (page 200) lists a hundred sites in alphabetical order with spider statistics, site scores and site descriptions, whilst appendix II (page 213) lists the top forty-nine sites in rank order, and includes species lists for each of these. The lists for gardens are given for their curiosity interest rather than for their ranking, and the low-ranking Lount Meadow (3) appears for comparison with the other two Lount grassland sites.

A final judgement on any site should be made only after careful analysis of all the facts available, of which the site scores are only a guide. However, Buddon Wood stands out clearly as the best site for spiders in the county, and this assessment is discussed in more detail below.

Woodland

Leicestershire is one of the least-wooded counties in England. With the exception of Buddon and Swithland Woods this prime habitat is very much underworked, and it speaks well for the quality of the county's woodland that encouraging lists have been obtained with relatively little collecting. Systematic study of the east Leicestershire woods, Burbage and Sheepy Woods, Cloud, Asplin and Pasture Woods, Oakley and Piper Woods and Martinshaw Wood would be worthwhile. It would appear that *Haplodrassus silvestris*, *Micrommata virescens*, *Ballus chalybeius*, *Mastigusa macrophthalma*, *Tetragnatha pinicola*, *Pachygnatha listeri*, *Gibbaranea gibbosa*, *Cyclosa conica*, *Walckenaeria incisa*, *Saloca diceros*, *Asthenargus paganus*, *Lepthyphantes obscurus*, *Lepthyphantes tenebricola* and *Lepthyphantes midas* are good indicator species of ancient woodland in Leicestershire, and that the more of these that occur at any one site the greater will be the stability and age of that ecosystem.

Climax woodland contains the widest variety of micro-habitats of any ecosystem represented in Leicestershire, and therefore prime sites would be expected to yield the highest number of spider species. This is shown to be the case with Buddon Wood – ancient birch/oak woodland on acid siliceous/clayey soils with wet flushes within the wood and adjoining herb-rich wet meadows – where 191 species are recorded, the highest total for any site in the county. It is known that many invertebrate species were lost to the wood following clear-felling and burning before the first spider survey was carried out in the 1970s. It is also known that a high percentage of the species recorded (appendix IIf) has survived the massive quarrying operations which have recently destroyed over 60% of the wood. This woodland was a favourite collecting ground of the Victorian entomologists whose work, and that of subsequent collectors, has shown Buddon Wood to have been one of the best woodland sites for invertebrates in Britain. The spider list speaks for itself and the ranking of the wood, based on the variety of its specialist and rare species, places it in a class of its own. In view of the continuing destruction of this site it is considered appropriate to quote in full the personal reminiscences of naturalist Peter H. Gamble (pers.comm. 3.6.95), who knew this woodland prior to wartime clear felling.

> "My earliest recollections of Buddon Wood, in the 1930s, were of its large size and sense of mystery – even after very many visits I still seemed to keep finding precious new spots – and each visit produced something new. I suppose the outstanding things were:—
>
> 1) The vast numbers of red wood ants and their great communal 'nests'.
> 2) The large amount of dead wood – most of the oaks carried much dead wood in their upper branches and trunks.
> 3) The fine old silver birches many of which were peppered with the nesting holes of great spotted woodpeckers.
> 4) The magnificent display of bluebells in May, acres of the woodland floor being painted deep blue, especially along the western side of the wood, and their glorious scent detectable from the public footpath two narrow meadows distant. Similarly the spectacular display of wood anemones along the eastern side of the wood in April.
> 5) The large population of green woodpeckers feeding and breeding in the wood, nesting-holes sometimes being located low down in the oaks allowing wonderful opportunities for close observation.
> 6) The large population of bats – hundreds, including large numbers of noctules, disporting themselves over the wood and Swithland Reservoir on warm summer evenings.

7) The large population of badgers, c.10 setts, including a large sett in a splendid position near the summit of the hill.
8) The large numbers of cuckoos and tree pipits, their songs an ever present feature during May and June. Alas, both are now something of a rarity in the area.
9) The magnificent show of orchids: late marsh and common spotted, plus numerous hybrids, on the floor of the Old Cocklow Quarry. Also dense colonies of bee orchid and numerous moss, liverwort and lichen species.

"The felling of Buddon's woodland during the war years of the early 1940s was a great tragedy, and left me with a feeling of utter despair and helplessness. However, such is the wonderful resilience of Nature that natural regeneration was soon at work transforming the bare ground with a green mantle of seedling birch and sallows and re-sprouting oaks and hazel stumps. For a number of years the scrubby conditions proved ideal for breeding warblers – willow, sedge, grasshopper and whitethroat – and during the 1940s–50s up to four pairs of nightjar bred here. Many an evening in June and July was spent listening to their churring song and watching their courtship and display flying over the hill tops whilst the lights twinkled in the valley below. Also, during the summers of 1943-45, nightingale sang and nested on the eastern edge of the wood near the Cocklow Quarry.

"Although, when viewed from the Charnwood Forest hills, Buddon Hill looked like an elongated dome (like a sleeping dinosaur!), in fact it was made up of numerous hillocks and ridges with many little valleys and hollows with wet ground with sphagnum and other bog mosses. To climb up to the summit in the evening time was to experience considerable changes in temperature, and around dusk map-winged swift and gold swift moths were often plentiful in season, flying just above the bracken.

"From what I remember of the woodland prior to felling I would estimate the age of the standard oaks to have been between 100 and 200 years but, as most were growing on shallow soils over rock, they were probably older than they appeared. Many coppiced trees such as the small-leaved limes *Tilia cordata* were obviously older than this and, as this particular species grew, and still grows in surviving areas of woodland, in discrete colonies, individual trees – probably connected by their root systems – could represent ancient growths many hundreds of years old. Some old hornbeam *Carpinus betulus* trees grew in the wood, and I understand that after being felled these were purchased by Messrs. M. Wright & Sons, the hosiery and webbing manufacturers at Quorn, to use for making shuttles. Bearing in mind that almost all the oaks on Buddon Wood were, and still are, sessile oaks *Quercus petraea*, it seems unlikely that the oaks present here were the result of planting and reafforestation, but would have been far more likely the result of natural regeneration. Unfortunately when the wood was clear-felled during the last Great War quite a lot of sycamore *Acer pseudoplatanus* seeded into it, especially opposite the sycamore plantation off the Wood Lane on the eastern side of the wood.

"The red wood ants were an especially important feature of the old wood and their 'nests' were scattered across the whole area, some individual 'nests' being of great size. The ants were formerly so plentiful – they must have numbered countless millions – and I still find it difficult to accept that within a few decades they had disappeared so completely. Before the felling of Buddon Wood, I can remember seeing red wood ants' 'nests' outside the confines of the wood itself, in more open wooded parts of Quorn House Park, and even on the Brazil Island in the middle of Swithland Reservoir. During the aerial nuptial flights of this species vast numbers of these insects must have ranged over Buddon and its environs on certain warm summer days.

"During the war years the wood was much used by the American Airborne troops (stationed nearby on the Quorn House Park Camp) for manoeuvres, often with live ammunition, and the shell holes which became water filled were colonised by dragonflies and damselflies, including the striking broad-bodied chaser. The wood was, of course, noted for its rare invertebrates and a small distinctive day flying moth, the argent and sable, was observed flying here in the spring shortly following the clear felling. I do not think it has been seen in Leicestershire since this time. Sadly this must be true of many of the wood's special invertebrates.

"Following the felling of Buddon Wood rapid regeneration took place and after some years the wealth of scrub, bracken, bramble and seedling birches made progression through the wood difficult. Presumably for this reason – to help shooting parties – the Earl of Lanesborough and members of his Estate staff tried burning as a method of controlling the growth in certain parts of the wood. Unfortunately for the wood, and its flora and fauna, this proved a disastrous policy, as the fire penetrated deep down into great accumulations of humus and kept breaking out again when fanned by the breeze, long after it was thought to have been quelled. At one time the fire burned for about six months before it eventually subsided and a large area – from memory at least a third of the total area – was gutted. This undoubtedly had a particularly harmful effect on the wood's ecology and may have been a major factor in the loss of the red wood ant.

"However, as the years progressed and the trees grew in size and stature, I remember being filled with hope, for Buddon was well on the way to becoming a fine exciting wood again, and species that were feared to have been lost were rediscovered. Thus the purple hairstreak was found to be still present in numbers on the western edge of the wood and many pairs of woodcock bred within the wood's confines. Moreover, some species such as the wood warbler, previously not recorded, colonised several areas with now tall close growing birches.

"Sadly, all hopes were abruptly dashed with the eventual realisation that the whole site was to be sacrificed for the extraction of 'granite' – to become reputedly the largest hole in Europe. Needless to say, I have still not come to terms with the destruction of so much I had grown to love, for surely no activity of mankind is more totally destructive of ecosystems than modern quarrying activities!"

The Forest Vision – a New National Forest

In conclusion, and as we contemplate the pressures upon our surviving semi-natural habitats, it is good to be able to look forward with some degree of optimism, not only to the active conservation of our remaining precious wildlife sites, but to the creation of a major new National Forest. This concept was pioneered by the Countryside Commission in 1987, and received Government backing in 1991 to prepare a strategy, which was published in 1994 (Countryside Commission, 1994). It is intended to stimulate economic regeneration in an area of high unemployment. The new Forest is to link the ancient forests of Needwood in the west and Charnwood in the east, covering an area of 502 sq km (200 sq miles) and spanning three Midland counties: Leicestershire, Derbyshire and Staffordshire. Of these, Leicestershire is to have the largest area, embracing Charnwood Forest and the Ashby coalfield plateau, including Markfield, Ibstock, Coalville, Whitwick, Measham and Ashby-de-la-Zouch.

At present, existing woodland covers only 6.1% of the Forest area, of which 36% is broadleaved, 9% coniferous and 55% mixed. Ancient woodland accounts for 46% of all woods and is concentrated mainly in Charnwood Forest, the Calke Uplands and Needwood. The intention is to increase the total woodland to 33%, involving some 135 sq km of new planting – of which 70% is planned in the first 10 years. This multi-purpose woodland will combine large scale commercial plantings and farm woodland, with recreational development and restoration of industrial derelict land. Primarily a commercial venture centred on sustainable timber cropping and tourism, the management of the new forest should create many new long term jobs and local leisure-related industries. The National Forest Company (established on 1st April 1995) will be working closely with English Nature, English Heritage and local authorities. It is intended that habitat creation will be taken seriously, in an effort to restore some of the damage to the environment and losses of flora and fauna through industrial exploitation and urban expansion over the last two centuries. This aproach will have to override profit-generation and recreational demand, allocating certain areas for low level land use with limited biotic pressures, and ensuring sensitive management of areas set aside for wild life. Altogether, the scheme presents a healthy challenge for the future!

APPENDIX I

Site Details

Appendix 1 is in two parts. Part 1 lists the top 100 sites, according to rank. The ranking score is determined on the number and status of local and rare species (Local= 1, Na= 4, Nb= 6, RDB3 & RDBK= 8, RDB2= 12). Part 2, the gazetteer, gives information about 125 main sites.

List of 100 Sites with spider statistics and site scores

Location	Map Ref	Records	Species	Local	Rare	Score
Acresford Sandpit, Donisthorpe	SK3013	123	45	9		9
Altar Stones, Markfield	SK4810	34	29	5	1RDB3	13
Ashby Canal	SK3311-SP3998	84	35	7		7
Aylestone Meadows, Leicester	SK5701	487	76	13	1Nb	17
Bardon Hill	SK4613	906	147	34	1RDB3	42
Barrow Gravel Pits	SK5616	75	34	10		10
Beacon Hill, Woodhouse	SK5014 & SK5114	58	44	6	1RDB3	14
Benscliffe Wood, Newtown Linford	SK5112	74	37	5		5
Blackbrook Reservoir, Shepshed	SK4517	154	48	9		9
Botcheston Bog, Desford	SK4804	53	34	3		3
Bradgate Park	SK5310	347	108	15	1RDB3 1RDBK	31
Brand; The, Woodhouse	SK5313	31	23	4		4
Brazil Wood, Swithland	SK5513	18	18	3		3
Brown's Hill Quarry, Ab Kettleby	SK7423	16	16	4		4
Buddon Wood, Quorn	SK5515	1394	191	54	2Nb 1RDB3	70
Buddon Brook Marsh, Quorn	SK5515	68	43	2		2
Burbage Common, Hinckley	SP4495	69	49	4		4
Burbage Wood, Hinckley	SP4594	72	44	7	1Nb	11
Burrough Hill, Somerby	SK7612	48	45	3		3
Cademan Moor, Grace Dieu	SK4317	252	78	17		17
CEGB derelict site, Leicester	SK5802	203	51	9		9
Charnwood Lodge N.R., Charley	SK4615	952	162	36	1Nb	40
Cloud Wood, Breedon-on-the-Hill	SK4121	55	39	5		5
Coalfield West opencast site	SK3812 & SK3911	148	45	2	1Nb	6
Cooper's Plantation, Croxton Kerrial	SK8627	24	23	3		3
Cropston Reservoir	SK5410	104	49	9		9
Croxton Park, Croxton Kerrial	SK8227	68	34	5		5
Donington Park, Castle Donington	SK4126	310	63	7	1RDB3 1RDB2	27
Drift, The	SK8628-SK8726	190	56	10		10
Drybrook Wood, Charley	SK4516	74	38	7		7
Dunton Bassett	SP5489	112	39	7		7
Eyebrook Reservoir	SP8595	74	43	11		11
Garden & house (1) Leicester	SK6001	110	49	5		5
Garden & house (2) Leicester	SK6105	31	23	1		1
Garden & house (3) Leicester	SK6205	86	74	9		9
Garden & house (4) Loughborough	SK5218	181	59	6		6
Garden & house (5) Loughborough	SK5217	63	43	3		3
Geeston Quarry, Ketton	SK9803	1134	114	28	2Nb	36
Grand Union Canal	SP6196-SP6989	25	20	4		4
Great Bowden Pit	SP7489	67	31	3		3
Great Merrible Wood, Great Easton	SP8396	60	36	5	1Nb	9
Groby Pool	SK5208	198	63	9	1Nb	13
Harby Hills	SK7627	49	29	6	1Nb	10
High Sharpley, Whitwick	SK4417	314	75	17	1RDB3	25
Hill Hole, Markfield	SK4810	29	23	3	1RDB3	11
Holwell Mouth, Ab Kettleby	SK7224	12	10	3		3
Holwell North Quarry, Ab Kettleby	SK7423	38	33	3		3

Location	Map Ref	Records	Species	Local	Rare	Score
Ives Head, Shepshed	SK4717	19	19	1	1RDB3	9
Kendall's Meadow, Sutton Cheney	SP3998	74	28	3		3
Ketton Quarry	SK9705	287	120	28	3Nb 1Na	46
King Lud's Entrenchments, Sproxton	SK8627	253	77	18	1Nb	22
Knighton Spinney, Leicester	SK6000	82	41	6	1Nb	10
Knipton Reservoir	SK8130	36	21	3		3
Launde Big Wood	SK7803	75	46	7	1Nb	11
Lea Wood, Ulverscroft	SK5011	29	24	2		2
Leicester Cattle Market site	SK5802	66	32	6		6
Lockington Marsh	SK4830	12	12	2		2
Lockington Meadows	SK4829-SK4930	20	14	5		5
Loughborough Meadows	SK5321	330	62	13		13
Loughborough Outwoods	SK5116	153	65	10		10
Lount Grassland (1) reseeded	SK3918 & SK3919	187	46	10		10
Lount Meadow (2) Coleorton SSSI	SK3819	175	54	9		9
Lount Meadow (3) Worthington SSSI	SK3919	66	30	3		3
Luffenham Heath Golf Course	SK9502	238	88	23	2Nb 1Na	37
Martinshaw Wood, Ratby	SK5107	33	32	6	1Nb	10
Moira Junction, Ashby Woulds	SK3015	204	68	15	1Nb	19
Nailstone Wiggs	SK4208	14	14	4		4
Narborough Bog	SP5497	305	72	10	2Nb	18
Newfield Heath, Ashby Woulds	SK3115	200	72	14		14
Newton Burgoland Marshes, Swepstone	SK3808	93	43	3		3
North Brook, Rutland	SK9413-SK9508	16	10	2		2
Owston Wood	SK7906	147	62	10	2Nb	18
Pickworth Great Wood	SK9815	82	54	10	1Nb	14
Pignut Spinney Marsh, Loughborough	SK5217	65	28	3		3
Puddledyke, Newtown Linford	SK5411	57	42	3		3
River Chater	SK8004-SK8403	23	14	3		3
River Eye	SK7718-SK7918	15	9	2		2
River Sence Marsh, Bardon	SK4512	24	21	1		1
River Soar (north)	SK4930-SK5913	809	92	21		21
River Trent	SK4127-SK4429	80	40	8		8
River Welland	SP8996	17	16	2		2
River Wreake	SK6716-SK6817	38	24	1		1
Rutland Water	SK80/90	121	56	8		8
Saddington Reservoir	SP6691	146	39	6		6
Seaton Meadow	SP9198	83	31	4		4
Shacklewell Hollow, Empingham	SK9707	15	13	3	1Nb	7
Sharnford Meadow	SP4991	17	15	2		2
Sheet Hedges Wood, Newtown Linford	SK5208	87	46	5		5
Skeffington Wood	SK7503	75	46	4		4
St Mary's Allotments, Leicester	SK5802	121	36	5	1Nb	9
Stonesby Quarry N.R., Sproxton	SK8125	206	63	13	1Nb	17
Stoneywell Wood, Ulverscroft	SK4911	125	60	12		12
Swithland Reservoir	SK5514	550	119	26	1Nb	30
Swithland Wood	SK5312	317	114	22		22
Thistleton Gullet	SK8918	19	16	3		3
Twenty Acre Piece, Burton-on-the-Wolds	SK6421	9	9	1	1Nb	5
Ulverscroft N.R.	SK4912	388	107	15		15
Wanlip Gravel Pits, Syston	SK6011	35	20	4		4
Watermead Country Park, Leicester	SK6011	237	55	8	1Nb	12
Wymondham Rough N.R.	SK8317	5	5	2		2

Gazetteer of Main Sites

Acresford Sandpit, Donisthorpe SK3013
Disused sandpit which has its eastern half partly filled with a large mound of fly-ash. The site is sparsely covered by vegetation and there are small pockets of broom and birch scrub. The western perimeter has small cliffs within the boundary, to a height of 10 metres, making the western half of the site a sun trap. There is informal access to the whole site, with a footpath around the boundaries.

Altar Stones, Markfield SK4810
An outcrop of slate-agglomerate at a height of 213m, situated between the former main road and the new A50. The outcrop is surrounded by remnant heath-grassland with some oak and gorse scrub. It was presented to the County Council by Miss J.A.F.Elgood in 1949 and is now a Local Nature Reserve with open access at all times. This is a Regionally Important Geological Site (RIGS).

Ashby Canal SK3311 to SP3998
From Snarestone to south of Ambion Wood, most of the canal is a SSSI and contains rich plant communities that typify eutrophic water conditions. There is a wide variety of submerged and floating vegetation which, together with the emergent and riparian plants, provides habitats for a diverse invertebrate fauna. The canal is used extensively for pleasure boating and for fishing, with all parts of it being accessible via the towpath.

Aylestone Meadows, Leicester SK5701
Flood meadows to the south of Leicester alongside the River Soar and Grand Union Canal with patches of marsh, bare mud, sedge beds, damp herb-rich grassland and rough areas. Now a Local Nature Reserve and riverside park, which also incorporates part of the dismantled Great Central Railway line, a section of the River Biam, the abandoned Evesham Road Allotments, and subsided football pitches on a previous sandpit landfill site. Many trees have been planted in the area, and several wetland habitats and a network of public paths created. The area is owned mostly by Leicester City Council, with open access at all times.

Bardon Hill, Charnwood SK4613
A prominent feature of the landscape, this hill – at 278m – is the highest point in Leicestershire. The Pre-Cambrian rocks on its northern and western flanks have been quarried away for roadstone, but the profile has recently been restored with overburden infill. There is only a remnant of the once extensive moorland around the summit, and this is being invaded by oak and birch scrub. A small pocket of the former oakwood remains on the southern flank, but much of the southern and eastern slopes have been planted with conifers. The moorland around the rocky summit of the hill is part of the SSSI which also includes the rock exposures in the main quarry. There is access to the summit along a public footpath.

Barkby Thorpe Park, Barkby SK6309
Open parkland grazed by cattle, with mature oak trees, on the south side of Barkby village.

Barnsdale Wood, Exton SK9108
Ancient damp oakwood on clay with an open canopy, carpeted with bluebells in the spring, situated on the north bank of Rutland Water between Burley Wood and Whitwell. The eastern end of the wood was partly cleared, ploughed and replanted with conifers in 1955.

Barrow Gravel Pits, Barrow upon Soar SK5616
A complex of open water, scrub, flood meadows and woodland within a bend of the River Soar, with a rich flora and fauna characteristic of the River Soar flood plain. The disused gravel pits and surrounding small hay meadows with tall hedges have been colonised by an interesting variety of plants and breeding birds. The islands have dense stands of sallows and osiers, and reed beds fringe some of the pools. Part of the Quorn and Mountsorrel bypass was built through the northern part of the site in the mid 1990s. The area is a SSSI and is a privately owned leisure park.

Beacon Hill, Woodhouse SK5014 & SK5114
One of the largest expanses of semi-natural habitat in the county, the Beacon rises to 248m and covers an area of over 80 ha. The rocky open moorland summit gives way to bracken-dominated scrubby acid-grassland on its eastern flank, with mature birch and oak wood lower down. There is an extensive conifer plantation on the northern slope. Although the eastern flank has been drained, there are wet flushes in the woodland and two ponds that still retain botanical interest. Remains of a Bronze Age hill fort have been found on its south-eastern flank. The area, acquired in 1946 as a public open space, is a country park and a SSSI. It is managed by Leicestershire County Council, and is heavily used by the public for leisure and recreation.

Benscliffe Wood, Newtown Linford SK5112
Much of Benscliffe Hill, which rises to 238m, was formerly covered with sessile oak woodland, but most of it was felled between 1914 and 1950. Larch and pine were planted commercially on about 57 ha, but there are some residual areas of bracken-dominated deciduous woodland. Several outcrops of felsitic-agglomerate occur along the ridge, two of which are SSSIs established for their lichen interest. The areas of heath around the outcrops have been largely planted with conifers. The wood is privately owned and has been used for commercial timber production. A study of the area was published by the LNC in 1990.

Bitteswell Aerodrome SP5085
Abandoned airfield with rough grassland.

Blackbrook Reservoir, Shepshed 38.6 ha SK4517
Standing on the northern fringes of the Charnwood Forest, the present reservoir occupies the site of the former header-reservoir built to feed the 18th century Charnwood Forest Canal. There is documentary evidence indicating a large lake on this site before the first reservoir. The present reservoir, which was completed in 1906, is a mesotrophic water body with plant communities unusual for the East Midlands – a mixture of northern and southern species. The reservoir is owned and managed by Severn Trent Water plc. for water storage and fishing. The site is a SSSI.

Bloody Oaks Quarry, Tickencote   SK970108
This 1.46 ha site is a SSSI owned by the LRTNC. It is the relict of a small, shallow, limestone pit that has been partly restored, and was shown as a quarry on the First Edition 25" Ordnance Survey Map (1884). There are good areas of rich limestone grassland, with patches of thorn-dominated scrub and rough grassland. Management work on the reserve is concentrating on preserving the limestone flora by removal of invading scrub and coarse grasses. There is a local rabbit population, and sheep-grazing is being considered on part of the site.

Botcheston Bog, Desford 3.19 ha   SK4804
A spring-fed marsh, which is a SSSI, at the side of Thornton Brook on alluvial soils over Keuper marl. A layer of peat has been laid down with a pH of around 6.95 and the bog supports a wetland plant community, including bogbean, now unusual in the Midlands. Beside the brook is an alder grove. A second stream crosses this cattle-grazed area, which is privately owned.

Bradgate Park, Charnwood   SK5310
An ancient deer park, and formerly the home of the Greys of Groby. It was purchased by Mr Charles Bennion in 1928 and given to the people of Leicestershire. The park is managed by the County Council as a country park, although some areas are kept private to provide seclusion for the deer. High dry-stone walls surround the park enclosing 344 ha of bracken-dominated moorland, with rocky hilltops rising to 206m. Several wooded enclosures were planted in the 19th century for game cover. Drainage channels cut deep into the peaty soil and some patches of wet heath remain. The River Lin runs through the park to feed Cropston Reservoir. There are several ponds within the park, and a small lake inside the Bradgate House ruins enclosure. A prominent feature is the number of ancient pollarded oaks, many of them hollow and some known to be over 300 years old.

The Brand, Woodhouse   SK5313
Part of the Swithland Wood SSSI, this area was quarried for slate until the early 1800s. The quarries are now water-filled and surrounded by oak woodland, scrub and some grassy heath. The area is notable for its ferns and lichens. The Brand, owned by the Martin family since 1892, has been largely undisturbed, with limited access to the public on regular Open Days.

Brazil Wood, Swithland 3 ha   SK5513
Now an island in Swithland Reservoir, the wood is connected to the shore by the Great Central Railway line that crosses the reservoir. The island is all that is left of an extensive ancient woodland; the terrain is rocky and still covered with oak woodland, reminiscent of the old Buddon Wood.

Breedon Hill   SK4023
A large part of the hill has been quarried for limestone, but the west side remains as common land covered by calcareous grassland with some areas of thick thorn scrub. There is a tidy churchyard at the top of the hill.

Brown's Hill Quarry, Ab Kettleby 2 ha   SK7423
This former ironstone quarry is owned by the LRTNC who manage it as a nature reserve. Quarrying terminated in 1957 and the area is now a Regionally Important Geological Site. The south-west of the area is mixed woodland planted in the 1930s. The slopes and screes have been colonised by herb-rich grassland and some scrub, which is controlled to maintain the site's open grassy aspect. There are small damp areas, and the quarry supports a rich variety of plants and invertibrates.

Buddon Wood, Quorn   SK5515
This ancient woodland was clear-felled in the 1940s and allowed to regenerate. The underlying rock is a hard granitic material in demand for roadstone, and much of the wood has been destroyed by the present quarrying operations which began in the 1970s. The remaining woodland forms a fringe (not more than 300m wide), around three sides of the quarry, and this remnant amounts to less than a quarter of the original woodland. Despite the wood's reduced size it is still a SSSI and is regarded as one of the best remnants of ancient woodland in the East Midlands. It has an impressive list of plant species and is the best site in Leicestershire for spiders. Moths and other invertebrates have been recorded here for many years. The wood is sessile oak/birch with small-leaved lime and alder around the wet flushes. Invading sycamore threatens the native tree species.

Buddon Brook Marsh, Quorn   SK5515
The marsh lies at the side of Buddon Brook, which is a continuation of the River Lin, flowing north towards Quorn and the River Soar, and linking the Soar Valley with Charnwood Forest. These unimproved marshy meadows, adjacent to Buddon Wood, are privately owned and are heavily grazed by cattle. A leat stream ran along the east side of the site to feed the water mill at Quorn. There is an area of marshy alder-woodland bordering Buddon Wood and the winding brook is overhung by alders. The wetter patches have tall marsh vegetation and the large hedges and blackthorn scrub provide good cover for wildlife. The meadows are included in the Buddon Wood and Swithland Reservoir SSSI. Alder scrub invading the meadows has been cleared and the tenant manages the area to conserve wildlife. A public footpath runs along part of the brook.

Burbage Common, Hinckley   SP4495
A public open space, situated two miles north-east of Hinckley and managed, together with Burbage Wood, Sheepy Wood and Aston Firs, by Hinckley & Bosworth Borough Council as a country park. Part of the open common is maintained as a golf course, whilst much of the rough dry heath-grassland is invaded by gorse and oak scrub. There is also a little marshy ground and a pond of recent origin. Another part is mown regularly for general recreation.

Burbage Wood, Hinckley   SP4594
An ancient oak/ash wood with a rich woodland flora including occasional pendulous sedge and a good shrub layer containing field maple & hazel. A very small part of the southern central section appears to have been continuously wooded, whilst the rest of the wood is on ridge & furrow. Coppicing was reintroduced in the 1980s. Aston Firs, an adjacent wood of similar composition, and Burbage Wood, stand on boulder clay and together are a SSSI. They are managed as a nature reserve with public access.

Burrough Hill, Somerby 33.2 ha   SK7612
A marlstone hill, 213m high, to the north of Burrough-on-the-Hill, which is the site of an Iron Age hill fort, the ramparts of which still remain. The ramparts possess areas of herb-rich grassland which is grazed by sheep and rabbits, whilst the flanks of the hill are colonised in parts by gorse and elder scrub. The hill is privately owned by the Ernest Cook Trust who lease it to the County Council as a public open space.

Cademan Moor, Grace Dieu   SK4317
This large area of acidic grassland, with wet flushes containing *sphagnum*, is being invaded by birch, oak, gorse and thorn scrub. Some scrub clearance has taken place at the southern end of the moor, and there is a pond at the lower northern end, which often dries out in summer. The moor is bordered by Cademan Wood on three sides. It is privately owned and is part of the Grace Dieu and High Sharpley SSSI.

CEGB derelict site, Rawdykes, Leicester   SK5802
Site of the former coal-fired power station which was demolished in the mid 1980s. The power station was built in 1922 (although electricity has been generated on the site since 1894) and located adjacent to the River Soar, whose waters it used in its cooling system. At the time the site was surveyed in 1992, it was used as a car storage area. Historically, this site has been associated with man for at least two thousand years, with part of the Raw Dyke – an old entrenchment dating back to Roman times – on its south eastern boundary. The vegetation is predominantly ruderal, with small areas of tall herb and invading birch scrub, the seemingly bare areas across the site have been colonised by lichen heath. The many years of coal storage and burning on the site has acidified the soil, providing an interesting pocket of heathland in a neutral area. The site is within the area of the urban renewal scheme, and there are plans to build a marina and create a mown-grass recreation area.

Charnwood Lodge Nature Reserve   SK4615
The best and most extensive example of moorland in Leicestershire, containing the largest stands of heather and bilberry in the county. The 227 ha reserve, which is a SSSI, was left to the LRTNC by Miss C.E.Clarke and is managed as an undisturbed wildlife sanctuary with limited access. There is a series of Pre-Cambrian rocky outcrops and ridges of geological importance, rising to about 250m on Timberwood Hill. Here the moorland flourishes, although bracken presents a problem. The acid heathland is grazed during the summer months to help keep the scrub and coarse grasses in check. In the lower areas there are broadleaved and conifer plantations, wet pasture, several ponds, marshy areas with *sphagnum* and a small reservoir.

Cloud Wood, Breedon-on-the-Hill 32.37 ha SK4121
An ancient woodland site, this area was felled in the 1940s and natural regeneration has been allowed to take place. Situated next to a working limestone quarry, the wood stands on Keuper marl and boulder clay over carboniferous limestone. The wood extended originally over the limestone on the upper slopes of Cloud Hill, but this unique part has now been quarried away. The variety of soil conditions and drainage combined with the alkalinity, caused by dumping of quarry waste, has given rise to a very rich ground and shrub flora, for which the wood is given SSSI status. It has a long history of coppicing, which ceased in the early years of the 20th century. The Wood was given to LRTNC in 1993 by Breedon Plc., the quarry owners, and is now a nature reserve open to Trust members.

Coalfield West coal opencast site.
                          SK3812 & SK3911
Part of an extensive area of intensively managed farmland which has been worked by opencast techniques for the extraction of shallow coal seams since 1945. The Coalfield West site extends into three parishes: Heather, Normanton and Ravenstone. The deeper coal seams were worked traditionally from deep mine shafts at Heather in the 19th century and, latterly, shallow seams were extracted by opencast over a small area in Ravenstone Parish during the late 1940s. A recent planning application to re-excavate the Ravenstone site and take the coal from under the Normanton and Heather sites was not granted. Spiders were collected in 1991 from three restored sites, then down to ley, and from three other sites on unmined land, comprising hedgerows, improved grassland and arable.

Cooper's Plantation, Croxton Kerrial   SK8627
Part of the SSSI embracing King Lud's Entrenchments and The Drift. The plantation is a mixture of broadleaf and mixed woodland, with scrub developing on limestone grassland at its eastern end.

Cropston Brickpit, see Puddledyke

Cropston Reservoir, Charnwood   SK5410
The reservoir covers about 55 ha lying on the southern fringes of Charnwood Forest adjacent to Bradgate Park, and is part of that SSSI. There are extensive reed beds, areas of willow scrub and bare mud habitats. An unusual marginal flora has developed containing species traditionally having a more northerly distribution, as well as several nationally scarce species. The reservoir, which is owned by Severn Trent Water Plc., was built in 1870 and is stocked for fishing. There is no public access.

Croxton Park, Croxton Kerrial   SK8227
This former medieval deer park, owned by the Belvoir Estate, is a SSSI. The unimproved grassland with ancient oaks and hawthorns, some hollow, is grazed by sheep and cattle. It is partly on Northampton Sand with the valleys eroding down into Upper Lias clay. The sands give a neutral to acid grassland but some areas are affected by the Lincolnshire limestone under the clay. There are lakes with marshy areas and the valley sides are partially covered with scrub, which is often dense. The limestone walls, which surround part of the park and the old oak plantations, are a feature of the park, and are excellent for lichens.

Donington Park, Castle Donington 37.5 ha
                                        SK4126
This privately owned former medieval deer park, with its old oaks and pasture woodland, still retains its herd of fallow and red deer. The slopes that face north, towards the River Trent, have been invaded by bracken

which is now under control, but ragwort threatens to take over the extensive deer-grazed pasture. Part of the deer park is a SSSI. Recent survey work on dead wood beetles has prompted a project to re-enclose some former parkland – now an arable field – containing several old oaks, and return it to pasture woodland. Two Red Data Book spiders, both associated with the ancient oaks, have been recorded from this site.

The Drift　　　　　　　SK8628 to SK8726
An ancient pre-Roman trackway forms the county boundary with Lincolnshire for 14km. The width of the track varies and there are areas of species-rich limestone grassland, which are bordered on each side by either hedgerows or deciduous woodland. The site is partly a SSSI, with management aimed at keeping the track and grassland open. The track is a green lane and footpath with open access at all times. Boundary changes along this section of The Drift in 1986 moved the county boundary from the centre of the track to the Leicestershire side, so that now this area is in Lincolnshire. This has been done to facilitate management under one county authority.

Drybrook Wood, Charley　　　　　　　SK4516
An area of broadleaved woodland with a ground flora dominated by bracken and brambles. The western third is more open with occasional trees and shrubs and at the southern end are large rocky outcrops with lichens, heather and bilberry. It is privately owned.

Dunton Bassett Quarry　　　　　　　SP5489
Owned by Bruntingthorpe Gravels, this site has been quarried for sand and gravel since the early 1980s. A pool developed with reedswamp which was excellent for amphibians and invertebrates. Other habitats include willow carr and some rough grassland, but half of the quarry has been used as a landfill site.

Eyebrook Reservoir 178.07 ha　　　SP8595
This reservoir was built privately by Stewarts and Lloyds in 1940, to supply their growing steel works at Corby in Northamptonshire. Besides the vast area of open water there is a reasonable emergent and aquatic flora with small areas of sedge beds. There is a conifer plantation on the eastern margin, with the other reservoir margins having sheep grazed semi-improved pasture. The reservoir, which is a SSSI and an important habitat for wintering wildfowl, is now owned by British Steel and is stocked for trout fishing.

Farnham's Wood, Quorn　　　　　　　SK5615
Wooded parkland adjoining Buddon Wood, on part of the Quorn medieval field system. There are blocks of woodland with 100-year-old plus pedunculate oak trees, and invading sycamore. The open grassland is let for grazing (mostly sheep), but contains a good number of scattered lime and oak trees, many of which are over 200 years old. Throughout the park, shed limbs of trees are allowed to decay naturally, providing an important habitat for invertebrates. The two hilly areas, on rocky ground with heath grassland, have been landscaped and planted with rhododendrons and conifers.

Fort Henry Lakes, Horn　　　　　　　SK9412
Lakes on the Horn Brook, Rutland.

Fosse Meadows Nature Park, Sharnford
58.68 ha　　　　　　　SP4891
An area comprising seven parcels of permanent grassland, including a herb-rich meadow, surrounded by hedgerows and ditches. A small tributary of the upper reaches of the River Soar runs along the Park's northern boundary. The site is being developed for informal recreation, and several belts of trees have been planted and two ponds created. There is a small car park on the southern side of the Park which is open to the public at all times. The site is owned and managed by Blaby District Council.

Garden 1, Knighton Church Road, Leicester
　　　　　　　SK6001
Semi-detached house and garden situated within the post-war residential development between Evington Golf Course and Knighton Park, south-east Leicester. The garden is fairly open with flower beds, a large area of mown grass, the occasional tree and hedges along two sides.

Garden 2, The Portwey, Leicester　　SK6105
Post-1945 semi-detached house built on farm-land at Humberstone, off the A47 Uppingham Road. The garden has large areas of mown grass and flower beds with occasional shrubs, trees and hedges. It has an open aspect, overlooking allotment gardens to the rear and is close to the open spaces of Humberstone Park, Humberstone Grange and the Towers Hospital, giving a suburban aspect.

Garden 3, Scraptoft Lane, Leicester　SK6205
The garden is situated on the east side of Leicester, about three miles from the centre of the city. The post-war detached house and garden are surrounded by a five-foot high timber panel fence, with a mortared stone wall either side of the main entrance, and a garage at the end of the back garden. The front garden contains evergreen bushes and trees, surrounding herbaceous borders, and a small pond close to the house. The back garden is more open, and is centred on a medium sized lawn with a fringe of flower beds and herbaceous borders sloping down towards the house. There is a small greenhouse in one corner of the terraced garden which has small drystone retaining walls. The perimeter has a range of deciduous shrubs and trees that generally screen the garden from adjacent properties.

Garden 4, Outwoods Drive, Loughborough SK5218
Semi-detached house built 1934 with separate timber/asbestos garage, toolshed and greenhouse, on southern edge of town. It is surrounded by school playing fields, parkland and nearby, an extensive abandoned clay quarry. The long garden was laid out formally with a small vegetable plot, three lawns separated by shrubberies and flower beds. An ornamental hedge was planted along one side near the house, and an ancient parish boundary hedge runs across the bottom of the garden. On either side are contiguous orchards and mature birch, rowan and conifer trees. There is a permanent compost heap and woodpile.

Garden 5, Bramcote Road, Loughborough SK5217
Detached house with integral garage and timber outbuildings, built mid-1970s on ridge & furrow farmland at northern edge of Charnwood Forest, below Loughborough Outwoods and Beacon Hill. Small landscaped garden, naturalised, with limestone and sandstone rockeries and raised ericaceous beds, a shrubbery, lawn, two ponds, paved areas and a patio, enclosed by timber fencing on three sides, two of which are covered by trained climbing shrubs.

Garendon Park, Loughborough SK494199
Marshy ground beside Oxley gutter in Hermitage Spinney. The Park was once part of the estates of Garendon Abbey, but is now intensively farmed. Recent plantations replace the old park woodland.

Geeston Quarry, Ketton SK9803
This is a former limestone quarry, abandoned in 1950. At present parts are being used to store firewood and coal, and there are bee-hives in one corner. One of the former uses of the quarry was as a run for poultry and pigs, so that when the Rutland Flora was published (Messenger, 1971) the vegetation was thought to be of little botanical interest. Today, the quarry holds an interesting limestone flora, that includes orchids as well as alien and garden escapes growing on a large area of dumped soil. The quarry is completely surrounded by a thicket of thorn scrub, with several further small patches within the quarry itself. The actual quarry has sparse vegetation, with rabbits helping to keep down most of any invading scrub. The site is privately owned and still contains outstanding pre-war rights to continue quarrying the area.

Grantham Canal SK7329 to SK7935
Opened in 1797, this canal was closed to traffic in 1929 and finally abandoned during the 1960s. The stretch between Harby and Redmile was designated a SSSI, as typifying the plant communities growing in a nutrient-rich slow-flowing water body. Diversity of the emergent, marginal and aquatic floras is great and the invertebrate life reflects this; floating mats of vegetation are also a feature and are unique in the area. The towpath makes the canal accessible to all, but the rich habitat is threatened by the proposed resumption of navigation and the consequent disturbance by dredging and clearing of vegetation. The LRTNC has an access agreement with the British Waterways Board over this stretch, but is not involved with its management.

Grand Union Canal SP6196-SP6989
This navigation was completed in 1814 and is still used by leisure craft. The water is of relatively high alkalinity and the more shallow margins have rich swamp and tall-fen communities. Its riparian flora is also well developed, and the open water supports nine species of pondweed with a diverse aquatic invertebrate fauna. Bridges, locks, wharves and a tunnel provide sheltered habitats for various forms of wildlife. This canal is a popular site for visitors in the summer and has been developed as a tourist attraction. The stretch from Kilby Bridge to Foxton Locks is a SSSI and is owned by the British Waterways Board, who allow the LRTNC to use this stretch for educational purposes.

Great Bowden Borrowpit 2.61 ha SP7489
Lying between Great Bowden and Thorpe Langton, the pit was excavated from the Lower Lias clay during the building of the railway line to Peterborough in 1884. Since then the pit has become water-filled and has developed an interesting acid bog flora, with cotton grass and *sphagnum*. Reedmace has invaded the southern corner and threatens to spread across the site unless kept in check. Dense thorn scrub covers the banks of the pit. The site, which is a SSSI, is owned by British Rail and leased to a local farmer whose stock graze the surrounding pasture. LRTNC has a management agreement with the tenant and the reserve is open to Trust members.

Great Merrible Wood, Gt. Easton 12 ha SP8396
This ancient woodland is a SSSI and is owned by the LRTNC. It was managed by coppicing for 120 years until the Second World War. Main tree species are ash, field maple and hazel, which have been coppiced, with oak standards. The wood is on calcareous boulder clay and has drainage ditches, a pond and streams. Part of the wood shows underlying ridge and furrow. The rides are wide and the shrub and herb layers are well developed and rich. It is Trust policy to continue to manage the wood by coppice rotation.

Groby Pool SK5208
This is the largest expanse of semi-natural water in the county, believed to have been created around the 12th century. Its actual size has varied throughout history and the present lake, at about 13 ha, covers half the area of its greatest size. It contains rich marginal and aquatic vegetation with several fine reed-beds which grade into wet alder woodland on the western margin. In the centre of the lake is a man-made island. The site is now owned by Amalgamated Roadstone Corporation Plc and has been managed as a nature reserve, with public access to the north-eastern side of the lake. It is an important site for wintering wildfowl, and the lake, together with surrounding woodland and some grassland, is a SSSI.

Gypsy Lane Brickpit, Leicester SK6106
Former brickpit and works abandoned in 1991. In 1992 there were large areas of bare ground with a ruderal flora developing on the mounds of overburden. The surrounding sides of the clay workings had growing on them tall-herb and thorn scrub. Since the pit was excavated below the natural water table it is necessary to keep the workings dry by pumping. There are plans to landfill the site for possible development.

Harby Hills, N.E. Leicestershire SK7627
Lying close to the bottom of a steep marlstone scarp, this cattle-grazed, undulating, slightly acidic grassland contains banks of thorn and gorse scrub. There is a marshy area fed by one of the many springs that rise from the escarpment, whose steeper parts are wooded. Although the woodland is not believed to be ancient in origin, its flora is interesting. This SSSI, which covers 18.45 ha, is privately owned but there is a public footpath running through it.

High Sharpley, Whitwick SK4417
Set around a rocky Pre-Cambrian outcrop, this is one of the few remaining examples of moorland which was

once widespread on Charnwood Forest. Dry acidic grassland on the higher areas, with heather and bilberry, gives way to open birch/oak woodland on the slopes, with the more open areas dominated by bracken. The area is part of the Grace Dieu and High Sharpley SSSI, in all about 89.3 ha. It is privately owned and managed for shooting, and public access is no longer permitted.

Hill Hole, Markfield　　　　　　　　　SK4810
This former granite quarry closed during the 1920s and the deeper part, with sheer quarry faces, has become filled with water. The surrounding higher ground contains rocky outcrops and spoil heaps, partially covered with heathy grassland and gorse scrub. Acid grassland is a feature of the surrounding area, but the previously excavated levels are sparsely vegetated and closely cropped by rabbits. The site is still owned by the quarry company, Tarmac Roadstone Central, and though public access is not encouraged, the area is widely used by local people.

Holwell Mouth, Ab Kettleby　　　　　　SK7224
This site contains the spring source of the River Smite, which has cut a valley into the Middle and Lower Lias marlstone scarp. Although the area is only 16.31 ha it contains a diversity of habitats, including a marshy area around the spring and an area of neutral grassland. In the north and on the upper slopes of the valley, ash woodland predominates with a hawthorn and hazel shrub layer, and a profusion of bluebells. In the south-east there is an area of bracken. The site is privately owned and is a SSSI.

Holwell North Quarry, Ab Kettleby　　SK7423
This small 2 ha ironstone quarry was worked up to 1917 and again from 1943 to 1963. It is a Regionally Important Geological Site (RIGS). Much of the quarry was infilled and farmed but the exposed face was left, with some spoil heaps, and is leased to LRTNC who manage it as a nature reserve. Herb-rich grassland has developed with some scrub. Trust members have access to the site but parts of it are liable to subsidence.

Ives Head, Shepshed　　　　　　　　　SK4717
A prominent hill-feature on the northern edge of Charnwood Forest, this Regionally Important Geological Site (RIGS) rises to 201m, with heathy rock-strewn grassland surrounding a Pre-Cambrian rock outcrop. There is a considerable amount of gorse scrub on the lower southern flank. The site, which is privately owned, is grazed and has no public access.

Kendall's Meadow, Sutton Cheney　　　SP3998
A meadow of 2.7 ha partly on alluvial soils and partly on clay, some of which shows ridge and furrow. The flora is particularly rich and diverse. It is privately owned and is cut for hay in the traditional manner. Designated as a SSSI, it is considered to be an exceptional example of this type of grassland community.

Ketton Quarry, Rutland　　　　　　　　SK9705
This is an area of several former limestone quarries that have been allowed to develop naturally to form a herb-rich limestone grassland. The vegetation here has been studied for more than a century. Part of the site is fenced and grazed, by sheep and donkeys, to help control the scrub which is encroaching into the open grassland. Other parts are cleared by hand to maintain the mosaic of habitats and there is a stable population of rabbits. The north-western part of the site is deciduous plantation, dominated by beech, with some self-seeded sycamore woodland together with thorn and gorse scrub. The site is owned by Castles Cement who still work the adjacent quarry and it is managed as a nature reserve by a local voluntary group. It is designated as a SSSI for its geological interest as well as for its floristic significance.

King Lud's Entrenchments, Sproxton　SK8627
An ancient earthwork, on the north boundary of a former American airfield, consisting of a double bank and ditch. This is part of a larger site owned by the Buckminster Estate, and was formerly under a management agreement with the LRTNC. The site is predominantly rough calcareous grassland, typical of communities on oolitic limestone, and contains the concrete bases of war-time Nissen huts. It is being invaded by thorn scrub and sycamore woodland. Adjacent is the mixed woodland of Cooper's Plantation (see above). The two sites are part of a SSSI.

Knighton Spinney, Leicester　　　　　　SK6000
Now owned by Leicester City Council, this oak/ash spinney was part of a larger area planted in the 1840s, as a timber crop, by Sir Edmond Craddock Hartopp, then lord of Knighton Manor. The spinney was acquired by LCC during the early 1950s as part of Knighton Park. The wood has a good structure with a hawthorn, field maple and holly shrub layer, and a ground flora that includes wood anemone and wood sorrel. Sycamore has invaded the southern part of the spinney, but is being removed gradually. The wood is being opened up to create different habitats, including some coppice. It is used as an educational resource for local schools, with access to the public being limited to two open days a month, throughout the year.

Knipton Reservoir, Croxton Kerrial　　SK8130
This small reservoir is in a shallow valley on the Belvoir Estate; it is fed by the River Devon from the south-west and a stream from the north. The open water is silting up in the two southern arms and the water is shallow here with floating vegetation. There are marshy areas to the south of the reservoir and a belt of willow and thorn scrub. Deciduous and mixed woodland borders the west and north sides with a small area on the east. Reed-beds fringe the reservoir and there is a heronry nearby. The woodland is managed for shooting, with no public access.

Launde Big Wood 41.23 ha　　　　　　SK7803
A large ancient woodland, mainly of the ash/maple type, which had most of its larger timber removed during the First World War. Medieval woodbanks run along the northern and southern boundaries. The wood stands mainly on boulder clay and Upper Lias clays with patches of glacial sands and gravels. There is thick undergrowth, dominated by hazel in places, with a rich ground flora and wide rides throughout the wood. There are some small streams flowing through the wood, and in the north, adjacent to the boundary, is a large area of bracken. The wood, which is a SSSI, is owned by the Diocese of Leicester and leased to the Forestry Commission.

Launde Park 106 ha SK7904
Parkland surrounding Launde Abbey, which is on the site of the former Augustinian Launde Priory, giving a long association with human habitation. The soils formed on boulder clay and Upper Lias clays have given rise to neutral grassland which is scattered with mature oaks and grazed by sheep and cattle. This pasture is classed as semi-improved. There is a small area of mixed plantation around a reservoir, several ponds and some marshy areas.

Lea Wood, Ulverscroft SK5011
This site was a meadow in 1287 when it formed part of the Lea Assart – an enclosed area of farmland with a bank and fence to separate it from common land. Part of it has been a wood since at least the mid 1700s but it was clear-felled between 1927 and 1957, and the oak woodland replaced with conifers and beech. A few broadleaved trees remain along the rides and around the edge of the wood. Since the wood is managed for timber, undergrowth is discouraged, but there is some bramble and bracken, with a little more diversity along the rides. A wetter area to the east of the site has a dense growth of alder with birch, hazel and willow. The wood is privately owned and there is no public access.

Leicester Cattle Market site SK5802
The cattle market was built in 1871. After the market was demolished in 1991 the concrete bases of the former buildings and animal pens were left, and a ruderal and tall-herb flora sprang up between them. The site is due to be developed for recreational use.

Lockington Meadows SK4829-SK4930
Wet meadows in the flood plain of the Rivers Soar and Trent which include Lockington Marsh SSSI. The meadows are a mosaic of wetland habitats with a large area of willow carr and some wet woodland. To the south are marshy areas with shallow pools, swamp and marshy grassland, the best of which are designated sites covering 9.2 ha. The marsh is grazed and is a rich area for invertebrates. In the north-west is a larger pool, fringed with reed-swamp.

Loughborough Meadows 63.5 ha SK5321
An area of flood meadow lying adjacent to the River Soar, most of which is traditionally managed for hay. It is cut in June or July, then grazed until the end of summer, with some smaller fields being used for grazing horses. The meadows have multiple ownership with complex commoners rights. Part is owned by the LRTNC and used as a nature reserve with access by permit only. The Summerpool Brook runs through the site and there is an oxbow lake. Recent flood prevention work on the adjacent river bank may affect the frequency of inundation. The meadows have a very rich flora and are a SSSI.

Loughborough Outwoods c. 81 ha SK5116
Ancient woodland standing on Pre-Cambrian and Triassic rocks. Known to be at least 500 years old, the woods are now planted partly with conifers, but these are being replaced gradually with native broadleaf trees. There is a wide variety of stand types from wet alder to dry oak/birch woodland. Streams flow through the wood and there are a number of high rocky outcrops. The ground flora contains an interesting variety of plants that are unusual for the county. In 1947, the wood was given to the people of Loughborough, "for their enjoyment", by Mr & Mrs A.Moss and Mr G.Bowler. The site is a SSSI and is managed as a country park by Charnwood Borough Council in consultation with LRTNC through the Charnwood Wildlife Project. Heavy public use has eroded some areas which have been fenced to allow them to recover.

Lount Grassland 1 Re-seeded sites:
Coleorton – SK3918
Worthington – SK3919
These are two areas of new, slightly acid, grassland on disturbed substrates, which have only a short history. One is a re-profiled colliery spoil tip in Coleorton Parish, and the other a restored opencast mine in Worthington Parish. The spoil tip ceased to be used for colliery waste in the 1960s, at which time spontaneous combustion in the clay/coal body of the tip was still evident. Much of the burnt shale was subsequently sold for civil engineering projects, and the remains of the tip were landscaped in the late 1980s and re-seeded with a wild flower mix. The site is now a local amenity area with ponds, grassy picnic areas and a wild flower meadow. Similarly, the opencast site was restored at the same time by replacement of the original overburden and topsoil, prior to grading, re-seeding and reversion to agricultural use.

Lount Meadow 2 Coleorton SSSI SK3819
A mosaic of slightly acid/neutral rough pasture with ant hills and banks of thorn scrub on the site of ancient coal pits. There is a marshy area in the north of the site surrounding a shallow seasonal pool. Management of the site is based on removal of an annual hay crop and control of the invading scrub. The land is privately owned.

Lount Meadow 3 Worthington SSSI SK3919
An old slightly acid/neutral herb-rich meadow with an area of marsh in one corner. This meadow is privately owned.

Luffenham Heath Golf Course, Rutland SK9502
This site, covering 87.34 ha, includes the largest piece of semi-natural limestone grassland in Leicestershire. The diverse range of grassland and heath habitats is the result of the site's varied geology. The lowest parts of the golf course are grassland on limestone soils. In places this is overlain by an area of sand-based soils, which supports acid heath species. This in turn has areas of further depositions, these being clay which give rise to neutral grassland. The golf course has been laid out across undulating countryside with large roughs that include herb-rich grassland, gorse and thorn scrub. Some woodland borders the fairways. Of note are the pockets of acidic soils that contain tormentil and patches of heather. This site supports a rich invertebrate population, with many species uncommon in the county having been recorded. The golf club is privately run on behalf of its members and the whole of the site is a SSSI.

Martinshaw Wood, Ratby 103 ha SK5107
A large wood, probably of ancient origin, that was originally dominated by sessile oak. Most of it was felled and replanted with a mixture of conifers and deciduous trees between 1954 and 1968. There is a small part of the original wood left in the parish of Groby. The vegetation of the wide rides is interesting and there are woodbanks, old marl pits and wet areas with some *sphagnum*. The wood has been purchased from the Forestry Commission by the Woodland Trust and is managed to combine wildlife conservation, timber production and education. There is a network of paths provided for public access.

Moira Junction, Ashby Woulds SK3015
A former branch line and sidings that connected the now abandoned Donisthorpe colliery to British Railways main line. The branch line closed in 1975 and the rails were removed. At the time the site was surveyed in 1992, it was being developed as an open space for the local community. The council had put down cinder paths and excavated several small ponds, but most of the existing vegetation that included willow scrub, thorn and gorse banks, heather and heath grassland had been left. There were also a number of bare areas which had been planted with small trees.

Nailstone Wiggs SK4208
Former ancient woodland with very deep layers of leaf-mould, much reduced by coal mining activities, such that, only a 15m wide marginal strip of mainly oak and birch remains at the former colliery site. The scrub layer is thick in parts. There are several steep-sided deep pools used to collect slurry from the spoil heaps, and one shaded pool used for fishing, which has some marginal vegetation.

Narborough Bog 9.5 ha SP5497
This SSSI and LRTNC nature reserve lies adjacent to the River Soar and contains a range of habitats that include reedbed, willow carr and wet woodland. It has the only substantial peat deposit in the county, which is about 180cm deep, overlying alluvium and river gravels. Major engineering work on the River Soar has lowered the water table of the bog over the past 20 years; thus, the largest natural reedbed in the county is drying out and being invaded by meadow-sweet. The woodland consists mainly of mature willow and alder, with oak and ash in the drier parts. A railway line separates the main bog from the southern third of the reserve, which contains willow scrub and two damp meadows that are cut for hay. There are plans to pump water back on to the reserve and regenerate the reedbed.

Newfield Heath, Ashby Woulds SK3115
Site of the old Newfield Colliery, abandoned in the 1880s and left as rural derelict land. It has regenerated naturally, producing a mosaic of habitats. Scrubby oak/ash woodland predominates, with ponds and marshy areas in the southern part of the site. In the northern corner is a small field grazed by horses, adjacent to which is a modest area of heath grassland containing several large patches of heather. Management involves mainly the removal of invading scrub and trees from the heath and wetland areas.

Newton Burgoland Marshes, Swepstone SK3808
An area of neutral grassland and species-rich marsh, on alluvial soils and boulder clay, alongside the River Sence. The site, which extends to 7.7 ha, is bounded by large hedges especially on the western side, with a wide ditch crossing the marsh. The southern part of the SSSI tends to be drier. It is privately owned and used for grazing.

North Brook, Rutland SK9413-SK9508
The brook rises west of Greetham and flows south-east through the Fort Henry lakes at Horn, to join the River Gwash at Empingham. Adjoining it is a narrow mosaic of limestone grassland, marsh, woodland and other habitats.

North Luffenham Quarry SK9603
A shallow but extensive linear quarry face is all that remains of a once much larger limestone quarry. There are areas of rabbit-grazed herb-rich limestone grassland, but much of the quarry is covered with dense thorn and elder scrub, with patches of tall herb. Several bee-hives are situated at its southern end, with the northern end used to store manure from a local farm. The quarry is privately owned and is designated as a SSSI.

Owston Wood SK7906
This is the largest continuous area of semi-natural woodland in the county. The wood, which stands predominantly on boulder clay and can be very wet in winter, holds a wide variety of woodland tree species, with ash/maple the predominant stand type. It is privately owned and was leased to the Forestry Commission who started planting in the 1960s, with both deciduous and coniferous species, including beech, sycamore, pine and oak. The undergrowth is diverse and dominated by hazel in most areas, with a rich and varied herb layer. There are extensive rides, and the shrubs from both sides are allowed to overgrow the rides to provide pathways for the resident dormice. The wood may hold the oldest living organism in Leicestershire – a coppiced small-leaved lime stool – which is over 6m in diameter and may be 800 years old. To the east of Owston Wood, across a minor road, is Little Owston Wood which is similar in composition. Both woods are a SSSI and cover about 141 ha. There is a footpath through the wood.

Pickworth Great Wood SK9815
One of the largest remaining blocks of deciduous woodland in Leicestershire, lying on a hilltop of heavy clay soils where drainage is locally impeded. This ancient woodland SSSI was originally predominantly ash/maple with ash/elm in the wetter parts, and was coppiced for many years, but the Forestry Commission who own the wood, have carried out extensive planting of conifers, ash, beech, birch and oak over the last seven decades. The undergrowth is quite varied, ranging from bare ground to rich herb and shrub layers, and there are wide rides with metalled roads for timber extraction. Bird and invertebrate life is abundant and varied with good lists of uncommon species; there is also a large deer population which causes some damage to the young trees. The wood is managed for commercial timber production and shooting.

Pignut Spinney Marsh, Loughborough SK524176
Unimproved, damp, species-rich grassland, 1.7 ha, lying on the Beacon Hill leg of the Woodbrook about 450 m north of Moat House. It is owned by Charnwood Borough Council and managed as a local nature reserve. The marsh is a remnant of the once-extensive marshy ground bordering the streams draining the lower slopes of Beacon Hill, Hangingstone and the Outwoods. Recently it has suffered from a diversion of its several feeder springs and the planting of ornamental trees along the upper footpath, which have caused significant drying out of the habitat. There is a small wet area within the reserve, that is mown annually to encourage the establishment of the marsh flora. This is the last known station in the Beacon Hill area for bog pimpernel, *Anagallis tenella*. The main habitat now is rough grass, which should be left undisturbed for the invertebrate fauna.

Puddledyke, Newtown Linford SK5411
Often known as Cropston Brickpit, this is a water-filled clay pit reputed to have been dug in the 1860s to make bricks for the Cropston Reservoir dam. The open water has become reduced by excessive aquatic and emergent vegetation spreading from the island and margins, and some dredging is necessary to preserve the site as a lake. The surrounding grassland is unspoilt and is bordered by tall hedges making a sheltered site which is excellent for invertebrates. It is owned by Severn Trent Water Plc and is part of the Cropston Reservoir SSSI.

River Chater SK8004-SK8403
A fairly small but fast flowing river in a deep valley, with a stony and gravelly bed. As it flows east, the river develops more meanders and muddy pools, and the margins vary from steep banks with sparse vegetation to reedswamp and woodland. Where tributary streams join the river there are marshy areas. The stretch above Launde Park Wood is in the Chater Valley SSSI. The River Chater finally joins the River Welland south-west of Tinwell (TF0005).

River Eye, Melton Mowbray SK7718-SK7918
One of the last remaining semi-natural and unpolluted stretches of river in Leicestershire, that still has natural features including riffles, pools, small cliffs and meanders.
The geology of its catchment area is comprised mainly of Jurassic and glacial boulder clays, but there is some influence from the Jurassic Limestone to the east. The 8 km stretch above Melton Mowbray is a designated SSSI, and for much of this length its banks are covered with reedswamp. There are small areas of willow scrub which add to the variety of habitats; the aquatic vegetation is also very diverse. This river is a Nature Conservation Review (NCR) site. There is no public access to most of its length.

River Sence Marsh, Bardon SK4512
The western River Sence has its prime source on the eastern flank of Bardon Hill. It flows westward, then in a south-westerly direction to join the River Anker at Atherstone. In its embryonic state, the River Sence embraces two small marshes on the south side of Bardon Hill, the first at Old Rise Rocks, and the second on the Hugglescote side of the main Leicester to Coalville Road (A50). The Old Rise marsh (SK4612) has a well-developed structure, typified by a varied field-layer flora containing meadowsweet, whilst the second (SK4512) is dominated by reedmace, having a sparse muddy substrate. Old Rise marsh is incorporated into the Bardon Hill study, and references to the River Sence marsh apply to the second site.

River Soar (north – Lockington to Rothley)
SK4930-SK5913
The River Soar to the north of Leicester is a slow moving river, flowing over a largely muddy and sandy substrate with a wide range of aquatic and emergent vegetation. It is used extensively for navigation and in places is canalised. From the Trent to Leicester the River Soar's water quality is Grade B-C, on a national grading system of A-F (where A is very good, and F is very poor). There are gravel pits and flood meadows along the banks, with oxbows, damp woodland and osier beds in places. The old willows and bankside vegetation provide a habitat and highway for many species of wetland spiders. Pitfall trapping of some of the exposed mud banks and sandbars to the south of Loughborough has produced some interesting species, such as *Erigone longipalpis*, *Erigone arctica* and *Prinerigone vagans*. There is easy access along the towpaths.

River Trent, N.W.Leicestershire SK4127-SK4429
For 12 km the river forms the county boundary with Nottinghamshire and there are extensive flood meadows at Lockington where the Soar joins the Trent. The River Trent is fairly fast flowing in places and has produced several large shingle banks and cut-offs where meanders have become isolated. It is not navigable upstream of Sawley, and the Trent and Mersey Canal, on the north side of the river, continues the navigation westwards. The river is owned and managed by British Waterways Board with access along the towpath and public footpaths. Where it borders Leicestershire, the Trent is classified as a Grade C river.

River Welland, Thorpe-by-Water SP8996
The River Welland, which rises at Sibbertoft, Northamptonshire, enters the county south-east of Husbands Bosworth (SP655833). For most of its association with the shire it forms the county boundary between Leicestershire and Northamptonshire until leaving the county south-east of Tinwell (TF018060). Within Leicestershire, the river has been straightened and deepened in its eastern portion, with its surrounding catchment area having been improved agriculturally. Despite these drainage schemes, the river still manages to flood large areas of farmland occasionally during the winter period. The only remaining piece of unimproved grassland along the River Welland is at Seaton (see Seaton Meadows). The river's water quality is GradeB-C, on the national grading system.

River Wreake, Rotherby SK6716-SK6817
The river is slow flowing and meandering in this stretch, and was straightened in the 1790s for navigation, leaving several cut-offs and marshy areas with tussock sedge. These marshes have been grazed but are otherwise undisturbed and have a good flora. The Wailes, at Frisby-on-the-Wreake, is part of the Frisby Marsh SSSI.

Rutland Water   SK80/90
This is one of the largest man-made water bodies in Britain. It covers 1255 ha, is over 34m deep and has a shoreline of 41 km. The reservoir was completed in 1977 and in addition to water storage it is used for water sports and fishing. There is a cycle track around the entire perimeter with picnic areas and, at the western end, two nature reserves. The reserves, which are managed by the LRTNC, contain a range of habitats including lagoons, islands, mudflats, reedswamp, marsh, old meadows, scrub and woodland. There are hides for public use and access is by day or season permit. Most of the other margins of the reservoir are semi-improved sheep pasture. The reservoir, which is a SSSI and a Ramsar site, is an important wintering area for wildfowl. It is owned by Anglian Water plc.

Saddington Reservoir   SP6691
This body of water was built in the 1790s on Lower Lias clays as a feeder for the Grand Union Canal and is owned by British Waterways. It has a variable water level with the margins consisting of willow carr, swampy sedge beds and bare mud with moss cover in places. The marginal invertebrate life is varied and interesting and there is some aquatic vegetation. The small marina is used mainly for dinghy sailing.

Seaton Meadow   SP9198
This traditionally-managed hay meadow is on alluvial soils near the River Welland and has escaped drainage despite the flood alleviation measures on the River. The area is 11.43 ha and contains a wide variety of grassland species. The site has some drier slopes as well as extensive wet flushes and smaller wet hollows. It is a SSSI.

Shacklewell Hollow, Tinwell   SK9707
An area of neutral marsh on peat and alluvial soils along a tributary of the River Gwash. The stream cuts into Northamptonshire Sands and Jurassic Lincolnshire Limestone, which gives rise to some calcareous flushes on the valley sides. There is a mosaic of habitats including a series of ponds, alder woodland, calcareous grassland, marsh and boggy pasture. The 4 ha site is privately owned and is a SSSI.

Sharnford Meadows,
see Fosse Meadows Nature Park

Sheet Hedges Wood, Newtown Linford   SK5208
An ancient woodland site covering 25.57 ha which, although much reduced by quarrying, is a SSSI. There are no old trees and it is thought that the saleable timber was felled in the early part of this century. The basic woodland type is ash/maple with pedunculate oak on clay soils, consisting of boulder clay and Keuper Marl. Sycamore has invaded the northern part, and in the south on wetter soils alder dominates. The shrub layer is very thick in parts and a more open area in the centre is bracken covered. The ground flora is rich and varied. A deep ditch marks the western boundary. Sheet Hedges Wood is owned by Leicestershire County Council; there is no public access, although a public footpath runs along part of the boundary.

Six Hills, see Twenty Acre Piece

Skeffington Wood, Tilton   SK7503
Ancient woodland on the north bank of the Eye Brook. The soils are varied, with clays, glacial sand and gravel, and are mainly heavy and calcareous. The original woodland was probably ash/hazel with maple/ash but it has been extensively planted with deciduous and coniferous species, including stands of oak. The hazel shrub layer was once coppiced and the ground flora under the broad-leaved trees is interesting. Small streams run through the wood and there is some alder in the wetter parts. The wood is now privately owned but still managed by the Forestry Commission. It is part of the Leighfield Forest SSSI.

St Mary's Allotments, Leicester   SK5802
Former allotments that were officially closed in 1992. At the time of the survey later that year, most of the plots were overgrown with ruderal vegetation or rough grass, brambles with wild garden crops, fruit trees and a little invading scrub. The area is owned by the City Council who intend to develop the site under an urban renewal scheme, with plans that include industrial units and a mown area for recreation.

Stonesby Quarry Nature Reserve   SK8125
This reserve covers an area of 3.6 ha and is the southern section of a worked-out quarry on Lincolnshire Limestone. Here, a rich limestone flora has developed on the thin soil. Much of the northern section has been used as a refuse tip. During the Second World War soil was imported to part of the site for the growing of crops, and some of this has been scraped off to leave bare limestone rock for plants to recolonise. There are several areas of hawthorn scrub and rank grass around the perimeter, and across the centre of the reserve. The area, which is a SSSI, is owned by the Diocese of Leicester and leased to the LRTNC, with access by permit only. Bees have been kept on the site for many years.

Stoneywell Wood, Ulverscroft   SK4911
One of the few remaining unspoilt sessile oak woods in the county with trees between 100 and 150 years old. Other tree species include birch and mature rowan, with a shrub layer of hazel, hawthorn and aspen. Some replanting has taken place. The herb layer is diverse. There is a pond on the eastern side. The wood is part of the mosaic of habitats that make up the Ulverscroft Valley SSSI and is privately owned with no public access.

Swithland Reservoir 66 ha   SK5514
Completed in 1896, this reservoir is fed by the overflow stream from Cropston Reservoir and a stream that runs through Swithland village, on which stood the old water mill. The reservoir, which lies on Keuper Marl, is owned by Severn Trent Water Plc and is subject to large fluctuations in water level. Parts of the reservoir are becoming very shallow due to silt deposits. The marginal tall-fen communities are extensive but there are muddy and stony shores which provide good habitat for wintering waterfowl and waders. There are some areas of willow and alder carr and wet woodland, and the inflow end is a dense reedswamp. The reservoir is part of the Buddon Wood – Swithland Reservoir SSSI.

Swithland Wood SK5312

This ancient woodland of about 70 ha stands on acidic loamy soils, derived mainly from marls and sandstones of the Triassic period overlying Pre-Cambrian rocks. The woodland is dominated by sessile oak, birch and small-leaved lime with some of the best stands of oak/lime and alder in the county. There is an open shrub layer, dominated by hazel with a herb-rich ground flora in its northern part. The wood contains a small area of unimproved grassland that is occasionally grazed. Within the wood are two abandoned slate quarries which are water-filled, one being used by a diving club. Weathered spoil heaps of broken slate provide an important scree habitat. Small streams run in shallow valleys and there are several wet hollows. In the south of the wood a stream runs on alluvial soils, supporting a stretch of alder woodland. The wood, which is a SSSI, was bought from Mr W. Gimson by the Rotary Club of Leicester and, in line with Gimson's wishes, was given to the people of Leicestershire in 1931. It is now managed as a country park by the County Council, with open access at all times. A part of the east side of the wood is privately owned and the shrub layer here is very thick.

Thistleton Gullet SK8918

A former ironstone quarry in Jurassic limestone owned by British Steel Pension Fund. Most of the quarry has been backfilled leaving a gully with rocky cliffs on the north and slopes of clay and scree to the south. There is a long pool at the bottom of the gully, with a shore partly of rock-strewn mud, with some emergent vegetation and reedswamp overhung by willows. At the east end of the site, farm refuse has been dumped making the water body nutrient rich. On the west side is thick willow scrub.

Tugby Wood SK7602

An ancient wood of about 18 ha which was felled during the First World War; it has been allowed to regenerate and has had a small amount of planting. The original wood was thought to be maple/ash or ash/hazel, but willow, birch and sycamore have invaded. Conifers have been planted in the east of the wood together with some deciduous species. The rides are broad and have a rich flora, and the general ground flora of the woodland is good. Two streams run through the site and join the Eye Brook to the north-east of the wood. A pond has been created by damming one of the streams. The wood is part of the Leighfield Forest SSSI and is privately owned with no public access.

Twenty Acre Piece, Burton-on-the-Wolds SK6421

An 8 ha parcel within a larger area of common land at Six Hills, the site of an ancient cross-roads, with poorly drained acidic soils derived from glacial boulder clay. The site contained remnants of old *Calluna* heath until a fire in 1967 destroyed this. Today, the site is a mosaic of damp acid grassland, scrub and secondary woodland; there is also a small pond in one corner. It is owned by the Duke of Somerset and leased to the LRTNC as a nature reserve, with access to members. The reserve is a SSSI by virtue of its acid grassland, which is usually cut and cleared annually. The drift geology around Six Hills is complicated by chalky boulder clay deposits, resulting in patches of calcicolous plants in an otherwise heathy environment.

Ulverscroft Nature Reserve 60 ha SK4912

A large area of mature and young plantation, heathland and grassland, streams and ponds. The lower parts of the reserve are on Keuper Marl, with an outcrop of Syenite and Pre-Cambrian rocks forming the higher ground to the south-west. Poultney Wood is on an ancient woodland site which has been replanted. The plantations are mostly of oak with some beech and conifers. The moorland and the small fields in the centre of the reserve have both been invaded by scrub since grazing ceased in the 1950s, so that the moorland is now dominated by bracken. In the valley are marsh and damp meadows with an area of wet alder woodland. Part of the reserve is owned by the National Trust and part by the LRTNC who manage it as a nature reserve with access for members. Most of the reserve lies in the Ulverscroft Valley SSSI.

Wanlip Gravel Pits, Syston SK6011

North of the Wanlip to Syston road is a series of flooded gravel pits – a continuation of the gravel workings that form Watermead Park – with reed and sedge beds and scrub around the margins. There is some rough grassland between the pits. The area is still being worked for gravel and there is no public access.

Watermead Country Park, Leicester SK6011

A complex of flooded gravel pits that have been developed for recreational use and nature conservation on the former site of unimproved flood meadows. The complex stretches from the northern corner of the city boundary, northwards to the village of Wanlip. There is a range of habitats including open water, sedge and reed beds, inundation meadow, willow carr and wet woodland. Also the River Soar and the Grand Union Canal meander through the park. This area has been popular with naturalists since gravel extraction began in the early 1950s, when it was known as Birstall or Wanlip Gravel Pits. It was not until 1990 that the County Council declared it a country park, with public access at all times, since when a number of car parks have been built.

Welby Osier Beds, Asfordby SK719210

Former dense wet woodland containing alder, ash and oak, with extensive bryophyte cover. It borders a swampy area around a stream supporting very large willows. The site has now been incorporated into the industrial surface of the new Asfordby Mine.

Wymondham Rough Nature Reserve SK8317

A site of 12.5 ha near the Leicester to Peterborough railway line. It contains a variety of habitats including old canal, ponds, species-rich neutral grassland, broadleaved and coniferous woodland and marsh. boulder clay, Lower Lias clay and alluvium have produced a calcareous clay soil. Part of the reserve, which includes unimproved grassland, is a SSSI and the whole of the site is owned by LRTNC, with access for members. The grassland is managed for hay and the woodlands are being selectively felled and replanted to improve their species range.

APPENDIX II

Forty-nine Key Sites in the County, and their Spiders

Species lists of spiders for 49 Leicestershire sites, grouped in 7 main habitat categories, are given in the following tables. The sites are not strictly comparable, but have been assembled so that the species lists will have characteristic affinities. The main purpose of the tables is to present species lists for as many Leicestershire sites as possible, over a wide range of habitats.

Table a) Calcareous and Heath Grassland p.215
Geeston Quarry, Ketton
Ketton Quarry
Luffenham Heath Golf Course
The Drift, Croxton Kerrial/Sproxton
King Lud's Entrenchments, Sproxton
Harby Hills
Stonesby Quarry

Table b) Coal Measures Heath Grassland p.219
Newfield Heath, old colliery site, Ashby Woulds
Moira Junction, Ashby Woulds
Lount 1, re-seeded opencast sites
Lount 2, Coleorton, heathy grassland
Lount 3, Worthington, marshy meadow
Coalfield West, restored opencast site
Acresford Sandpit, Donisthorpe

Table c) Moorland and Parkland p.222
Bardon Hill
High Sharpley & Gun Hill, Whitwick
Charnwood Lodge N.R., Charley
Cademan Moor, Grace Dieu
Ulverscroft N.R.
Bradgate Park
Donington Park
Beacon Hill, Woodhouse

Table d) Riparian Habitats p.226
River Trent, including Lockington Marsh
River Soar
Narborough Bog
Aylestone Flood Meadows
Watermead Riverside Park, Wanlip
Barrow Gravel Pits
Loughborough Meadows

Table e) Reservoirs and Open Water p.229
Swithland Reservoir
Cropston Reservoir
Blackbrook Reservoir
Groby Pool
Rutland Water
Eyebrook Reservoir
Saddington Reservoir

Table f) Woodland p.232
Buddon Wood, including Buddon Marsh, Quorn
Swithland Wood
Loughborough Outwoods
Stoneywell Wood, Ulverscroft
Burbage Wood
Owston Wood
Launde Big Wood
Pickworth Great Wood

Table g) Garden and House Spiders p.236
Knighton Church Road, Leicester
The Portwey, Leicester
Scraptoft Lane, Leicester
Outwoods Drive, Loughborough
Bramcote Road, Loughborough

Schedule of the 49 sites for which species lists are given in the following tables, presented in rank order as determined by the species rarity score.

	Table	Records	Species	Local	Nb	Na	Red Data Book	Score
Buddon Wood	f)	1394	191	54	2		IRDB3	70
Ketton Quarry	a)	287	120	28	3	1		46
Bardon Hill	c)	906	147	34			IRDB3	42
Charnwood Lodge N.R.	c)	952	162	36	1			40
Luffenham Heath	a)	238	88	23	2	1		37
Geeston Quarry	a)	1134	114	28	2			36
Bradgate Park	c)	347	108	15			IRDB3 IRDBK	31
Swithland Reservoir	e)	550	119	26	1			30
Donington Park	c)	310	63	7			IRDB3 IRDB2	27
High Sharpley	c)	314	75	17			IRDB3	25
King Lud's Entrenchments	a)	253	77	18	1			22
Swithland Wood	f)	317	114	22				22
River Soar	d)	809	92	21				21
Moira Junction	b)	204	68	15	1			19
Owston Wood	f)	147	62	10	2			18
Narborough Bog	d)	305	72	10	2			18
Aylestone Meadows	d)	487	76	13	1			17
Stonesby Quarry	a)	206	63	13	1			17
Cademan Moor	c)	252	78	17				17
Ulverscroft N.R.	c)	388	107	15				15
Pickworth Great Wood	f)	82	54	10	1			14
Beacon Hill	c)	58	44	6			IRDB3	14
Newfield Heath	b)	200	72	14				14
Groby Pool	e)	198	63	9	1			13
Loughborough Meadows	d)	330	62	13				13
Watermead Country Park	d)	237	55	8	1			12
Stoneywell Wood	f)	125	60	12				12
Eyebrook Reservoir	e)	74	43	11				11
Burbage Wood	f)	72	44	7	1			11
Launde Big Wood	f)	75	46	7	1			11
Harby Hills	a)	49	29	6	1			10
Barrow Gravel Pits	d)	75	34	10				10
Lount Grassland (1)	b)	187	46	10				10
The Drift	a)	190	56	10				10
Loughborough Outwoods	f)	153	65	10				10
Acresford Sandpit	b)	123	45	9				9
Blackbrook Reservoir	e)	154	48	9				9
Cropston Reservoir	e)	104	49	9				9
Lount Meadow (2)	b)	175	54	9				9
Garden 3, Leicester	g)	86	74	9				9
River Trent	d)	80	40	8				8
Rutland Water	e)	121	56	8				8
Saddington Reservoir	e)	146	39	6				6
Coalfield West	b)	148	45	2	1			6
Garden 4, Loughborough	g)	181	59	6				6
Garden 1, Leicester	g)	110	49	5				5
Garden 5, Loughborough	g)	63	43	3				3
Lount Meadow (3)	b)	66	30	3				3
Garden 2, Leicester	g)	31	23	1				1

The symbols used in the following tables indicate the species status:
♦ Common species,   ★ Local species,   ★ Nationally notable or rare species.   (♂ dimorphic male.)

Table a) Calcareous and Heath Grassland

Spiders recorded from
Geeston Quarry; Ketton Quarry; Luffenham Heath Golf Course;
The Drift; King Lud's Entrenchments; Harby Hills; Stonesby Quarry.

		Geeston	Ketton	Luffenham	Drift	King Lud's	Harby	Stonesby
Amaurobius fenestralis		♦						
.... similis		♦	♦					
.... ferox		♦						
Dictyna arundinacea			♦	♦	♦	♦		
.... uncinata		♦	♦	♦				
.... latens	L			★				
Lathys humilis	L		★	★				
Dysdera erythrina		♦	♦					
.... crocata		♦	♦					♦
Segestria senoculata		♦			♦			
Drassodes lapidosus		♦	♦					♦
.... cupreus		♦	♦	♦		♦	♦	♦
Haplodrassus signifer	L	★						
Zelotes latreillei	L	★	★		★	★	★	★
Drassyllus pusillus	L	★			★	★	★	
Micaria pulicaria		♦	♦		♦	♦		♦
Clubiona reclusa			♦	♦	♦			♦
.... pallidula				♦				
.... terrestris		♦						
.... lutescens			♦	♦				
.... comta		♦	♦	♦				
.... brevipes				♦				
.... diversa	L					★		★
Cheiracanthium virescens	L	★						
Agroeca proxima		♦				♦		
Phrurolithus festivus		♦	♦			♦	♦	
Zora spinimana		♦	♦	♦		♦		
Anyphaena accentuata			♦	♦				
Diaea dorsata	L			★				
Xysticus cristatus		♦	♦		♦	♦	♦	♦
.... audax			♦	♦				
.... erraticus	L		★					
.... ulmi	L		★	★		★		
Ozyptila scabricula	Nb	★						
.... sanctuaria	L	★	★					
.... praticola	L	★						
.... trux					♦	♦		
.... atomaria			♦					
.... brevipes	L		★	★	★			
Philodromus dispar			♦	♦				
.... aureolus			♦	♦				
.... cespitum		♦	♦	♦				
.... collinus	Nb		★	★				
Tibellus oblongus			♦	♦	♦			
Salticus scenicus			♦					♦
.... cingulatus						♦		
Heliophanus flavipes			♦	♦		♦		♦
Euophrys frontalis			♦	♦	♦			
.... aequipes	L	★	★					
Sitticus pubescens	L							★
Evarcha falcata			♦					
Pardosa agrestis	Nb	★						
.... palustris		♦	♦		♦	♦	♦	♦

APPENDIX II

Calcareous and Heath Grassland

Species		Geeston	Ketton	Luffenham	Drift	King Lud's	Harby	Stonesby
Pardosa pullata		♦	♦	♦	♦	♦	♦	♦
.... *prativaga*		♦	♦	♦	♦	♦	♦	♦
.... *amentata*		♦	♦	♦	♦	♦	♦	♦
.... *nigriceps*		♦	♦	♦	♦	♦	♦	♦
Alopecosa pulverulenta		♦	♦		♦	♦	♦	♦
.... *barbipes*		♦						
Trochosa ruricola		♦	♦	♦	♦	♦	♦	♦
.... *robusta*	Nb		★					
.... *terricola*		♦	♦		♦	♦	♦	♦
Arctosa perita	L	★						
Pirata hygrophilus						♦		
.... *uliginosus*	L		★		★	★		★
Pisaura mirabilis		♦	♦	♦	♦	♦		♦
Agelena labyrinthica		♦	♦					
Textrix denticulata		♦						♦
Tegenaria gigantea		♦						
.... *agrestis*	L	★						
Coelotes atropos							♦	♦
Cicurina cicur	L	★	★					
Hahnia montana			♦			♦		
.... *nava*	L	★	★			★	★	★
Ero cambridgei			♦			♦		
.... *furcata*		♦	♦			♦		
Episinus angulatus	L	★	★					
Crustulina guttata	L					★		
Steatoda bipunctata			♦	♦				
Anelosimus vittatus		♦	♦	♦				
Achaearanea lunata	L			★				
Theridion sisyphium		♦	♦	♦				
.... *impressum*	L	★	★	★				
.... *varians*			♦	♦				
.... *simile*				♦				
.... *mystaceum*		♦		♦				
.... *tinctum*	L		★	★				
.... *bimaculatum*		♦	♦	♦	♦	♦		
.... *pallens*		♦	♦	♦				
Enoplognatha ovata		♦	♦	♦				♦
.... *thoracica*	L	★	★			★	★	★
Robertus lividus					♦			
.... *neglectus*	L			★				
Tetragnatha extensa			♦	♦				
.... *pinicola*	Nb		★	★				
.... *montana*			♦	♦				
.... *obtusa*	L		★					
.... *nigrita*	L		★	★				
Pachygnatha clercki							♦	♦
.... *degeeri*		♦	♦	♦	♦	♦	♦	♦
Metellina segmentata		♦	♦	♦	♦			♦
.... *mengei*		♦	♦	♦	♦	♦		
.... *merianae*				♦				
Gibbaranea gibbosa	L			★				
Araneus diadematus		♦	♦	♦				
.... *quadratus*				♦				
Larinioides cornutus				♦				
.... *patagiatus*	L			★				
Nuctenea umbratica								♦
Agalenatea redii	L		★					
Araniella cucurbitina		♦	♦	♦				
.... *opistographa*	L		★	★				

Calcareous and Heath Grassland

		Geeston	Ketton	Luffenham	Drift	King Lud's	Harby	Stonesby
Hypsosinga pygmaea	L		★	★	★	★		
Cercidia prominens	L		★	★				
Cyclosa conica	L			★				
Ceratinella brevipes							♦	
.... brevis		♦	♦			♦		
.... scabrosa	L	★	★		★	★		
Walckenaeria acuminata		♦	♦		♦	♦		♦
.... antica		♦	♦		♦	♦		
.... atrotibialis	L		★		★	★		★
.... incisa	Nb					★		
.... dysderoides	L				★	★		
.... unicornis						♦		
.... cuspidata		♦						
Dicymbium nigrum			♦	♦	♦			♦
Entelecara acuminata				♦				
.... erythropus	L	★						
Hylyphantes graminicola	L	★	★	★				
Gongylidium rufipes		♦	♦	♦			♦	
Dismodicus bifrons			♦	♦		♦		♦
Hypomma cornutum			♦	♦				
Gonatium rubens		♦	♦	♦	♦	♦		
Maso sundevalli		♦	♦		♦			
.... gallicus	Na		★	★				
Peponocranium ludicrum		♦	♦					
Pocadicnemis juncea		♦	♦	♦	♦	♦		♦
Oedothorax fuscus		♦	♦	♦	♦			♦
.... agrestis	L							★
.... retusus			♦		♦		♦	♦
.... apicatus	L	★		★	★			
Pelecopsis parallela	L	★				★		
Silometopus reussi	L	★						
Cnephalocotes obscurus	L	★				★		
Ceratinopsis stativa	L					★		★
Tiso vagans								♦
Troxochrus scabriculus	L	★				★		★
Thyreosthenius parasiticus	L	★						
Monocephalus fuscipes		♦	♦		♦	♦		♦
Saloca diceros	Nb						★	
Gongylidiellum vivum								♦
.... latebricola	L		★					
Micrargus herbigradus					♦	♦	♦	♦
.... subaequalis	L		★	★			★	★
Erigonella hiemalis		♦	♦	♦	♦			♦
Savignia frontata		♦		♦				
Diplocephalus cristatus		♦			♦			♦
.... permixtus			♦			♦		
.... latifrons		♦	♦	♦	♦	♦		♦
.... picinus		♦						
Panamomops sulcifrons	L	★	★			★		
Typhochrestus digitatus	L	★			★			
Milleriana inerrans	L			★				
Erigone dentipalpis		♦	♦	♦	♦	♦	♦	♦
.... atra		♦	♦	♦		♦		♦
.... arctica	L	★						
Porrhomma pygmaeum		♦		♦				
.... microphthalmum	L	★	★	★				★
.... errans	Nb							★
Meioneta rurestris		♦	♦	♦	♦	♦		♦
.... saxatilis		♦	♦		♦	♦		♦

APPENDIX II

Calcareous and Heath Grassland

		Geeston	Ketton	Luffenham	Drift	King Lud's	Harby	Stonesby
Microneta viaria		♦			♦			
Centromerus sylvaticus		♦	♦		♦	♦		
.... *prudens*						♦		
.... *dilutus*		♦	♦	♦				
Centromerita bicolor		♦			♦			
.... *concinna*		♦				♦		
Saaristoa abnormis			♦			♦		♦
Bathyphantes gracilis		♦	♦	♦	♦	♦	♦	♦
.... *parvulus*			♦		♦	♦		♦
Kaestneria dorsalis				♦				
.... *pullata*				♦	♦	♦		
Diplostyla concolor		♦	♦		♦	♦	♦	♦
Poeciloneta variegata	L			★				
Tapinopa longidens						♦		
Stemonyphantes lineatus		♦						♦
Lepthyphantes obscurus	L			★				
.... *tenuis*		♦	♦	♦	♦	♦	♦	♦
.... *zimmermanni*		♦	♦					
.... *mengei*						♦		
.... *flavipes*		♦						
.... *ericaeus*		♦	♦		♦	♦		♦
.... *pallidus*	L	★	★		★	★		★
Linyphia triangularis		♦	♦	♦				
.... *hortensis*		♦	♦	♦		♦		
.... *montana*		♦	♦	♦				
.... *clathrata*		♦	♦			♦		
.... *peltata*			♦	♦				♦
Microlinyphia pusilla					♦	♦		♦

	Geeston	Ketton	Luffenham	Drift	King Lud's	Harby	Stonesby
TOTALS (species)	114	120	88	56	77	29	63
LOCAL species	28	28	23	10	18	6	13
NOTABLE Nb	2	3	2	0	1	1	1
NOTABLE Na	0	1	1	0	0	0	0
SITE STATUS (rarity score)	36	46	37	10	22	10	17

Table b) Coal Measures Heath Grassland

Spiders recorded from
Newfield Heath Old Colliery Site; Moira Junction;
Lount 1 – re-seeded opencast sites; Lount 2 – Coleorton SSSI heathy grassland;
Lount 3 – Worthington SSSI marshy meadow; Coalfield West; Acresford Sandpit.

Species	Status	Newfield	Moira	Lount 1	Lount 2	Lount 3	Coalfield	Acresford
Amaurobius fenestralis				♦		♦		
Dictyna arundinacea		♦	♦		♦			
Oonops pulcher		♦					♦	
Drassodes cupreus		♦	♦	♦				
Haplodrassus signifer	L	★	★	★				
Zelotes latreillei	L	★	★					★
.... *apricorum*	L	★						
Micaria pulicaria		♦	♦					♦
Clubiona reclusa				♦	♦	♦		
.... *neglecta*	L			★	★			★
.... *diversa*	L	★	★					
Cheiracanthium virescens	L		★					
Agroeca proxima		♦		♦				
Phrurolithus festivus		♦	♦					♦
Xysticus cristatus		♦	♦	♦	♦	♦	♦	♦
Ozyptila trux							♦	
Philodromus cespitum		♦	♦					
Tibellus oblongus								♦
Euophrys frontalis		♦	♦					♦
.... *aequipes*	L							★
Pardosa palustris		♦	♦	♦	♦	♦	♦	♦
.... *pullata*		♦	♦	♦	♦	♦	♦	♦
.... *prativaga*		♦	♦	♦	♦	♦	♦	♦
.... *amentata*		♦		♦	♦	♦	♦	♦
.... *lugubris*						♦		
Alopecosa pulverulenta		♦	♦	♦	♦	♦		♦
Trochosa ruricola		♦	♦	♦	♦	♦		♦
.... *terricola*		♦	♦	♦	♦	♦	♦	
Arctosa perita	L	★	★					★
Pirata piraticus	L				♦			
.... *hygrophilus*		♦		♦	♦			
Pisaura mirabilis					♦			♦
Argyroneta aquatica	L	★						
Textrix denticulata			♦					
Tegenaria gigantea			♦					
.... *agrestis*	L							★
Cicurina cicur	L			★				
Hahnia montana		♦						
.... *nava*	L		★					
Ero cambridgei		♦	♦					
.... *furcata*							♦	
Euryopis flavomaculata	L		★					
Anelosimus vittatus			♦					
Achaearanea simulans	Nb		★					
Theridion sisyphium		♦	♦		♦			♦
.... *bimaculatum*		♦	♦	♦	♦			♦
.... *pallens*								♦
Enoplognatha ovata		♦	♦		♦		♦	
.... *thoracica*	L	★	★	★	★			
Robertus lividus		♦					♦	♦
Pholcomma gibbum		♦						
Tetragnatha extensa		♦						

Coal Measures Heath Grassland

Species		Newfield	Moira	Lount 1	Lount 2	Lount 3	Coalfield	Acresford
Tetragnatha montana		♦	♦					
.... *nigrita*	L		★					
Pachygnatha clercki			♦	♦	♦	♦	♦	
.... *degeeri*		♦	♦	♦	♦	♦	♦	♦
Metellina mengei		♦						
Nuctenea umbratica			♦					
Araniella opistographa	L		★					
Ceratinella brevipes					♦			
.... *brevis*		♦	♦	♦	♦		♦	
Walckenaeria acuminata		♦	♦			♦	♦	
.... *antica*		♦	♦		♦			♦
.... *cucullata*		♦						
.... *dysderoides*	L	★						
.... *nudipalpis*				♦	♦	♦	♦	
.... *furcillata*	L				★			
.... *unicornis*							♦	
.... *vigilax*	L			★				
Dicymbium nigrum	L	♦	♦	♦	♦		♦	♦
.... *tibiale*	L				★			
Gnathonarium dentatum					♦			
Dismodicus bifrons			♦					
Hypomma bituberculatum					♦			
Gonatium rubens		♦	♦				♦	
Maso sundevalli		♦	♦				♦	
Pocadicnemis pumila								♦
.... *juncea*		♦	♦	♦	♦	♦	♦	♦
Oedothorax gibbosus					♦			
(.... *tuberosus*)					♂			
.... *fuscus*				♦	♦	♦	♦	♦
.... *retusus*				♦	♦	♦	♦	♦
.... *apicatus*	L			★				
Pelecopsis parallela	L							★
Silometopus elegans	L	★			★			
Cnephalocotes obscurus	L	★	★		★			★
Ceratinopsis stativa	L				★			
Evansia merens	L	★						
Tiso vagans		♦	♦		♦			
Tapinocyba praecox	L	★	★					
Monocephalus fuscipes		♦	♦	♦			♦	♦
Lophomma punctatum	L				★			
Gongylidiellum vivum		♦			♦		♦	
Micrargus herbigradus		♦	♦	♦			♦	♦
.... *subaequalis*	L		★		★			★
Savignia frontata					♦	♦		
Diplocephalus permixtus					♦	♦		
.... *latifrons*			♦					
.... *picinus*		♦	♦					
Araeoncus humilis	L			★				
Milleriana inerrans	L			★			★	
Erigone dentipalpis		♦	♦	♦	♦	♦	♦	♦
.... *atra*		♦	♦	♦	♦	♦	♦	♦
Leptorhoptrum robustum	L				★			
Ostearius melanopygius							♦	♦
Porrhomma pygmaeum				♦	♦		♦	
.... *campbelli*	L	★						
.... *microphthalmum*	L			★	★	★		
Agyneta conigera							♦	
.... *decora*	L							★
.... *ramosa*	L		★					

Coal Measures Heath Grassland

		Newfield	Moira	Lount 1	Lount 2	Lount 3	Coalfield	Acresford
Meioneta rurestris			♦	♦			♦	♦
.... *saxatilis*		♦	♦		♦	♦		
.... *beata*	L			★				
Microneta viaria		♦	♦	♦			♦	
Centromerus sylvaticus		♦	♦				♦	♦
.... *dilutus*		♦	♦					
Tallusia experta					♦			
Centromerita bicolor				♦	♦	♦	♦	
.... *concinna*		♦	♦	♦				
Saaristoa abnormis		♦						
Bathyphantes approximatus					♦			
.... *gracilis*		♦	♦	♦	♦	♦	♦	♦
.... *parvulus*		♦	♦	♦	♦	♦	♦	♦
Diplostyla concolor		♦	♦		♦	♦	♦	♦
Stemonyphantes lineatus		♦	♦					♦
Lepthyphantes tenuis		♦	♦	♦	♦	♦	♦	♦
.... *zimmermanni*							♦	
.... *ericaeus*		♦	♦		♦	♦	♦	♦
.... *pallidus*	L	★	★				★	
.... *insignis*	Nb						★	
Linyphia triangularis		♦						
.... *clathrata*		♦	♦		♦		♦	
Microlinyphia pusilla		♦	♦					
TOTALS (species)		72	68	46	54	30	45	45
LOCAL species		14	15	10	9	3	2	9
NOTABLE Nb		0	1	0	0	0	1	0
SITE STATUS (rarity score)		14	19	10	9	3	6	9

APPENDIX II

Table c) Moorland and Parkland

Spiders recorded from
Bardon Hill; High Sharpley and Gun Hill; Charnwood Lodge Nature Reserve;
Cademan Moor; Ulverscroft N.R.; Bradgate Park; Donington Park; Beacon Hill

		Bardon	Sharpley	C.L.N.R.	Cade.	Ulvers.	Bradgate	Don.	Beacon
Amaurobius fenestralis		♦	♦	♦		♦	♦	♦	♦
.... *similis*		♦		♦			♦		
.... *ferox*		♦							
Dictyna arundinacea					♦	♦			
Oonops pulcher		♦	♦	♦			♦	♦	♦
Dysdera erythrina							♦		
Harpactea hombergi	L	★	★	★	★		★	★	★
Segestria senoculata		♦		♦			♦	♦	
Drassodes lapidosus		♦							
.... *cupreus*		♦	♦	♦		♦	♦		♦
Haplodrassus signifer	L	★	★	★	★				
Zelotes latreillei	L	★	★	★	★				★
Micaria pulicaria		♦	♦	♦					♦
Clubiona corticalis				♦			♦	♦	
.... *reclusa*		♦	♦	♦		♦	♦	♦	
.... *stagnatilis*		♦							
.... *terrestris*		♦		♦		♦	♦		
.... *lutescens*				♦		♦	♦		
.... *comta*		♦		♦		♦	♦	♦	
.... *brevipes*				♦		♦	♦		
.... *trivialis*			♦	♦					
.... *diversa*	L	★	★	★	★	★	★		
Agroeca proxima		♦	♦	♦	♦	♦			
Zora spinimana			♦	♦		♦	♦		♦
Xysticus cristatus		♦	♦	♦	♦	♦	♦		♦
.... *erraticus*	L		★						
Philodromus aureolus		♦		♦				♦	
.... *cespitum*				♦					♦
Tibellus oblongus				♦		♦	♦		
Salticus scenicus				♦			♦		
.... *cingulatus*		♦				♦			
Heliophanus flavipes							♦		
Neon reticulatus		♦		♦		♦	♦	♦	♦
Euophrys frontalis		♦	♦	♦		♦	♦		
.... *erratica*	L	★		★			★	★	★
.... *aequipes*	L		★						
Sitticus pubescens	L	★		★					
Pardosa palustris		♦	♦	♦					♦
.... *pullata*		♦	♦	♦	♦	♦	♦	♦	♦
.... *amentata*		♦	♦	♦		♦	♦		
.... *nigriceps*		♦	♦	♦	♦	♦	♦		
.... *lugubris*		♦	♦	♦	♦	♦	♦		
Alopecosa pulverulenta		♦	♦	♦	♦	♦	♦	♦	
Trochosa ruricola							♦	♦	
.... *terricola*		♦	♦	♦	♦	♦	♦	♦	
Pirata piraticus		♦		♦		♦			
.... *hygrophilus*		♦		♦		♦			
.... *latitans*	L					★	★		
Pisaura mirabilis				♦			♦		
Argyroneta aquatica	L			★					
Textrix denticulata		♦	♦	♦		♦	♦		
Tegenaria domestica		♦		♦					
Coelotes atropos		♦	♦	♦	♦	♦	♦		♦

Moorland and Parkland

Species		Bardon	Sharpley	C.L.N.R.	Cade.	Ulvers.	Bradgate	Don.	Beacon
Cryphoeca silvicola		♦		♦		♦			♦
Mastigusa macrophthalma	RDB3	★	★				★	★	★
Antistea elegans	L				★	★			
Hahnia montana		♦		♦		♦			
Ero cambridgei		♦				♦	♦	♦	
.... *furcata*			♦	♦			♦	♦	♦
Steatoda bipunctata		♦		♦			♦	♦	
Anelosimus vittatus		♦		♦		♦			
Theridion sisyphium		♦		♦		♦			♦
.... *impressum*	L			★					
.... *varians*		♦		♦		♦			
.... *mystaceum*		♦		♦			♦	♦	♦
.... *bimaculatum*		♦			♦	♦	♦	♦	
.... *pallens*		♦		♦		♦	♦	♦	
Enoplognatha ovata		♦		♦	♦	♦	♦	♦	
.... *thoracica*	L				★				
Robertus lividus		♦	♦	♦	♦	♦	♦	♦	♦
Pholcomma gibbum		♦	♦	♦		♦	♦	♦	
Theonoe minutissima	L	★	★	★		★			
Nesticus cellulanus	L	★		★					★
Tetragnatha extensa		♦		♦	♦	♦			
.... *montana*				♦					
.... *obtusa*	L						★		
Pachygnatha clercki		♦		♦	♦	♦	♦		
.... *degeeri*		♦	♦	♦	♦	♦	♦	♦	♦
Metellina segmentata		♦	♦	♦	♦	♦	♦	♦	♦
.... *mengei*		♦		♦		♦			♦
.... *merianae*		♦		♦			♦	♦	♦
Zygiella x-notata		♦		♦			♦	♦	
.... *atrica*		♦		♦			♦	♦	
Araneus diadematus		♦	♦			♦		♦	
.... *quadratus*					♦				
Larinioides cornutus				♦		♦			
Nuctenea umbratica		♦					♦	♦	
Atea sturmi	L			★					
Araniella cucurbitina				♦					
.... *opistographa*	L			★					
Ceratinella brevipes					♦				
.... *brevis*		♦	♦	♦	♦	♦	♦		♦
Walckenaeria acuminata		♦	♦	♦	♦	♦	♦	♦	♦
.... *antica*		♦	♦	♦	♦			♦	
.... *cucullata*		♦		♦	♦	♦	♦		
.... *atrotibialis*	L					★			
.... *capito*	L			★					
.... *dysderoides*	L	★	★	★	★				
.... *nudipalpis*		♦	♦	♦	♦	♦			♦
.... *furcillata*	L	★	★	★					
.... *unicornis*		♦	♦	♦		♦			
.... *cuspidata*		♦				♦	♦		
.... *vigilax*	L			★	★				
Dicymbium nigrum				♦	♦	♦	♦	♦	
.... *tibiale*	L			★	★				
Entelecara acuminata						♦			
.... *erythropus*	L	★		★		★		★	
Gnathonarium dentatum				♦			♦		
Dismodicus bifrons		♦	♦	♦	♦	♦			
Hypomma bituberculatum		♦		♦					
.... *cornutum*							♦		
Metopobactrus prominulus	L	★	★						

APPENDIX II

Moorland and Parkland

		Bardon	Sharpley	C.L.N.R.	Cade.	Ulvers.	Bradgate	Don.	Beacon
Baryphyma trifrons	L					★			
Gonatium rubens		♦	♦	♦	♦	♦	♦		
.... *rubellum*				♦				♦	
Maso sundevalli		♦		♦				♦	
Peponocranium ludicrum		♦	♦	♦	♦				
Pocadicnemis pumila		♦	♦	♦	♦	♦			
.... *juncea*		♦			♦	♦			
Oedothorax gibbosus		♦		♦	♦	♦			
(.... *tuberosus*)		♂		♂	♂	♂			
.... *fuscus*		♦		♦		♦	♦	♦	
.... *agrestis*	L	★							
.... *retusus*		♦		♦	♦		♦	♦	
.... *apicatus*	L							★	
Silometopus elegans	L	★	★	★	★				
Cnephalocotes obscurus	L	★	★	★		★	★		★
Evansia merens	L	★	★	★					
Tiso vagans		♦	♦	♦		♦	♦	♦	
Troxochrus scabriculus	L							★	
Tapinocyba praecox	L		★	★	★		★		
.... *pallens*	L	★							
.... *insecta*	L					★			
Microctenonyx subitaneus	L			★					
Thyreosthenius parasiticus	L	★		★			★		
Monocephalus fuscipes		♦	♦	♦	♦	♦	♦		
.... *castaneipes*	L			★					
Lophomma punctatum	L	★		★	★	★	★		
Gongylidiellum vivum		♦	♦	♦	♦	♦			
Micrargus herbigradus		♦	♦	♦	♦	♦	♦		♦
.... *apertus*	L	★		★					
.... *subaequalis*	L	★					★		★
Erigonella hiemalis							♦		
.... *ignobilis*	L						★		
Savignia frontata		♦	♦	♦		♦	♦	♦	♦
Diplocephalus cristatus				♦					
.... *permixtus*		♦		♦		♦	♦		
.... *latifrons*		♦		♦		♦	♦		
.... *picinus*		♦				♦	♦		
Milleriana inerrans	L							★	
Erigone dentipalpis		♦		♦	♦	♦	♦	♦	♦
.... *atra*		♦	♦	♦	♦	♦	♦	♦	
Prinerigone vagans	L	★		★					
Leptorhoptrum robustum	L			★	★				
Drepanotylus uncatus	L			★					
Asthenargus paganus	L	★				★			
Ostearius melanopygius				♦				♦	
Porrhomma pygmaeum		♦	♦			♦	♦	♦	
.... *pallidum*	L	★		★		★			
.... *microphthalmum*	L	★		★			★		
.... *egeria*	L	★							
Agyneta subtilis				♦				♦	
.... *conigera*			♦	♦					
.... *decora*	L	★		★			★		
.... *cauta*	L	★							
Meioneta rurestris		♦		♦		♦		♦	
.... *saxatilis*					♦	♦		♦	
Microneta viaria		♦	♦	♦	♦	♦			
Centromerus sylvaticus		♦	♦	♦	♦	♦			
.... *prudens*		♦	♦	♦			♦		♦
.... *dilutus*		♦	♦	♦	♦		♦		

		Moorland and Parkland							
		Bardon	Sharpley	C.L.N.R.	Cade.	Ulvers.	Bradgate	Don.	Beacon
Tallusia experta						♦			
Centromerita bicolor		♦	♦	♦	♦			♦	
.... *concinna*		♦	♦	♦				♦	♦
Saaristoa abnormis		♦	♦	♦	♦	♦			
.... *firma*	L	★							
Macrargus rufus		♦	♦	♦	♦	♦	♦		♦
Bathyphantes approximatus		♦		♦			♦		
.... *gracilis*		♦	♦	♦		♦	♦	♦	
.... *parvulus*		♦	♦	♦		♦	♦		
.... *nigrinus*		♦				♦	♦	♦	
Kaestneria pullata		♦		♦	♦	♦	♦		
Diplostyla concolor		♦		♦			♦	♦	
Poeciloneta variegata	L	★	★	★		★			
Drapetisca socialis		♦		♦		♦	♦		
Tapinopa longidens				♦					♦
Labulla thoracica				♦	♦	♦		♦	
Stemonyphantes lineatus		♦	♦	♦			♦	♦	♦
Bolyphantes luteolus		♦		♦			♦		
.... *alticeps*	L	★		★	★				
Lepthyphantes leprosus		♦		♦					
.... *minutus*				♦		♦	♦	♦	
.... *alacris*		♦		♦	♦	♦			
.... *obscurus*	L	★	★	★	★	★	★		
.... *tenuis*		♦	♦	♦	♦	♦	♦	♦	♦
.... *zimmermanni*		♦	♦	♦	♦	♦	♦	♦	♦
.... *cristatus*	L			★					
.... *mengei*		♦	♦	♦	♦	♦	♦		♦
.... *beckeri*	RDBK						★		
.... *flavipes*		♦		♦			♦		
.... *tenebricola*	L	★							
.... *ericaeus*		♦	♦	♦	♦	♦	♦	♦	♦
.... *pallidus*	L	★	★	★		★	★	★	
.... *insignis*	Nb			★					
.... *midas*	RDB2							★	
Helophora insignis		♦		♦		♦			
Linyphia triangularis		♦		♦	♦	♦	♦		♦
.... *hortensis*				♦			♦		
.... *montana*		♦		♦		♦	♦	♦	
.... *clathrata*		♦		♦	♦	♦	♦	♦	♦
.... *peltata*		♦		♦		♦			
Microlinyphia pusilla		♦	♦	♦	♦	♦	♦		
TOTALS (species)		147	75	162	78	107	108	63	44
LOCAL species		34	17	36	17	15	15	7	6
NOTABLE Nb		0	0	1	0	0	0	0	0
RDB3		1	1	0	0	0	1	1	1
RDB2		0	0	0	0	0	0	1	0
RDBK		0	0	0	0	0	1	0	0
SITE STATUS (rarity score)		42	25	40	17	15	31	27	14

Table d) Riparian Habitats

Spiders recorded from
the River Trent including Lockington Marsh; the River Soar; Narborough Bog;
Aylestone Flood Meadows; Watermead Riverside Park, Leicester;
Barrow Gravel Pits; Loughborough Meadows

	R.Trent	R.Soar	Nar.Bog	Aylestone	Watermd	Barrow	Lough.Mead.
Amaurobius fenestralis		♦					
Dictyna arundinacea		♦		♦	♦		
.... uncinata			♦				
Oonops pulcher		♦					
Segestria senoculata		♦					
Zelotes latreilli	L						★
Micaria pulicaria		♦			♦		♦
Clubiona reclusa		♦	♦	♦	♦	♦	
.... pallidula		♦					
.... phragmitis	L	★	★	★		★	★
.... terrestris		♦					
.... lutescens	♦	♦	♦		♦		
.... brevipes							♦
Phrurolithus festivus				♦			
Xysticus cristatus			♦	♦			♦
Ozyptila praticola	L				★		
Philodromus aureolus			♦	♦			
.... cespitum			♦	♦			
Tibellus oblongus		♦					
Salticus cingulatus		♦					
Pardosa palustris		♦		♦			♦
.... pullata		♦	♦	♦			♦
.... prativaga		♦	♦	♦	♦		
.... amentata		♦	♦	♦	♦	♦	♦
Alopecosa pulverulenta			♦	♦	♦		♦
Trochosa ruricola		♦	♦	♦	♦	♦	♦
.... terricola			♦	♦			
Pirata piraticus	♦	♦		♦	♦	♦	♦
.... hygrophilus			♦				
Pisaura mirabilis			♦	♦	♦	♦	♦
Tegenaria silvestris	L		★				
Antistea elegans	L	★					
Hahnia nava	L			★			★
Ero cambridgei			♦	♦	♦		♦
Anelosimus vittatus				♦			
Theridion pictum	L	★					
.... bimaculatum		♦		♦	♦	♦	
.... pallens			♦		♦	♦	
Enoplognatha ovata			♦	♦	♦	♦	
Robertus lividus			♦	♦		♦	
Tetragnatha extensa	♦	♦				♦	
.... pinicola	Nb		★				
.... montana			♦		♦	♦	
.... nigrita	L			★		★	
Pachygnatha clercki		♦	♦	♦	♦	♦	♦
.... degeeri			♦		♦		♦
Metellina segmentata		♦	♦		♦		
.... mengei				♦		♦	♦
Zygiella x-notata			♦				♦
Larinioides cornutus		♦	♦	♦		♦	♦
.... sclopetarius	L		★				★
Nuctenea umbratica				♦			

Riparian Habitats

Species	R.Trent	R.Soar	Nar.Bog	Aylestone	Watermd	Barrow	Lough.Mead.
Araniella cucurbitina							♦
.... *opistographa*				♦			
Ceratinella brevipes		♦		♦			♦
.... *brevis*			♦				
.... *scabrosa*	L		★				
Walckenaeria acuminata		♦	♦				
.... *antica*		♦		♦	♦		♦
.... *dysderoides*	L	★					
.... *nudipalpis*		♦	♦		♦		♦
.... *unicornis*		♦	♦	♦			♦
.... *cuspidata*		♦	♦	♦	♦		
Dicymbium nigrum		♦	♦	♦	♦	♦	♦
Gnathonarium dentatum	♦	♦		♦	♦	♦	
Tmeticus affinis	L	★	★			★	
Gongylidium rufipes	♦	♦	♦	♦	♦		
Dismodicus bifrons		♦	♦	♦			
Hypomma bituberculatum	♦	♦	♦	♦	♦		♦
Baryphyma pratense	L	★	★	★	★	★	
Gonatium rubens		♦	♦				♦
.... *rubellum*	♦						
Maso sundevalli		♦					
Pocadicnemis pumila				♦			
.... *juncea*		♦	♦	♦	♦		♦
Oedothorax gibbosus	♦	♦	♦	♦	♦	♦	♦
(.... *tuberosus*)	♂	♂	♂	♂			
.... *fuscus*	♦	♦		♦	♦	♦	♦
.... *agrestis*	L	★					
.... *retusus*		♦	♦	♦	♦	♦	♦
.... *apicatus*	L	★					
Silometopus reussi	L	★					★
Cnephalocotes obscurus	L			★			★
Tiso vagans		♦		♦			
Troxochrus scabriculus	L	★		★			
Microctenonyx subitaneus	L						★
Monocephalus fuscipes		♦		♦	♦		
Lophomma punctatum	L	★	★	★	★	★	
Saloca diceros	Nb		★				
Gongylidiellum vivum		♦		♦	♦		♦
Micrargus herbigradus		♦	♦	♦	♦		♦
.... *subaequalis*	L	★		★			★
Erigonella ignobilis	L		★				
Savignia frontata		♦	♦	♦	♦		♦
Diplocephalus cristatus	♦	♦			♦		
.... *permixtus*	♦	♦				♦	
.... *latifrons*		♦	♦	♦	♦		
.... *picinus*		♦	♦				♦
Araeoncus humilis	L	★					
Panamomops sulcifrons	L	★		★	★		★
Milleriana inerrans	L	★		★			
Erigone dentipalpis	♦	♦	♦	♦		♦	♦
.... *atra*	♦	♦	♦	♦		♦	♦
.... *arctica*	L	★	★				
.... *longipalpis*	L	★					
Prinerigone vagans	L	★				★	
Leptorhoptrum robustum	L	★	★	★	★	★	★
Halorates distinctus	L	★	★		★	★	★
Ostearius melanopygius		♦					
Porrhomma pygmaeum		♦				♦	
.... *campbelli*	L		★				

APPENDIX II

Riparian Habitats

		R.Trent	R.Soar	Nar.Bog	Aylestone	Watermd	Barrow	Lough.Mead.
Porrhomma microphthalmum	L			★				★
.... *errans*	Nb				★	★		
Meioneta rurestris			♦	♦	♦			♦
.... *saxatilis*			♦	♦	♦			
.... *beata*	L							★
Microneta viaria			♦	♦				
Centromerus sylvaticus				♦				♦
Tallusia experta		♦	♦	♦	♦	♦		♦
Centromerita bicolor				♦	♦	♦		♦
.... *concinna*								♦
Saaristoa abnormis			♦	♦				
Bathyphantes approximatus		♦	♦	♦	♦		♦	♦
.... *gracilis*		♦	♦	♦	♦	♦	♦	♦
.... *parvulus*			♦	♦	♦	♦		
.... *nigrinus*			♦	♦	♦			
Kaestneria pullata				♦	♦	♦		♦
Diplostyla concolor			♦	♦	♦		♦	♦
Floronia bucculenta	L						★	
Stemonyphantes lineatus					♦	♦		♦
Lepthyphantes leprosus								♦
.... *tenuis*		♦	♦	♦	♦	♦	♦	♦
.... *zimmermanni*		♦	♦	♦	♦	♦		♦
.... *flavipes*		♦	♦					♦
.... *ericaeus*			♦	♦	♦	♦		♦
.... *pallidus*	L		★	★	★	★		
Linyphia triangularis		♦		♦				
.... *hortensis*				♦				
.... *montana*			♦				♦	
.... *clathrata*			♦	♦	♦	♦	♦	
Microlinyphia impigra	L				★		★	
Allomengea vidua	L		★		★		★	★
TOTALS (species)		40	92	72	76	55	34	62
LOCAL species		8	21	10	13	8	10	13
NOTABLE Nb		0	0	2	1	1	0	0
SITE STATUS (rarity score)		8	21	18	17	12	10	13

Table e) Reservoirs and Open Water

Spiders recorded from
Swithland Reservoir; Cropston Reservoir; Blackbrook Reservoir; Groby Pool;
Rutland Water; Eyebrook Reservoir; Saddington Reservoir.

		Swith.R.	Crop.R.	Blkbrk R.	Groby P.	Rutl.Wat.	Eyebrk.R.	Saddington
Amaurobius fenestralis		♦		♦	♦			
.... *similis*			♦		♦			
.... *ferox*		♦	♦					
Dictyna arundinacea		♦				♦		
.... *uncinata*		♦					♦	
Lathys humilis	L						★	
Oonops pulcher		♦			♦			
Harpactea hombergi	L	★						
Segestria senoculata					♦	♦	♦	
Scotophaeus blackwalli			♦					
Micaria pulicaria						♦		
Clubiona corticalis			♦					
.... *reclusa*		♦						
.... *pallidula*						♦		
.... *phragmitis*	L	★	★	★	★	★	★	★
.... *lutescens*		♦				♦		♦
.... *brevipes*					♦			
Phrurolithus festivus		♦						
Xysticus cristatus		♦	♦					
.... *ulmi*	L		★					
Philodromus dispar		♦				♦		
.... *aureolus*		♦					♦	
Tibellus oblongus		♦			♦			
Salticus scenicus		♦				♦		
.... *cingulatus*								♦
Euophrys frontalis					♦			
.... *aequipes*	L		★					
Evarcha falcata		♦						
Pardosa palustris						♦		
.... *pullata*		♦	♦			♦	♦	
.... *prativaga*		♦	♦			♦	♦	♦
.... *amentata*		♦	♦	♦	♦	♦		♦
.... *nigriceps*						♦		
.... *lugubris*		♦	♦					
Trochosa ruricola		♦	♦	♦		♦	♦	♦
Pirata piraticus		♦	♦	♦	♦	♦	♦	♦
.... *hygrophilus*		♦	♦			♦		
Pisaura mirabilis		♦				♦		
Textrix denticulata					♦			
Antistea elegans	L	★	★	★	★			
Hahnia montana				♦				
.... *helveola*	L			★				
Ero cambridgei		♦			♦	♦		♦
Steatoda bipunctata							♦	
Anelosimus vittatus		♦					♦	
Achaearanea lunata	L					★		
Theridion sisyphium		♦			♦		♦	
.... *pictum*	L	★						★
.... *varians*		♦					♦	
.... *mystaceum*		♦						
.... *tinctum*	L	★						
.... *bimaculatum*		♦						
.... *pallens*		♦	♦					

Reservoirs and Open Water

Species		Swith.R.	Crop.R.	Blkbrk R.	Groby P.	Rutl.Wat.	Eyebrk.R.	Saddington
Enoplognatha ovata		♦				♦	♦	
Robertus lividus		♦						
Pholcomma gibbum		♦						
Nesticus cellulanus	L	★		★				
Tetragnatha extensa		♦	♦		♦	♦	♦	♦
.... *montana*		♦			♦	♦	♦	♦
.... *obtusa*	L	★	★				★	
.... *nigrita*	L	★					★	
.... *striata*	Nb				★			
Pachygnatha clercki		♦	♦	♦	♦		♦	♦
.... *degeeri*		♦	♦		♦	♦		
Metellina segmentata		♦			♦		♦	♦
.... *mengei*		♦					♦	
.... *merianae*		♦		♦			♦	♦
Meta menardi	L	★		★				
Zygiella x-notata		♦			♦	♦		
.... *atrica*		♦						
Gibbaranea gibbosa	L						★	
Araneus diadematus		♦			♦	♦	♦	
Larinioides cornutus		♦			♦	♦	♦	♦
.... *sclopetarius*	L	★		★		★	★	
.... *patagiatus*	L	★						
Nuctenea umbratica		♦	♦		♦			
Araniella opistographa	L	★						
Ceratinella brevis					♦			♦
Walckenaeria acuminata		♦						
.... *incisa*	Nb	★						
.... *nudipalpis*		♦	♦	♦	♦			
.... *unicornis*		♦		♦	♦	♦		
.... *cuspidata*					♦			
Dicymbium nigrum					♦	♦		
Hylyphantes graminicola	L	★					★	
Gnathonarium dentatum		♦	♦	♦	♦	♦		♦
Tmeticus affinis	L	★			★			
Gongylidium rufipes		♦			♦			♦
Dismodicus bifrons		♦	♦		♦			
Hypomma bituberculatum		♦	♦	♦	♦	♦		♦
.... *cornutum*		♦					♦	
Gonatium rubens		♦				♦		
Maso sundevalli		♦		♦	♦			
Pocadicnemis pumila		♦						
.... *juncea*		♦				♦		♦
Oedothorax gibbosus		♦	♦	♦	♦	♦		♦
(.... *tuberosus*)		♂		♂	♂	♂		♂
.... *fuscus*		♦	♦	♦		♦		♦
.... *agrestis*	L		★					
.... *retusus*		♦	♦	♦		♦		♦
.... *apicatus*	L	★				★	★	
Cnephalocotes obscurus	L	★		★				
Monocephalus fuscipes		♦		♦				
Lophomma punctatum	L	★	★	★	★			
Gongylidiellum vivum		♦			♦	♦		
Micrargus herbigradus		♦	♦					
Erigonella ignobilis	L	★						
Savignia frontata		♦	♦		♦			♦
Diplocephalus cristatus		♦		♦	♦	♦		
.... *permixtus*		♦	♦	♦	♦			
.... *latifrons*		♦						
.... *picinus*		♦						♦

Reservoirs and Open Water

Species	L	Swith.R.	Crop.R.	Blkbrk R.	Groby P.	Rutl.Wat.	Eyebrk.R.	Saddington
Araeoncus humilis	L	★						★
Erigone dentipalpis		♦	♦	♦	♦	♦		♦
.... *atra*		♦	♦		♦	♦	♦	
.... *arctica*	L					★		
Prinerigone vagans	L	★	★	★	★	★	★	
Leptorhoptrum robustum	L	★						
Drepanotylus uncatus	L				★			★
Ostearius melanopygius					♦	♦		
Porrhomma pygmaeum		♦	♦	♦	♦	♦	♦	♦
.... *microphthalmum*	L	★				★		
Agyneta ramosa	L	★						
Meioneta rurestris		♦	♦	♦		♦	♦	♦
.... *saxatilis*		♦						
Centromerus sylvaticus		♦						
.... *dilutus*				♦				
Tallusia experta					♦			
Saaristoa abnormis		♦			♦			
Bathyphantes approximatus		♦	♦	♦		♦		♦
.... *gracilis*		♦	♦	♦	♦	♦	♦	♦
.... *parvulus*		♦	♦					
.... *nigrinus*		♦		♦	♦			
Kaestneria dorsalis		♦					♦	
.... *pullata*		♦	♦	♦	♦	♦		♦
Diplostyla concolor			♦	♦	♦	♦	♦	♦
Drapetisca socialis		♦						
Floronia bucculenta	L	★						
Labulla thoracica							♦	
Stemonyphantes lineatus					♦			
Lepthyphantes alacris		♦						
.... *obscurus*	L	★					★	
.... *tenuis*		♦	♦	♦	♦	♦	♦	♦
.... *zimmermanni*		♦		♦	♦			♦
.... *cristatus*	L				★			
.... *mengei*		♦			♦		♦	
.... *ericaeus*		♦		♦	♦			
.... *pallidus*	L	★			★		★	
Helophora insignis		♦						
Linyphia triangularis		♦						
.... *hortensis*		♦						
.... *montana*		♦	♦	♦		♦	♦	♦
.... *clathrata*		♦	♦	♦	♦	♦	♦	
.... *peltata*		♦	♦		♦			
Microlinyphia pusilla					♦			
.... *impigra*	L					★		★
Allomengea vidua	L	★			★			★
TOTALS (species)		119	49	48	63	56	43	39
LOCAL species		26	9	9	9	8	11	6
NOTABLE Nb		1	0	0	1	0	0	0
SITE STATUS (rarity score)		30	9	9	13	8	11	6

APPENDIX II

Table f) Woodland

Spiders recorded from
Buddon Wood; Swithland Wood; Loughborough Outwoods; Stoneywell Wood;
Burbage Wood; Owston Wood; Launde Big Wood; Pickworth Great Wood.

Species	Status	Budd.W.	Swith.W.	Outwoods	Stoneyw.	Burbage W.	Owst W.	Lnde B.W.	Pick.G.W.
Amaurobius fenestralis		♦	♦	♦	♦		♦		
.... *similis*			♦						
Dictyna arundinacea		♦	♦						♦
.... *uncinata*		♦							
Lathys humilis	L		★			★			★
Oonops pulcher		♦	♦	♦					
Dysdera erythrina		♦	♦						♦
Harpactea hombergi	L	★	★	★					
Segestria senoculata		♦	♦						
Drassodes cupreus		♦							
Haplodrassus silvestris	Nb	★							
Zelotes latreillei	L	★							
.... *apricorum*	L		★						
Micaria pulicaria		♦							
Clubiona corticalis		♦							
.... *reclusa*		♦	♦					♦	
.... *pallidula*		♦							
.... *terrestris*		♦	♦	♦	♦	♦			
.... *lutescens*		♦	♦			♦	♦	♦	
.... *comta*		♦	♦	♦	♦	♦	♦	♦	♦
.... *brevipes*		♦	♦	♦	♦	♦			♦
.... *diversa*		♦							
Agroeca brunnea	L	★		★					
.... *proxima*		♦							
Phrurolithus festivus		♦							
Zora spinimana		♦	♦					♦	
Anyphaena accentuata		♦	♦			♦	♦	♦	♦
Diaea dorsata	L								★
Misumena vatia							♦		♦
Xysticus cristatus		♦	♦					♦	
.... *ulmi*	L							★	★
Ozyptila praticola	L	★							
.... *trux*				♦					
Philodromus dispar		♦	♦	♦					
.... *aureolus*		♦				♦		♦	♦
.... *cespitum*		♦							
.... *collinus*	Nb								★
Tibellus oblongus		♦	♦					♦	
Salticus scenicus		♦	♦						
Heliophanus cupreus			♦						
Ballus chalybeius	L	★							
Neon reticulatus		♦	♦			♦		♦	
Euophrys frontalis		♦	♦						
.... *erratica*	L	★							
Evarcha falcata		♦							
Pardosa pullata		♦	♦						♦
.... *prativaga*		♦	♦					♦	♦
.... *amentata*		♦	♦			♦		♦	♦
.... *nigriceps*		♦							
.... *lugubris*		♦	♦		♦	♦		♦	♦
Alopecosa pulverulenta		♦	♦						
Trochosa ruricola			♦						
.... *terricola*		♦							♦
Pirata piraticus		♦							

		Woodland							
		Budd.W.	Swith.W.	Outwoods	Stoneyw.	Burbage W.	Owst W.	Lnde B.W.	Pick.G.W.
Pirata hygrophilus		♦	♦		♦			♦	♦
Pisaura mirabilis		♦	♦	♦			♦	♦	♦
Textrix denticulata			♦	♦					
Tegenaria gigantea		♦							♦
.... *agrestis*	L	★							
.... *domestica*			♦						
.... *silvestris*	L	★							
Coelotes atropos		♦	♦	♦			♦	♦	♦
Cryphoeca silvicola					♦		♦		
Mastigusa macrophthalma	RDB3	★							
Antistea elegans	L	★			★				
Hahnia montana		♦	♦	♦	♦				
.... *helveola*	L	★	★	★	★				
Ero cambridgei		♦			♦				
.... *furcata*		♦	♦				♦		
Episinus angulatus	L	★							
Euryopis flavomaculata	L	★							
Anelosimus vittatus		♦		♦		♦	♦	♦	♦
Achaearanea lunata	L	★			★				
Theridion sisyphium		♦	♦	♦		♦		♦	♦
.... *pictum*	L	★							
.... *varians*		♦			♦				♦
.... *simile*		♦							
.... *mystaceum*		♦	♦		♦	♦			
.... *tinctum*	L	★							★
.... *bimaculatum*		♦			♦		♦	♦	
.... *pallens*		♦	♦	♦	♦	♦	♦	♦	♦
Enoplognatha ovata		♦	♦	♦		♦	♦		♦
.... *thoracica*	L	★							
Robertus lividus		♦	♦	♦			♦		♦
.... *neglectus*	L						★		
Pholcomma gibbum		♦							
Theonoe minutissima	L	★			★				
Nesticus cellulanus	L	★	★						
Tetragnatha extensa		♦					♦	♦	
.... *pinicola*	Nb						★	★	
.... *montana*		♦	♦			♦	♦	♦	♦
.... *obtusa*	L			★					
.... *nigrita*	L	★							
Pachygnatha clercki		♦	♦	♦		♦	♦		
.... *listeri*	L		★			★	★	★	
.... *degeeri*		♦	♦	♦		♦		♦	♦
Metellina segmentata		♦	♦			♦			
.... *mengei*		♦	♦	♦	♦	♦	♦	♦	♦
.... *merianae*		♦	♦	♦			♦		♦
Zygiella x-notata		♦							
.... *atrica*		♦							
Gibbaranea gibbosa	L	★							
Araneus diadematus		♦	♦	♦	♦			♦	
.... *quadratus*		♦							
.... *marmoreus*	L						★		
Larinioides cornutus		♦							♦
.... *sclopetarius*	L	★							
.... *patagiatus*	L	★							
Nuctenea umbratica		♦	♦	♦					
Atea sturmi	L	★							
Araniella cucurbitina		♦	♦				♦	♦	♦
.... *opistographa*	L	★		★		★		★	★
Cyclosa conica	L	★				★	★		★
Ceratinella brevipes		♦							

APPENDIX II

Woodland

Species	Status	Budd.W.	Swith.W.	Outwoods	Stoneyw.	Burbage W.	Owst W.	Lnde B.W.	Pick.G.W.
Ceratinella brevis		♦	♦		♦				
Walckenaeria acuminata		♦	♦	♦	♦		♦		
.... *cucullata*		♦	♦						
.... *atrotibialis*	L	★							
.... *incisa*	Nb	★							
.... *dysderoides*	L	★	★	★	★				
.... *nudipalpis*		♦			♦				
.... *furcillata*	L	★							
.... *unicornis*		♦			♦				
.... *cuspidata*		♦							
Dicymbium nigrum		♦					♦		
.... *tibiale*	L	★					★		
Entelecara acuminata									♦
Moebelia penicillata					♦				
Hylyphantes graminicola	L	★							
Gnathonarium dentatum		♦			♦				
Gongylidium rufipes		♦	♦	♦	♦	♦	♦	♦	♦
Dismodicus bifrons		♦	♦		♦		♦		
Hypomma bituberculatum		♦			♦				
.... *cornutum*		♦	♦						♦
Baryphyma pratense	L	★							
Gonatium rubens		♦							
.... *rubellum*		♦		♦			♦		
Maso sundevalli		♦			♦	♦	♦	♦	
Pocadicnemis pumila		♦						♦	
.... *juncea*		♦				♦			
Oedothorax gibbosus		♦	♦	♦	♦		♦		
(.... *tuberosus*)		♂	♂	♂	♂				
.... *fuscus*						♦	♦		
.... *agrestis*	L	★	★						
.... *retusus*		♦	♦			♦			
Pelecopsis nemoralis	L		★						
Cnephalocotes obscurus	L	★	★						
Tiso vagans		♦							
Tapinocyba praecox	L	★							
.... *pallens*	L	★							
.... *insecta*	L	★	★						
Thyreosthenius parasiticus	L	★		★	★				
.... *biovatus*	L	★							
Monocephalus fuscipes		♦	♦	♦	♦	♦	♦		
Lophomma punctatum	L	★	★		★				
Saloca diceros	Nb					★	★		
Gongylidiellum vivum		♦	♦						
Micrargus herbigradus		♦	♦	♦	♦	♦	♦		
.... *apertus*	L	★	★						
.... *subaequalis*	L	★							
Erigonella hiemalis		♦					♦		
Savignia frontata		♦	♦	♦	♦				
Diplocephalus cristatus			♦						
.... *permixtus*		♦	♦	♦	♦		♦		
.... *latifrons*		♦	♦	♦	♦		♦		♦
.... *picinus*		♦	♦	♦	♦		♦		
Araeoncus humilis	L		★						
Erigone dentipalpis		♦				♦			♦
.... *atra*		♦	♦				♦		
Prinerigone vagans	L	★							
Leptorhoptrum robustum	L	★	★						
Asthenargus paganus	L				★				
Porrhomma pygmaeum		♦	♦	♦			♦		
.... *convexum*	L		★						

		Woodland							
		Budd.W.	Swith.W.	Outwoods	Stoneyw.	Burbage W.	Owst W.	Lnde B.W.	Pick.G.W.
Porrhomma pallidum	L	★	★	★	★				
.... *microphthalmum*	L	★	★				★	★	★
Agyneta subtilis						◆			
.... *conigera*			◆			◆		◆	
.... *ramosa*	L	★							
Meioneta innotabilis					◆				
.... *rurestris*		◆							
.... *saxatilis*		◆				◆			
Microneta viaria		◆	◆	◆	◆		◆		
Centromerus sylvaticus		◆							
.... *prudens*		◆	◆						
.... *dilutus*		◆	◆	◆	◆		◆		
Tallusia experta		◆	◆						
Centromerita bicolor			◆						
.... *concinna*		◆							
Saaristoa abnormis		◆							
.... *firma*	L	★			★		★		
Macrargus rufus			◆	◆	◆		◆		
Bathyphantes approximatus		◆	◆		◆				
.... *gracilis*		◆	◆	◆	◆	◆	◆	◆	◆
.... *parvulus*		◆						◆	
.... *nigrinus*		◆	◆	◆	◆	◆	◆	◆	◆
Kaestneria dorsalis		◆						◆	◆
.... *pullata*		◆	◆	◆					
Diplostyla concolor		◆	◆	◆			◆	◆	◆
Poeciloneta variegata	L	★			★				
Drapetisca socialis		◆	◆	◆					
Tapinopa longidens				◆					
Floronia bucculenta	L	★							
Labulla thoracica		◆							
Bolyphantes alticeps	L		★						
Lepthyphantes leprosus			◆						
.... *minutus*		◆	◆	◆	◆				
.... *alacris*		◆	◆	◆	◆	◆	◆		◆
.... *obscurus*	L	★	★	★	★		★	★	★
.... *tenuis*		◆	◆	◆	◆	◆	◆	◆	◆
.... *zimmermanni*		◆	◆	◆	◆	◆	◆	◆	◆
.... *cristatus*	L	★				★	★		
.... *mengei*		◆	◆	◆					
.... *flavipes*		◆	◆	◆					
.... *tenebricola*	L	★	★		★	★	★		★
.... *ericaeus*		◆	◆	◆		◆	◆		
.... *pallidus*	L	★	★	★			★	★	
Helophora insignis		◆	◆	◆					
Pityohyphantes phrygianus	L								★
Linyphia triangularis		◆	◆			◆			
.... *hortensis*		◆	◆		◆	◆	◆	◆	
.... *montana*		◆	◆		◆	◆	◆	◆	◆
.... *clathrata*		◆	◆		◆	◆	◆	◆	◆
.... *peltata*		◆	◆	◆	◆	◆	◆	◆	◆
Microlinyphia pusilla		◆	◆	◆					
Allomengea vidua	L	★							
TOTALS (species)		191	114	65	60	44	62	46	54
LOCAL species		54	22	10	12	7	10	7	10
NOTABLE Nb		2	0	0	0	1	2	1	1
NOTABLE Na		0	0	0	0	0	0	0	0
RDB3		1	0	0	0	0	0	0	0
SITE STATUS (rarity score)		70	22	10	12	11	18	11	14

APPENDIX II

Table g) Garden and House Spiders

Garden 1. Knighton Church Road, Leicester; Garden 2. The Portwey, Leicester; Garden 3. Scraptoft Lane, Leicester; Garden 4. Outwoods Drive, Loughborough; Garden 5. Bramcote Road, Loughborough.

		Garden 1	Garden 2	Garden 3	Garden 4	Garden 5
Amaurobius fenestralis				♦		♦
.... *similis*			♦	♦	♦	♦
.... *ferox*			♦	♦	♦	♦
Dictyna arundinacea				♦		
.... *uncinata*		♦	♦		♦	♦
Lathys humilis	L	★	★	★		
Oonops pulcher		♦		♦		
.... *domesticus*				♦	♦	♦
Harpactea hombergi	L			★		
Pholcus phalangioides						♦
Drassodes lapidosus				♦	♦	
Scotophaeus blackwalli		♦	♦	♦	♦	♦
Clubiona corticalis		♦		♦	♦	
.... *pallidula*		♦	♦	♦	♦	
.... *phragmitis*	L			★		
.... *terrestris*		♦		♦	♦	♦
.... *lutescens*		♦	♦	♦		
.... *comta*		♦	♦	♦		
.... *brevipes*				♦		
Misumena vatia				♦		
Xysticus cristatus		♦		♦	♦	♦
Ozyptila praticola	L	★		★		★
.... *atomaria*					♦	
Philodromus dispar		♦	♦	♦	♦	♦
.... *aureolus*		♦	♦	♦		
.... *cespitum*		♦		♦		
Tibellus oblongus				♦		
Salticus scenicus				♦	♦	♦
Heliophanus flavipes				♦		
Euophrys frontalis		♦				
.... *lanigera*						♦
Pardosa pullata		♦		♦	♦	♦
.... *prativaga*				♦		
.... *amentata*		♦	♦	♦	♦	♦
Alopecosa pulverulenta				♦		
Trochosa ruricola					♦	
Pisaura mirabilis		♦		♦		
Tegenaria gigantea		♦	♦	♦	♦	♦
.... *domestica*		♦	♦	♦	♦	♦
Ero furcata					♦	
Steatoda bipunctata		♦	♦	♦	♦	♦
Anelosimus vittatus		♦				
Theridion varians				♦		
.... *melanurum*			♦		♦	♦
.... *mystaceum*		♦		♦		
.... *tinctum*	L	★		★	★	
.... *bimaculatum*				♦	♦	
.... *pallens*		♦	♦	♦		
Enoplognatha ovata		♦		♦	♦	♦
Nesticus cellulanus	L					★
Tetragnatha extensa				♦		
.... *montana*				♦		♦
Pachygnatha clercki					♦	♦
.... *degeeri*				♦		♦

Garden and House Spiders

Species		Garden 1	Garden 2	Garden 3	Garden 4	Garden 5
Metellina segmentata			♦	♦	♦	♦
.... *mengei*				♦		
Zygiella x-notata		♦	♦	♦	♦	♦
Araneus diadematus		♦	♦	♦	♦	♦
Nuctenea umbratica			♦	♦	♦	♦
Araniella cucurbitina		♦				
Ceratinella brevis				♦		
Walckenaeria acuminata		♦				
.... *nudipalpis*						♦
Entelecara acuminata		♦		♦	♦	
.... *erythropus*	L	★			★	
Gongylidium rufipes		♦		♦		
Dismodicus bifrons				♦		
Hypomma cornutum				♦		
Pocadicnemis pumila					♦	
.... *juncea*		♦				
Oedothorax fuscus				♦	♦	
Silometopus reussi	L			★	★	
Tiso vagans				♦		
Microctenonyx subitaneus	L	★			★	
Micrargus herbigradus		♦		♦	♦	♦
.... *subaequalis*	L				★	
Savignia frontata					♦	
Diplocephalus cristatus					♦	♦
Lessertia dentichelis	L					★
Erigone dentipalpis				♦	♦	♦
.... *atra*		♦		♦	♦	♦
Ostearius melanopygius		♦			♦	
Porrhomma pygmaeum					♦	♦
.... *microphthalmum*	L			★		
Meioneta rurestris				♦	♦	♦
.... *saxatilis*		♦		♦		
Microneta viaria				♦		
Centromerus sylvaticus					♦	
.... *dilutus*				♦		
Centromerita bicolor		♦			♦	
Bathyphantes gracilis		♦		♦	♦	♦
Diplostyla concolor		♦		♦	♦	♦
Labulla thoracica					♦	
Stemonyphantes lineatus				♦	♦	♦
Lepthyphantes nebulosus	L			★	★	
.... *leprosus*		♦	♦	♦	♦	♦
.... *minutus*				♦		
.... *tenuis*		♦	♦	♦	♦	♦
.... *zimmermanni*					♦	
.... *mengei*				♦	♦	
.... *ericaeus*			♦			
.... *pallidus*	L				★	
Linyphia triangularis		♦			♦	♦
.... *hortensis*		♦				
.... *montana*		♦		♦	♦	
.... *clathrata*		♦		♦		♦
Microlinyphia pusilla				♦		
TOTAL (species)		49	23	74	59	43
LOCAL species		5	1	9	6	3
SITE STATUS (rarity score)		5	1	9	6	3

APPENDIX III

Natural History Societies

The natural history movement is an integrated network of institutions and people, harnessing the professional and amateur expertise of many thousands of individuals in a wide variety of disciplines. Whatever interest one may have in natural history, the exchange of ideas and the stimulation of fellow-enthusiasts are essential to enjoyment and development of our chosen subjects.

Five organisations are listed here, two national and three local, which cater for the specialist requirements of anyone interested in spiders. At national level, the *British Arachnological Society* caters for the serious arachnologist, whilst the *Spider Recording Scheme* is concerned primarily with field-work and recording. Within the county, the *Leicestershire Entomological Society* specialises in the study of invertebrates, whilst the *Lit.& Phil. Natural History Section* and *Loughborough Naturalists' Club* cover the whole range of natural history subjects. The *N.H. Section*, in the true traditions of the Society, is renowned for the quality and scope of its informative meetings, whereas the strength of the *L.N.C.* lies in its dedication to field-work and its publications.

The British Arachnological Society

The *British Arachnological Society* exists to promote the study of arachnids in Britain and around the world. Membership is open to all those interested in arachnids, and currently there are over 500 members worldwide, both amateur and professional, and nearly 100 subscribing institutions.

Members of the Society participate in identification courses, ecological surveys and field meetings. Lectures and conversazione are arranged at the time of the Annual General Meeting. The *Bulletin*, published three times a year in March, July and September, is supplied free to members, and contains papers on taxonomy, behaviour, life-histories, morphology, ecology and phenology of arachnids. Also, a Newsletter acts as the medium for the exchange of information on all arachnological matters, and includes short articles and notes, international news, book reviews, requests for advice and information, hints and tips, reports of activities and notices of forthcoming events. A valuable library of reprints of scientific papers, as well as books and photographic slides, is available, via a postal loan service, to members resident in Britain.

Details of membership of the Society may be obtained by sending a stamped addressed envelope to the Membership Treasurer, Mr Shaun H. Hexter, 71 Havant Road, London, E17 3JE.

The Spider Recording Scheme

The *Spider Recording Scheme* was launched in 1987, and aims to define the geographical distribution of each species of spider found in the British Isles; to record the spider fauna of selected sites of particular concern to nature conservation, and to provide opportunities to extend our knowledge of the biology of spiders, with special consideration of their habitats and seasonal occurrence.

It is intended that in 1997 (*i.e.* after ten years of field work and other types of recording) enough information will be available to allow the publication of provisional distribution maps based on the 10 km square of the Ordnance Survey National Grid. The complete Atlas, which will illustrate and, as far as possible, account for the distribution of spiders in the British Isles, will follow in 1999.

Further information on the Scheme may be obtained from the National Organiser, Mr David R. Nellist, 198A Park Street Lane, Park Street, St Albans, Herts., AL2 2AQ.

Leicestershire Entomological Society

The Society was founded in 1988 with the prime aims of bringing together all those, whether professional or amateur, interested in the study of insects and other invertebrates in and around Leicestershire (including Rutland), and furthering their conservation.

The Society publishes two Newsletters annually, which aim to keep members abreast of entomological news, conservation issues and literature, and also include short notes of field observations. Additionally, the Society produces Occasional Publications which contain more detailed articles and reports not appropriate for inclusion in the Newsletter. Occasional Publication No. 8 *Leicestershire Harvestmen* by Jon Daws was published in March 1994.

A winter programme of indoor talks and identification workshops is held by the Society, together with a varied programme of summer field visits. Whilst many of these meetings have included contributions from leading entomologists, the programme is balanced by contributions from local experts and members. The field meetings are aimed principally at increasing knowledge of the invertebrate fauna of local sites of interest, and provide a valuable opportunity to visit some of the more important locations in Leicestershire. A number of significant discoveries has been made in the course of such visits.

Anyone interested in joining the Society and making a positive contribution to its aims is invited to write to the Honorary Treasurer, Jane McPhail, 7 Station Close, Kirby Muxloe, Leicestershire LE9 2ES for further details.

Natural History Section of the Leicester Literary and Philosophical Society

The *Natural History Section* of the *Leicester Literary and Philosophical Society* was formed in 1849 to cater for members with a specialist interest in natural history. Today the Section organises a winter programme of 12 meetings which include talks by experts, both local and national, members' exhibition and slide evenings and social events. A summer programme of 15 field meetings, with experienced naturalists as leaders, takes members around sites of interest in the county, and includes a couple of coach trips farther afield. Newsletters are produced twice yearly and from time to time, joint day-schools are organised with the Adult Education Department of Leicester University.

Close links are maintained with the Leicestershire Museums Service, which the Lit. and Phil. helped to found in 1845. The winter meetings are held at the Museums Service H.Q. at The Rowans, College Street and the Assistant Keeper (Zoology) acts as the Section's Programme Secretary. Members have always helped with museum projects and surveys and the reports of the Section's field trips are filed in the Museum's databank providing useful information to assist the Museum in its work with the Planning Department to safeguard ecologically important sites in the county.

Anyone wishing to join the Natural History Section should contact the Hon Treasurer, Miss R. Ewen, 9 Midway Road, Leicester for details. Please note that Section membership does not entitle one to membership of the Parent Body, though Parent Body members can join the Section at a reduced rate.

Loughborough Naturalists' Club

The *Loughborough Naturalists' Club* was formed in 1960 and now has a membership of about 170. It is primarily a field club with a 'hands-on' approach to natural history, the emphasis being on observing and recording. The Club is fortunate in its number of expert naturalists covering most fields of interest, who are very willing to help any member in matters of identification.

Each quarter, members send in records and reports and these are collated and edited to produce a bulletin, *Heritage*, which is sent to all members. The records are then stored permanently by Leicestershire Museums Service where they form part of a valuable database which is consulted when questions of site evaluation arise. This gives point and purpose to individual naturalists to record their findings. Every member is encouraged to contribute to the quarterly records, though the degree of participation is left to the individual. The Club also publishes occasional special titles concerned with Leicestershire natural history, with an accent on in-depth surveys of areas of Charnwood Forest.

Indoor meetings are held monthly during the winter and field visits to sites of interest take place in summer. Monthly Wednesday morning walks are held throughout the year.

The membership Secretary will supply a prospectus and sample copies of Heritage and the Newsletter to anyone interested. Please contact:- Mr L. Hall, 47 Victoria Road, Woodhouse Eaves, Loughborough, Leicester LE12 8RF.

REFERENCES

Agar, W. (1890) 'Spiders'. *Trans.Leicester lit.phil.Soc.*, New Quarterly Series **2** (3), April, 114-122 (read before Section "D" 12.12.1889)

Arnold, G.A. & Crocker, J. (1967) '*Arctosa perita* (Latr.) from colliery spoil heaps in Warwickshire and Leicestershire'. *Bull.Br.Spider Study Group.* **35**, 7-8. July

Barfield, T. (1995) 'The flora of Valley Hay Meadows at Seaton, Rutland: Rutland Record No.15' pp.199-205, *Jn.Rut.loc.hist.rec.Soc.* Leicestershire County Council

Blackwall, J. (1850) 'Descriptions of some newly discovered Species of Araneida ...' *Ann.Mag.nat.Hist.* (2) **6**, p.342 [and (2) **10**, p.185 (1852)]

Blackwall, J. (1861; 1864) *A history of the spiders of Great Britain and Ireland.* Vol.I 1861, 1-174 pls i-xii, Vol.II 1864, 175-384 pls xiii-xxix. Ray Society, London

Bonnet, P. (1956) *Bibliographia araneorum* Tome II (2), p.1106 & note 174 p.1154 [chapter 3]

Bouskell, F. (1907) 'Arachnida - Spiders' pp.94-95, in: Page,W. [ed.] *The Victoria History of the County of Leicestershire.* **1**, Archibald Constable, London

Braendegaard, J. (1932) 'Catalogue of Danish Spiders' item 282, p.703, in: Nielsen,E. *The Biology of Spiders.* Vol 2, pp.725

Bratton, J.H. [ed.] (1991) *British Red Data Books No.3: Invertebrates other than Insects.* JNCC

Bristowe, W.S. (1939; 1941) *The comity of spiders.* Vols I & II. Ray Society, London

Bristowe, W.S. (1958) *The World of Spiders.* Collins New Naturalist Series, London, pp.304

Bristowe, W.S. (1963) 'Notes on Rare Spiders and Courtship as a clue to Relationships'. *Proc.ent.nat.Hist. Soc.S.Lond., 1962.* 184-190

Bryan, P.W. [ed.] (1933) *A Scientific Survey of Leicester and District.* Produced by the British Association for Science Annual Meeting held in Leicester, for which articles were contributed by A.R.Horwood, E.E.Lowe, W.E.Mayes and other members of the Literary and Philosophical Society and its Sections.

Caradine, E.L. (1994) *A Study of the Araneae and Opiliones fauna of wetland sites along the River Soar Valley, Leicester.* pp.149. MSc Thesis, University of Leicester

Caradine E., Page, S. & Bullock, J. (1995) 'Spider and Harvestmen communities of Urban Wetlands'. *Land Contamination & Reclamation.* **3** (2), 77-78

Cooke, J.A.L. & Lampel, G.P. (1953) '*Pirata uliginosus* Thor. (Araneae, Lycosidae) in Shropshire, and some further additions to the County Records of Spiders'. *J.Soc.Br.Ent.* **4** (8), 186-187

Countryside Commission (1994) *The National Forest - The Strategy.* CCP468, Countryside Commission, Cheltenham. pp.110

Crabbe, G. (1795) 'The natural history of the Vale of Belvoir' in: Nichols, J. *The history and antiquities of the county of Leicester.* **1** (1), cxci-ccviii. J. Nichols, London

Crocker, J. (1962a) 'Two rare spiders from Leicestershire'. *Bull.Flatford Mill Spider Group.* **15**, 3. July

Crocker, J. (1962b) 'An Initial Survey of the Spiders of Bradgate Park and Cropston Reservoir margins' pp.73-78, in: *Surveys of Leicestershire Natural History - No.1.* Loughborough Naturalists' Club, pp.108

Crocker, J. (1963) 'Spiders of Charnwood Heath'. *Bull.Flatford Mill Spider Group.* **18**, 4. April

Crocker, J. (1966) 'Some observations on the habitat requirements of *Dysdera* in Leicestershire'. *Bull.Br.Spider Study Group.* **32**, 4-6. October

Crocker, J. (1967) 'Hints, tips and gadgets for spider collectors - The Collecting Bag'. *Bull.Br.Spider Study Group.* **35**, 3-4, 4 figs, July

Crocker, J. (1969) 'Storing a Spider Reference Collection'. *Bull.Br.arachnol.Soc.* **1** (3), 44-47, July 1969

Crocker, J. (1973) 'The habitat of *Tetrilus macrophthalmus* (Kulczynski), in Leicestershire and Nottinghamshire (Araneae: Agelenidae)'. *Bull.Br.arachnol.Soc.* **2** (7), 117-123

Crocker, J. (1979) 'The Sherwood Forest Arachnid Survey'. *Newsl.Br.arachnol.Soc.* **25**, p.5

Crocker, J. [ed.] (1981) *Charnwood Forest: A Changing Landscape.* Loughborough Naturalists' Club & Sycamore Press Ltd. pp.184

Dalingwater, J.E. (1984) 'White Holes and Pink Dwarfs'. *Newsl.Br.arachnol.Soc.* **41**, 5-6. October

Dandy, J.C. (1969) *Watsonian Vice-counties of Great Britain.* Ray Society Publ. No.146, London

David, Carol (1989) *Vegetation History and Pollen Recruitment in a Lowland Lake Catchment; Groby Pool, Leicestershire.* Loughborough University of Technology, Dept.of Geography Working Paper No.4

Donisthorpe, H.St.J.K. (1927) 'Araneina (Spiders)' chapter XIII, in: *The Guests of British Ants.* 184-201. Routledge, London

Douglas, R.N. (1908) 'Invertebrates' pp.38-45, in: Page, W. [ed.] *The Victoria History of the County of Rutland.* Vol.I. Archibald Constable. London [Reprinted 1975]

Duffey, E. (1955) '*Lepthyphantes whymperi* F.O.Pickard-Cambridge (Araneae, Linyphiidae), in Charnwood Forest, Leicestershire, and other new county records'. *Entomologist's mon.Mag.* **91**, 236-237. October

Duffey, E. (1963) 'Ecological studies on the Spider Fauna of the Malham Tarn Area'. *Field Studies.* **1** (5), 1-23

Duffey, E. (1972) 'Ecological survey and the arachnologist'. *Bull.Br.arachnol.Soc.* **2** (5), 69-82

Duffey, E. (1993) 'A review of factors influencing the distribution of spiders with special reference to Britain'. *Memoirs of the Queensland Museum.* **33** (2), 497-502. Brisbane

Evans, I.M. (1963) 'Native or Immigrant? Some Spiders from Banana Warehouses, Leicester'. *Bull.Flatford Mill Spider Group.* **20**, 5, October

Evans, I.M. (1979) 'Spiders' (section 4.12) pp.203-215, in: Evans, I.M. [ed.] *North-east Leicestershire Coalfield. Report of a biological survey, 1978.* Leicestershire Museums, Leicester

Eversham, B.C. (1983) *Defining rare and notable species, with special reference to invertebrates: a discussion.* Invertebrate Site Register. Report No.49, pp.29. Nature Conservancy Council, London

Gilbertson, P. [ed] (1991) *Invertebrate Survey of Lockington Marshes SSSI and Surrounding Area.* Leicestershire Museums, Arts & Records Service, Leicester

Hadfield, Charles (1966) *The Canals of the East Midlands.* David & Charles (Publishers) Ltd, Newton Abbot [2nd edition 1970]

Horwood, A.R. (1907) 'Zoology: Invertebrata', pp.351-352, in: Nuttall, G.C. [ed.] *A guide to Leicester and District.* Leicester: Edward Shardlow (for British Association for the advancement of Science)

Jackson, A.R. (1913) 'On some new and obscure British spiders'. *Trans.Notting.Nat.Soc.* **60**, p.25

Jackson, A.R. (1932) 'On New and Rare British Spiders'. *Proc.Dorset Nat.Hist.F.Cl.* **53**, 200-214

Kulczynski, L. (1897) in: Chyzer, C. & Kulczyski, L. *Araneae Hungariae.* **2** (2), p.156. Budapest

Lehtinen, P.T. (1964) 'Additions to the spider fauna of Southern and Central Finland'. *Ann.Zool.Fenn.* **1**, 303-305

Lister, M. (1678) *Historiae Animalium Angliae Tres Tractatus.* Royal Society, London

Locket, G.H. & Millidge, A.F. (1951; 1953) *British Spiders.* Vols I & II. Ray Society, London

Locket, G.H., Millidge, A.F. & Merrett, P. (1974) *British Spiders.* Vol.III. Ray Society, London

Lott, D.A. & Daws, J. (1993) *Survey of Beetles in some grassland SSSIs in Leicestershire.* A Report to English Nature. Leicestershire Museums, Arts & Records Service, Leicester

Lott, F.B. (1935) *The Centenary Book of the Leicester Literary and Philosophical Society.* Leicester: Thornley. pp.263

Mackie, D.W. (1989) *How to begin the study of Spiders.* British Arachnological Society, pp.11

Menge, A. (1854) in: Koch, C.L. & Berendt, G.C. *Die im Bernstein befindlichen Crustaceen, Myriapoden, Arachniden und Apteren der Vorwelt.* **1** (2), pp.124 [p.8, p.94 (*n.g.*)]. Berlin

Merrett, P. (1964a) 'New County Records of British Spiders'. *Bull.Br.Spider Study Group.* **22**, 3-4

Merrett, P. (1964b) 'New County Records of British Spiders'. *Bull.Br.Spider Study Group.* **23**, 3-4

Merrett, P. (1964c) 'New County Records of British Spiders'. *Bull.Br.Spider Study Group.* **24**, 5-6

Merrett, P. (1965a) 'New County Records of British Spiders'. *Bull.Br.Spider Study Group.* **25**, 3-5

Merrett, P. (1965b) 'New County Records of British Spiders'. *Bull.Br.Spider Study Group.* **26**, 5-6

Merrett, P. (1967) 'New County Records of British Spiders'. *Bull.Br.Spider Study Group.* **34**, 5-6

Merrett, P. (1969) 'New County Records of British Spiders'. *Bull.Br.arachnol.Soc.* **1** (2), 19-21

Merrett, P. (1974) 'Distribution Maps of British Spiders' pp.132-285, in: Locket G.H., Millidge, A.F. & Merrett, P. *British Spiders.* Vol.III. Ray Society, London

Merrett, P. (1975) 'New County Records of British Spiders'. *Bull.Br.arachnol.Soc.* **3** (5), 140-141

Merrett, P. (1982) 'New County Records of British Spiders'. *Bull.Br.arachnol.Soc.* **5** (7), 332-336

Merrett, P. (1989) 'Twelve hundred new County Records'. *Bull.Br.arachnol.Soc.* **8** (1), 1-4

Merrett, P. (1990) *A Review of the Nationally Notable Spiders of Great Britain*. Report No.127, Nature Conservancy Council, Peterborough. pp.67

Merrett, P. (1995) 'Eighteen hundred new County Records'. *Bull.Br.arachnol.Soc.* **10** (1), 15-18

Merrett, P., Locket G.H. & Millidge A.F. (1985) 'A checklist of British spiders'. *Bull.Br.arachnol.Soc.* **6** (9), 381-403

Merrett, P. & Millidge A.F. (1992) 'Amendments to the checklist of British spiders'. *Bull.Br.arachnol.Soc.* **9** (1), 4-9

Messenger, Guy (1971) *Flora of Rutland*. Leicester Museums, Leicester

Mott, F.T. (1890) Quarterly Report of Section"D" (Biology) - Exhibits 15.1.1890. *Trans.Leicester lit.phil.Soc.* New Quarterly Series **2** (3), p.146. April

Nicholson,P.B. [ed.] (1975) *Wildlife Conservation in Charnwood Forest: Report by a Working Party*. Nature Conservancy Council, Midlands Region. pp.51

Nielsen, E. (1932) *The Biology of Spiders,* in 2 Vols, Vol.1 [English] summarising Vol.2 [Danish]; figs in Vol.2. Levin & Munksgaard, Copenhagen

Odenvall & Jarvi (1901) *Acta Soc.Fauna Flora fenn.* **20** (4), 1-12

Owen, J. (1991) 'Spiders' pp.325-330, in: *The Ecology of a Garden, The first Fifteen Years*. Cambridge University Press

Parker, J. & Harley, B. [eds] (1992) *Martin Lister's English Spiders 1678*. Harley Books, Colchester

Perring, F.H. & Walters, S.M. [eds] (1962) *Atlas of the British Flora*. B.S.B.I., pp.432

Pickard-Cambridge, O. (1881) *The Spiders of Dorset*. Part 2, 236-625. Sherborne

Pickard-Cambridge, O. (1900) *List of British & Irish Spiders*. Dorchester. pp.86

Pilawski, S. (1966) 'Preliminary investigations of Spiders in Swietokrzski Mountains'. *Proc.zool.Wroclaw Univ.* **2** (51), p.12

Primavesi, A.L. & Evans, P.A. [eds] (1988) *The Flora of Leicestershire*. Leicestershire Museums Publication No.89, pp.486

Ratcliffe, D.A. [ed] (1977) *A Nature Conservation Review: The selection of biological sites of national importance to nature conservation in Britain*. Vol.I. Published on behalf of the Nature Conservancy Council and the Natural Environment Research Council. Cambridge University Press, London

Roberts, M.J. (1985; 1987; 1993) *The Spiders of Great Britain and Ireland*. Vol.I 1985, pp.229; Vol.II 1987, pp.204; Vol.III 1985, pp.256; Supplement 1993, pp.16. Harley Books, Colchester

Roberts, M.J. (1995) Collins Field Guide: *Spiders of Britain and Northern Europe*. pp.383 Harper Collins, Glasgow

Rowley, F.R. (1897) Quarterly Report of Section "F" for Entomology. *Trans.Leicester lit.phil.Soc.* N.S. **4** (10), 435-437 [extract of paper "The Aranidae (Spiders) of Leicestershire" by G.B.Chalcraft, read before the Section 2nd July 1897]

Simon, E. (1884) *Les arachnides de France*. **5** (2), p.327

Simon, E. (1886) 'Descriptions de quelques especes nouvelles de la famille des Agelenidae'. *Ann.Soc.ent.Belg.* **30**, p.LX

Simon, E. (1937) *Arachn.de France*. **6** (5), p.1022-1023

Smith, C.J. (1987) 'The Spider Recording Scheme'. *Newsl.Br.arachnol.Soc.* **48**, 1-2

Squires, A.E. & Jeeves, M. (1994) *Leicestershire & Rutland Woodlands, Past & Present*. Kairos Press, Newtown Linford

Staveley, E.F. (1866) *British Spiders*. pp.280 + 16 pls. Lovell Reeve, London

Thorell, T. (1871) *Remarks on synonyms of European spiders*. Pt II, p.165. Uppsala

Wanless, F.R. (1973) 'The female Genitalia of British Spiders of the Genus *Lepthyphantes* (Linyphiidae) II'. *Bull.Br.arachnol.Soc.* **2** (7), 127-142

Wild, A.M. (1952) 'Additions to the known spider fauna of Glamorgan, Carmarthen, Monmouth and other counties in the British Isles'. *Rep.Trans.Cardiff Nat.Soc., 1948-50*. **80**, 8-11

Wunderlich, J. (1973) 'Weitere seltene und bisher unbekannte Arten sowie Anmerkungen zur Taxonomie und Synonymie (Arachnida: Araneae): Spinnenfauna Deutschlands, XV'. *Senckenberg. biol.* **54** (4/6), 405-428

Wunderlich, J. (1986a) *Fossile Spinnen in Bernstein und ihre lebenden Verwandten*. Quelle & Meyer, Wiesbaden

Wunderlich, J. (1986b) '"Lebende Fossilien" in Europa ?' *Fossilien* **4**, 177-181

INDEX

All spiders are indexed by species names first, with the generic name in brackets. The page numbers in bold indicate an entry in the Atlas section.

Ab Kettleby: 200, 203, 207
abnormis (Saaristoa): 28, **172**, 218, 221, 225, 228, 231, 235
accentuata (Anyphaena): 23, 35, 41, **92**, 215, 232
acid grassland: 21
Acresford Sandpit: 84, 86, 88, 104, 112, 149, 168, 194, 200, 202, 214, 219
acuminata (Entelecara): 26, 53, **140**, 217, 223, 234, 237
acuminata (Mastigusa): 70, 71
acuminata (Walckenaeria): 26, 43, **136**, 217, 220, 223, 227, 230, 234, 237
aequipes (Euophrys): 24, **102**, 194, 215, 219, 222, 229
affinis (Tmeticus): 26, **142**, 191, 194, 227, 230
Agar, William: 32, 33, 39, 62, 129
AGELENIDAE: 25, 110-114
agrestis (Oedothorax): 27, 54, **148**, 217, 224, 227, 230, 234
agrestis (Pardosa): 24, **103**, 190, 196, 215
agrestis (Tegenaria): 25, **112**, 194, 195, 216, 219, 233
alacris (Lepthyphantes): 28, **179**, 225, 231, 235
albidus (Philodromus): 24, 69, 77, **99**, 190, 196
Allexton Wood: 181, 190
alsine (Araneus): 30, 34, 35, 37, 41, 48, 49
Altar Stones: 114, 146, 200, 202
alticeps (Bolyphantes): 28, **178**, 193, 196, 225, 235
AMAUROBIIDAE: 22, 78
Ambion Wood: 36
amentata (Pardosa): 24, 35, 38, 41, **105**, 193, 216, 219, 222, 226, 229, 232, 236
ancient woodland: 16, 81, 92, 127, 134, 193, 197
angulatus (Araneus): 38-39
angulatus (Episinus): 25, 38, **116**, 216, 233
Anstey: 125
antica (Walckenaeria): 26, **136**, 217, 220, 223, 227
ANYPHAENIDAE: 23, 92
apertus (Micrargus): 22, 27, 69, **157**, 193, 224, 234
apicatus (Oedothorax): 27, **149**, 217, 220, 224, 227, 230
approximatus (Bathyphantes): 28, **173**, 221, 225, 228, 231, 235
apricorum (Zelotes): 23, 43, **85**, 219, 232
aquatica (Argyroneta): 24, 33, **110**, 219, 222
ARANEIDAE: 26, 129-134
arctica (Erigone): 27, **162**, 196, 217, 227, 231
ARGYRONETIDAE: 24, 110
arietina (mastigusa): 71
arundinacea (Dictyna): 22, **79**, 215, 219, 222, 226, 229, 232, 236
Asfordby: 47, 212
Ashby Canal: 12
Ashby Woulds: 77, 152, 201, 209
Ashby-de-la-Zouch: 11, 12, 13, 19, 102, 111, 199
Asplin and Pasture Woods: 197
atomaria (Ozyptila): 24, 52, **96**, 215, 236
atra (Erigone): 27, **162**, 193, 217, 220, 224, 227, 231, 234, 237
atrica (Tegenaria): 54
atrica (Zygiella): 26, 33, 34, 35, 39, 41, **129**, 223, 230, 233
atropos (Coelotes): 25, 35, 38, 41, 42, **113**, 216, 222, 233
atrotibialis (Walckenaeria): 26, **137**, 217, 223, 234
audax (Xysticus): 23, **94**, 215
aureolus (Philodromus): 24, 38, **97**, 215, 222, 226, 229, 232, 236
Aylestone: 35
Aylestone Holt: 181
Aylestone Meadows: 79, 135, 160, 200, 202, 214, 226
Aylestone power station: 21

Bagworth: 13
Ball, Harry: 60, 62
Baltic amber: 70, 71
banana spiders: 37, 44, 47

barbipes (Alopecosa): 24, 77, 106, 216
Bardon Hill: 9, 13, 32, 35, 36, 47, 50, 52, 67, 72, 77, 83, 84, 86, 87, 91, 100, 102, 103, 104, 114, 124, 136, 138, 144, 146, 148, 150, 151, 153, 154, 157, 164, 165, 167, 168, 171, 173, 178, 182, 183, 184, 193, 196, 200, 201, 202, 214, 222
Barkby: 202
Barkby Holt Wood: 87, 118, 125, 134
Barkby Thorpe Park: 98, 202
Barnack Hills and Holes: 10, 49, 134
Barnsdale Wood: 51, 92, 155, 182, 184, 202
Barrow Gravel Pits: 67, 142, 164, 177, 186, 200, 202, 214, 226
Barrow upon Soar: 148, 202
Barrow, W.H.: 37, 39
Beacon Hill: 43, 47, 67, 72, 84, 86, 92, 102, 114, 124, 136, 150, 171, 176, 193, 196, 200, 202, 214, 222
beata (Meioneta): 28, **170**, 221, 228
Beaumanor: 32, 67
beckeri (Lepthyphantes): 22, 28, 50, 55, 74, 75, 77, **181**, 193, 225
Bee, L.: 73
Belton: 78
Belvoir, Vale of: 9, 17, 32, 182
Benscliffe Wood: 46, 67, 92, 124, 165, 173, 193, 200, 202
Bescaby: 17
bicolor (Centromerita): 28, 35, 37, 41, **172**, 218, 221, 225, 228, 235, 237
bifrons (Dismodicus): 26, 53, **143**, 217, 220, 223, 227, 230, 234, 237
Biggs, L.S.: 37, 62
Billa Barra: 146, 150
bimaculatum (Theridion): 25, 38, **121**, 216, 219, 223, 226, 229, 233, 236
Biological Records Centre: 22, 47, 66, 68
biovatus (Thyreosthenius): 27, 30, 40, 41, 48, 77, **154**, 234
bipunctata (Steatoda): 25, 35, 41, **117**, 193, 216, 223, 229, 236
Birstall: 158, 178, 179
Bisbrooke: 78
bituberculatum (Hypomma): 26, **143**, 220, 223, 227, 230, 234
Bittesswell Aerodrome: 94, 202
Blaby: 33, 35
Blackbrook Reservoir: 19, 20
Blackwall, John: 31, 34, 35, 39, 42, 45, 48, 49
blackwalli (Scotophaeus): 23, 40, 41, **84**, 229, 236
blackwalli (Theridion): 25, 121
Blakeshay: 67
Bloody Oaks Quarry: 55, 94, 95, 190, 203
Boggild, O: 72
Bonner, Alec: 40, 62
Bosworth Duckery: 118
Bosworth Park: 170
Botcheston Bog: 20, 79, 109, 158, 200, 203
boulder clay: 13, 15
Bouskell, Frank: 32-36, 40, 48, 49, 62
Bradgate House: 74, 75
Bradgate Park: 17, 32, 36, 39, 40, 47, 50, 67, 72, 75, 77, 81, 86, 92, 100, 102, 105, 106, 109, 114, 135, 136, 145, 152, 154, 157, 158, 165, 166, 168, 171, 178, 181, 193, 196, 200, 203, 214, 222
Brand, the: 38, 43, 67, 77, 85, 92, 100, 116, 117, 148, 193, 196, 200, 203
Braunstone: 148
Brazil Wood: 153, 154, 198, 200, 203
Breedon Cloud Quarry: 91
Breedon Hill: 12, 132, 154, 169, 174, 203
Breedon-on-the-Hill: 200, 204
brevipes (Ceratinella): 26, **135**, 217, 220, 223, 227, 233
brevipes (Clubiona): 23, **89**, 215, 222, 226, 229, 232, 236

brevipes (Ozyptila): 24, 41, **97**, 215
brevis (Ceratinella): 26, **135**, 217, 220, 223, 227, 230, 234, 237
Briery Wood: 182
Bristowe, W.S.: 30, 38, 40-42, 51, 62, 71
British Arachnological Society: 22, 31, 44, 45, 48, 54, 66, 67, 68, 69, 73, 238
British Museum: 46, 73
British Spider Study Group: 31, 44, 45, 46, 67
broad specialists: 195
Broombriggs: 67
Brown's Hill Quarry: 109, 115, 122, 200, 203
brunnea (Agroeca): 23, 33, **91**, 193, 232
bucculenta (Floronia): 28, 42, **177**, 228, 231, 235
Buck Hill: 67
Buddon Brook: 177
Buddon Brook Marsh: 200, 203
Buddon Wood: 13, 16, 32, 35, 36, 37, 38, 39, 48, 49, 67, 77, 79, 81, 84, 86, 87, 89, 91, 92, 101, 102, 103, 105, 112, 114, 115, 116, 117, 118, 119, 120, 122, 124, 129, 131, 132, 134, 135, 136, 137, 138, 140, 144, 145, 148, 152, 153, 154, 157, 165, 168, 171, 173, 181, 182, 193, 196, 197-199, 200, 203, 214, 232
Burbage: 127
Burbage Common: 79, 104, 119, 200, 203
Burbage Wood: 77, 80, 118, 134, 155, 169, 181, 182, 197, 200, 203, 214, 232
Burley Wood: 16, 134
Burrough Hill: 12, 32, 80, 113, 157, 170, 200, 204
Burrough Hill Wood: 181
Burton-on-the-Wolds: 201, 212

C.E.G.B. Rawdykes: 108, 112, 115, 149, 188, 194, 200, 204
Cademan Moor: 79, 83, 84, 91, 122, 135, 136, 138, 139, 140, 146, 150, 152, 163, 178, 193, 200, 204, 214, 222
calcareous grassland: 195, 215-218
Calke Uplands: 199
cambridgei (Ero): 25, **115**, 216, 219, 223, 226, 229, 233
campbelli (Porrhomma): 28, **166**, 220, 227
Canals: 18-19, 87
Ashby: 19, 45, 110, 119, 148, 177, 200, 202
Charnwood: 19
Coventry: 19
Grand Junction: 18
Grand Union: 18, 19, 86, 88, 110, 142, 206
Grantham: 19, 110, 206
Leics & Northants Union: 19
Oakham: 19
Union: 18, 19
capito (Walckenaeria): 26, 77, **137**, 193, 196, 223
Caradine, Emma: 55, 60, 62, 68
carboniferous limestone: 13, 14
Carlton Curlieu: 160
carri (Lepthyphantes): 72
castaneipes (Monocephalus): 27, 77, **155**, 193, 196, 224
Castle Donington: 12, 200, 204
Catmose, Vale of: 12
cauta (Agyneta): 28, 77, **168**, 193, 224
cellulanus (Nesticus): 25, 52, 54, **124**, 223, 230, 233, 236
cespitum (Philodromus): 24, 41, 53, **98**, 215, 219, 222, 226, 232, 236
Chalcraft, G.B.: 32, 33-35, 40, 48, 49, 55, 62
chalybeius (Ballus): 24, 64, 77, **101**, 197, 232
Charley Hall: 32
Charnwood Forest: 9, 10, 12, 15, 16, 20, 21, 36, 43, 44, 45, 46, 47, 50, 53, 61, 66, 67, 70, 71, 72, 80, 81, 83, 89, 106, 107, 110, 113, 114, 123, 134, 136, 138, 141, 153, 165, 171, 176, 178, 188, 189, 190-193, 196, 199
Charnwood Forest Survey: 61, 63, 64, 67
Charnwood Lodge Nature Reserve: 21, 55, 67, 77, 83, 86, 89, 90, 92, 102, 103, 104, 110, 114,

243

119, 123, 124, 132, 136, 138, 139, 140, 145, 146, 150, 151, 152, 153, 154, 155, 156, 157, 165, 167, 168, 170, 171, 172, 173, 176, 178, 180, 181, 183, 193, 196, 200, 204, 214, 222
cicur (Cicurina): 25, **194**, 216, 219
cingulatus (Salticus): 24, **100**, 215, 222, 226, 229
clathrata (Linyphia): 29, 38, **185**, 218, 221, 225, 228, 231, 235, 237
Clements, H.A.B.: 61, 62
clercki (Pachygnatha): 26, 41, **126**, 216, 220, 223, 226, 230, 233, 236
Climate: 15
Cloud Wood: 118, 154, 197, 200, 204
CLUBIONIDAE: 23, 86-90
coal measures: 13, 14, 190, 219-221
Coalbourne Wood: 154
coalfield: 11, 12, 13, 45, 47, 61, 68
Coalfield West, Ravenstone: 54, 96, 161, 183, 188, 193, 200, 204, 214, 219
Coalville: 11, 199
Cocklow Quarry: 198
Coleorton: 12, 13, 19, 83, 86, 88, 91, 100, 101, 115, 138, 139, 140, 151, 208, 219
Coleorton Hall: 77, 82, 103, 113, 120, 160, 169, 172
collecting techniques: 63-65
collection books: 65, 66
colliery spoil heaps: 47
collinus (Philodromus): 24, 69, **98**, 190, 191, 215, 232
Comity of Spiders: 40, 41, 42
common species: 22
computer data storage: 65, 68
comta (Clubiona): 23, 41, **89**, 215, 222, 232, 236
concinna (Centromerita): 28, **172**, 218, 221, 225, 228, 235
concolor (Diplostyla): 28, 41, **175**, 193, 218, 221, 225, 228, 231, 235, 237
conica (Cyclosa): 26, 35, 41, **134**, 197, 217, 233
conigera (Agyneta): 28, **167**, 220, 224, 235
conifer woodland: 17
convexum (Porrhomma): 28, 54, 77, **165**, 234
Cooke, J.A.L.: 43, 62
Cooper's Plantation: 91, 95, 105, 200, 204
Coppice Leys Wood: 134, 157
coppice-with-standards: 16
cornutum (Hypomma): 26, 41, **143**, 217, 223, 230, 234, 237
cornutus (Larinioides): 26, 42, **131**, 216, 223, 226, 230, 233
corticalis (Clubiona): 23, 43, **86**, 222, 229, 232, 236
Cossington: 17
Cotes: 142, 162, 194
Cottesmore: 51
Countesthorpe: 146
Countryside Commission: 199
county records: 55-60
Crabbe, George: 31
cristatus (Diplocephalus): 27, **158**, 217, 224, 227, 230, 234, 237
cristatus (Lepthyphantes): 28, 74, **181**, 225, 231, 235
cristatus (Xysticus): 23, 35, 38, 41, **93**, 193, 215, 219, 222, 226, 229, 232, 236
crocata (Dysdera): 28, 40, 46, **81**, 215
Crocker, John: 44, 45-47, 49, 52, 54, 61, 62, 67, 69, 72, 74, 75
Crocker, M.G.: 38, 61, 62, 74, 75
Croft Hill: 13
Cropston: 143, 156
Cropston Brickpit: 204
Croxton Kerrial: 200, 204, 207
Croxton Park: 17, 98, 129, 200, 204
cucullata (Walckenaeria): 26, **136**, 220, 223, 234
cucurbitina (Araniella): 26, 34, 35, 37, 38, 41, 54, **133**, 216, 223, 227, 233, 237
cupreus (Drassodes): 23, 35, **83**, 215, 219, 222, 232
cupreus (Heliophanus): 24, 77, **100**, 232
cuspidata (Walckenaeria): 26, **139**, 217, 223, 227, 230, 234

Dalingwater, J.E.: 51
Daws, Jon: 44, 54-55, 61, 62, 64, 67

Dawson, J.E.: 62
decora (Agyneta): 28, **168**, 220, 224
degeeri (Pachygnatha): 26, 38, 41, **127**, 216, 220, 223, 226, 230, 233, 236
dentatum (Gnathonarium): 26, **142**, 220, 223, 227, 230, 234
dentichelis (Lessertia): 27, 52, **160**, 237
denticulata (Textrix): 25, 35, 37, 41, 53, **110**, 216, 219, 222, 229, 233
dentipalpis (Erigone): 27, **161**, 217, 220, 224, 227, 231, 234, 237
Desford: 200, 203
diadematus (Araneus): 26, 33, 35, 38, 41, **130**, 193, 216, 223, 230, 233, 237
diceros (Saloca): 27, **155**, 197, 217, 227, 234
DICTYNIDAE: 22, 79-80
Dietrick Vacuum Sampler: 64
digitatus (Typhochrestus): 27, 41, **161**, 190, 217
dilutus (Centromerus): 28, **171**, 218, 221, 224, 231, 235, 237
diorite: 13
dispar (Philodromus): 24, 38, 53, **97**, 215, 229, 232, 236
distinctus (Halorates): 28, **164**, 191, 194, 196, 227
distribution maps: 76-187
distribution of spiders: 194
diversa (Clubiona): 22, 23, **90**, 215, 219, 222, 232
DMAP for Windows: 76
Dobson, Stanley: 62, 68, 69
domestica (Tegenaria): 25, 35, 37, 41, 48, 52, **112**, 193, 222, 233, 236
domesticus (Oonops): 22, 42, 52, **80**, 236
Donington Park: 12, 17, 72, 73, 77, 86, 101, 102, 114, 152, 161, 167, 183, 196, 200, 204, 214
Donisthorpe: 12, 200, 202
Donisthorpe Colliery: 108
Donisthorpe, H.St.J.: 30, 33-34, 39, 62, 63, 71, 72
dorsalis (Kaestneria): 28, **175**, 218, 231, 235
dorsata (Diaea): 23, **93**, 190, 196, 215, 232
Drift, the: 79, 84, 85, 96, 97, 109, 133, 135, 136, 137, 157, 190, 200, 205, 214, 215
Drybrook Wood: 136, 145, 157, 164, 165, 200, 205
Duddington: 110
Duffey, Eric: 31, 43, 61, 62, 63, 68
Dunton Bassett: 149, 152, 168, 200
Dunton Bassett Quarry: 205
DYSDERIDAE: 23, 81
dysderoides (Walckenaeria): 26, 138, 217, 220, 223, 227, 234

east Leicestershire: 12-15, 21
egeria (Porrhomma): 28, 77, **167**, 193, 224
Egleton: 55, 88
elegans (Antistea): 25, **114**, 223, 226, 229, 233
elegans (Silometopus): 27, **150**, 220, 224
Ellistown: 51
Empingham: 201
Enderby: 82
English Nature: 188, 189, 199
Environment Agency: 18
Epping Forest: 61, 74
ericaeus (Lepthyphantes): 29, 43, **182**, 218, 221, 225, 228, 231, 235, 237
errans (Porrhomma): 28, **166**, 194, 217, 228
erratica (Euophrys): 24, **102**, 192, 222, 232
erraticus (Xysticus): 22, 23, 49, **94**, 215, 222
erythrina (Dysdera): 23, 46, **81**, 215, 222, 232
erythropus (Entelecara): 26, **141**, 217, 223, 237
Essendine: 77, 99, 103, 115
Evans, I.M.: 40, 42, 44-47, 53, 61, 62, 68, 69
Evesham Road Allotments: 115, 152, 161, 166
evolutionary cycle: 70
experta (Tallusia): 28, **171**, 221, 225, 228, 231, 235
extensa (Tetragnatha): 25, 35, 38, 41, 42, **124**, 216, 219, 223, 226, 230, 233, 236
Exton: 93, 98, 106, 202
Eye Kettleby Mill: 17

falcata (Evarcha): 24, 38, **103**, 215, 229, 232
Farnham's Wood: 182, 184, 205
fenestralis (Amaurobius): 22, 35, 41, **78**, 215, 219, 222, 226, 229, 232, 236

fera (Phoeutria): 44
ferox (Amaurobius): 22, 35, 36, 37, **78**, 193, 215, 222, 229, 236
festivus (Phrurolithus): 23, **91**, 215, 219, 226, 229, 232
field notebook: 65
fieldwork: 61
firma (Saaristoa): 28, **173**, 225, 235
Flatford Mill Spider Group: 31, 44, 45, 46
flavipes (Heliophanus): 24, 38, **100**, 215, 222, 236
flavipes (Lepthyphantes): 29, 74, **182**, 218, 225, 228, 235
flavomaculata (Euryopis): 25, 49, **116**, 194, 219, 233
flood meadows: 17
fog: 16
Fort Henry Lakes: 205
Fosse Meadows Nature Park: 205
Foxton: 19, 142
Foxwell, Derek: 44
Frisby-on-the-Wreake: 177
frontalis (Euophrys): 24, **101**, 215, 219, 222, 229, 232, 236
frontata (Savignia): 27, **158**, 217, 220, 224, 227, 230, 234, 237
furcata (Ero): 25, 42, **116**, 216, 219, 223, 233, 236
furcillata (Walckenaeria): 26, **138**, 220, 223, 234
fuscipes (Monocephalus): 27, **154**, 217, 220, 224, 227, 230, 234
fuscus (Oedothorax): 27, 43, 53, 54, **148**, 217, 220, 224, 227, 230, 234, 237

gallicus (Maso): 27, 51, **146**, 190, 191, 196, 217
Gamble, Peter: 61, 62, 197
Gardens: 52, 200, 205-206, 214, 236-237
Garden Spiders: 52-53, 236-237
Garendon Park: 206
Geeston Quarry: 62, 64, 77, 79, 81, 83, 84, 85, 86, 89, 92, 95, 100, 103, 106, 108, 110, 112, 113, 115, 116, 119, 122, 135, 136, 146, 149, 150, 152, 154, 157, 160, 161, 188, 190, 194, 196, 200, 206, 214, 215
generalist spiders: 195
geological influences: 195
geology: 13, 15
gibbosa (Cyclosa): 197
gibbosa (Gibbaranea): 26, **129**, 197, 216, 230, 233
gibbosus (Oedothorax): 27, 48, 51, **147**, 220, 224, 227, 230, 234
gibbosus f.tuberosus (Oedothorax): 27, 48, 51, **147**, 220, 224, 227, 230, 234
gibbum (Pholcomma): 25, **123**, 219, 223, 230, 233
gigantea (Tegenaria): 25, 33-35, 37, 41, 42, 53, 54, **111**, 216, 219, 233, 236
Glaston: 78
GNAPHOSIDAE: 23, 83-86
Goddard, D.G.: 61, 62
Grace Dieu: 46, 67, 92, 127, 148, 193, 200, 204
gracilis (Bathyphantes): 28, 41, 53, **174**, 193, 218, 221, 225, 228, 235, 237
graminicola (Hylyphantes): 26, 51, **141**, 217, 230, 234
Grange Wood: 36
grassland: 20
gravel pits: 87
Great Bowden Pit: 20, 54, 87, 110, 119, 200, 206
Great Casterton: 121
Great Easton: 16, 92, 103, 200, 206
Great Merrible Wood: 127, 135, 140, 145, 153, 155, 181, 190, 200, 206
Greater Cotswolds: 189
Green, Colin: 61
Groby: 32, 168
Groby Pool: 19, 46, 67, 77, 126, 142, 143, 163, 181, 194, 200, 206, 214, 229
Gumley: 88
Gun Hill: 72, 222
guttata (Crustulina): 25, 35, 41, **117**, 216
Gypsy Lane Brickpit: 112, 206

habitat diversity: 195
habitat evaluation: 188

HAHNIIDAE: 25, 114-115
Hambleton: 186
Hambridge: 18
Hangingstone: 67
Harby: 19
Harby Hills: 84, 85, 115, 122, 135, 155, 160, 190, 200, 206, 214, 215
Headly, C.B.: 33, 36, 49, 55, 62
heath grassland: 195
Heather: 13, 78
heathland: 21, 195
helveola (Hahnia): 25, **115**, 192, 193, 229, 233
herbigradus (Micrargus): 27, 53, **156**, 217, 220, 224, 227, 230, 234, 237
Heritage: 46
HETEROPODIDAE: 23, 92
Heyday Hays: 176
hiemalis (Erigonella): 27, **157**, 217, 224, 234
High Sharpley: 21, 43, 47, 72, 83, 89, 91, 92, 94, 114, 124, 138, 144, 146, 150, 151, 152, 171, 176, 180, 182, 193, 196, 200, 206, 214, 222
High Tor: 155
Hill Hole: 72, 114, 146, 178, 200, 207
Hinckley: 11, 16, 200, 203
Holly Hayes Wood: 154
Holocene: 15
Holwell Mouth: 109, 114, 200, 207
Holwell North Quarry: 109, 200, 207
hombergi (Harpactea): 22, 23, 38, **81**, 222, 229, 232, 236
Horn: 205
hortensis (Linyphia): 29, 41, **185**, 218, 225, 228, 231, 235, 237
Horwood, A.R.: 37, 40
Hose: 47, 145
Houghton-on-the-Hill: 12
house spiders: 52, 236-237
Humberstone Quarry: 86, 88, 112, 115
humilis (Araeoncus): 27, **160**, 220, 227, 231, 234
humilis (Lathys): 22, 52, **80**, 215, 229, 232, 236
Humphries Rowell Associates: 54, 61, 62
Husbands Bosworth: 17, 19
hygrophilus (Pirata): 24, **108**, 216, 219, 222, 226, 229, 233

Ibstock: 103, 199
identification aids: 51
ignobilis (Erigonella): 27, **158**, 224, 227, 230
Ikin, Helen: 54, 61, 62
impigra (Microlinyphia): 29, **186**, 228, 231
impressum (Theridion): 25, **119**, 216, 223
improved grassland: 12
incisa (Walckenaeria): 26, **137**, 197, 217, 230, 234
inerrans (Milleriana): 27, **161**, 194, 195, 217, 220, 224, 227
innotabilis (Meioneta): 28, 77, **169**, 235
insecta (Tapinocyba): 27, **153**, 224, 234
insignis (Helophora): 29, **184**, 225, 231, 235
insignis (Lepthyphantes): 29, **183**, 193, 194, 221, 225
Institute of Terrestrial Ecology: 66
inundation meadow: 11
ironstone: 12, 13
Ives Head: 47, 67, 72, 77, 104, 114, 171, 193, 201, 207

Jackson, A.R.: 39, 45
Jerrard, P.C.: 61, 62
John-o'-Gaunt Fox Covert: 103
Jones, P.E.: 73
juncea (Pocadicnemis): 27, 69, **147**, 217, 220, 224, 227, 230, 234, 237
Jurassic rocks: 13, 14

Kalamazoo Visible Binder system: 66, 68
Kaye, W.J.: 33, 34, 36, 62
Kegworth: 142, 148
Kendall's Meadow: 104, 135, 139, 160, 201, 207
Ketton: 13, 121, 200, 206
Ketton Quarry: 55, 77, 79, 80, 81, 92, 94, 95, 96, 99, 100, 103, 104, 107, 109, 110, 113, 115, 116, 119, 122, 125, 132, 133, 135, 136, 137, 146, 156, 157, 160, 190, 201, 207, 214, 215

Keuper marl: 13, 14
Kibworth: 34, 110
Kilby Bridge: 19
King Lud's Entrenchments: 85, 90, 91, 92, 95, 96, 100, 109, 115, 117, 122, 133, 135, 136, 137, 147, 149, 152, 157, 161, 171, 176, 181, 190, 201, 207, 214, 215
Kirby Muxloe: 102
Kirk, Christine: 55
Knighton: 236
Knighton Park: 205
Knighton Spinney: 80, 86, 87, 113, 118, 134, 201, 207
Knipton: 19

La Touche, A.A.: 69
labyrinthica (Agelena): 25, 38, **110**, 190, 196, 216
Lampel, G.P.: 43, 62
Lane, C.J.: 40
lanigera (Euophrys): 24, **102**, 195, 236
lanio (Xysticus): 24, 77, **94**
lapidosus (Drassodes): 23, 38, 41, **83**, 215, 222, 236
latebricola (Gongylidiellum): 27, 77, **156**, 190, 217
latens (Dictyna): 22, **79**, 215
latifrons (Diplocephalus): 27, 43, **159**, 217, 220, 224, 227, 230, 234
latitans (Pirata): 24, **109**, 222
latreillei (Zelotes): 23, 43, **84**, 194, 215, 219, 222, 226, 232
Launde Big Wood: 92, 95, 101, 106, 125, 127, 173, 190, 201, 207, 214, 232
Launde Park: 98, 208
Lea Meadows: 55
Lea Wood: 67, 95, 201, 208
Leicester: 10, 11, 37, 40, 42, 52, 55, 61, 77, 78, 80, 82, 86, 93, 100, 102, 103, 112, 115, 140, 150, 153, 200, 201, 202, 205, 206, 207, 211, 212, 236
Leicester Cattle Market: 86, 112, 139, 153, 161, 194, 201, 208
Leicester Literary and Philosophical Society: 31, 33, 39, 42, 44, 46, 61, 67, 239
Leicester Museum: 39, 43, 44, 67
Leicester University: 32, 42, 51, 55, 61, 62
Leicestershire Coalfield: 12
Leicestershire Entomological Society: 238
Leicestershire Fauna Survey: 32, 42, 43, 61
Leicestershire Museums Service: 40, 46, 51, 54, 55, 68
Leicestershire Reference Collection: 69
Leicestershire Spider Survey: 63
leprosus (Lepthyphantes): 22, 28, **179**, 225, 228, 235, 237
Lias: 13, 14
life strategies: 195
limestone: 12-14, 188-191
limestone grassland: 21
Lincolnshire Limestone: 14, 188-191, 196
lineatus (Stemonyphantes): 28, 42, **177**, 218, 221, 225, 228, 231, 237
LINYPHIIDAE: 26, 135-187
LIOCRANIDAE: 23, 91
listeri (Pachygnatha): 26, 38, **127**, 197, 233
Little Dalby: 78, 134
lividus (Robertus): 25, 38, 41, **123**, 216, 219, 223, 226, 230, 233
Living Fossils: 70
local species: 22
Locket, G.H.: 30, 45, 69, 71
Lockington Grounds: 9, 17
Lockington Marsh: 15, 17, 20, 54, 79, 144, 201, 226
Lockington Meadows: 100, 142, 162, 164, 201, 208
Loddington Reddish: 43, 106, 142, 159, 177, 190
Long Whatton: 97
Longcliffe: 67
longidens (Tapinopa): 28, **176**, 218, 225, 235
longipalpis (Erigone): 27, **162**, 194, 196, 227
Lott, Derek: 44, 54, 55, 61, 62, 64, 67
Loughborough: 11, 12, 18, 52, 78, 82, 86, 96, 102, 140, 141, 150, 153, 157, 160, 164, 181, 205, 206, 236

Loughborough Meadows: 20, 61, 64, 84, 115, 135, 148, 150, 153, 160, 170, 201, 208, 214, 226
Loughborough Naturalists' Club: 44, 45, 46, 61, 66, 67, 239
Loughborough University: 62
Lount: 54, 106, 138, 140, 150, 151, 160, 170, 188, 194, 197, 219
Lount Colliery: 122
Lount Grassland: 201, 208, 214
Lount Meadow: 135, 201, 208, 214
Lowe, E.E.: 37, 38, 39, 40, 55, 62
lowland heath: 12, 21
LRTNC: 55, 61, 67
ludicrum (Peponocranium): 27, 77, **146**, 217, 224
Luffenham Heath: 21, 79, 80, 87, 92, 93, 94, 95, 97, 98, 118, 119, 120, 122, 123, 125, 129, 131, 133, 134, 140, 146, 157, 161, 190, 201, 208, 214, 215
lugubris (Pardosa): 24, **106**, 219, 222, 229, 232
lunata (Achaearanea): 25, **118**, 216, 229, 233
luteolus (Bolyphantes): 28, **178**, 225
lutescens (Clubiona): 22, 23, **88**, 215, 222, 226, 229, 232, 236
Lutterworth: 11
LYCOSIDAE: 24, 103-109
Lyndon Wood: 118, 186

Mackie, D.W.: 61, 62, 69
macrophthalma (Mastigusa): 22, 25, 40, 41, 46, 47, 50, 52, 64, 70-72, 74, **114**, 192, 193, 196, 197, 223, 233
main spirit collection: 65
Malaise trap: 52, 54, 100, 143
Manchester University Museum: 68, 69
Market Bosworth: 36, 110
Market Harborough: 11, 40
Markfield: 32, 72, 88, 178, 199, 200, 202, 207
marmoreus (Araneus): 26, 30, 35, 41, 48, 76, 77, **130**, 233
Martinshaw Wood: 17, 80, 113, 118, 197, 201, 209
Mathias, John: 61
Mayes, W.E.: 39, 62
Mayes, W.P.: 40, 55, 62
Measham: 12, 13, 199
melanopygius (Ostearius): 28, **164**, 195, 220, 224, 227, 231, 237
melanurum (Theridion): 25, 42, 52, 54, 69, **120**, 236
Melton Mowbray: 11, 12, 17, 18, 19
menardi (Meta): 26, **128**, 230
mengei (Lepthyphantes): 28, 53, 74, **181**, 218, 225, 231, 235, 237
mengei (Metellina): 26, 53, **128**, 216, 220, 223, 226, 230, 233, 237
merens (Evansia): 27, **151**, 193, 220, 224
merianae (Metellina): 26, 35, 38, 41, **128**, 216, 223, 230, 233
Merrett, Peter: 22, 30, 45, 47, 49, 50, 51, 62, 66, 77
Merry Lees: 13
METIDAE: 26, 127-129
microphthalmum (Porrhomma): 28, 53, **166**, 217, 220, 224, 228, 231, 235, 237
midas (Lepthyphantes): 22, 29, 50, 51, 72, 73, 74, 77, **183**, 196, 197, 225
Middle England: 189
Millidge, A.F.: 30, 45, 71
MIMETIDAE: 25, 115-116
minutissima (Theonoe): 25, 64, **124**, 192, 193, 196, 223, 233
minutus (Lepthyphantes): 28, **179**, 225, 235, 237
mirabilis (Pisaura): 24, 35, 38, 41, 42, **109**, 216, 219, 222, 226, 229, 233, 236
mist: 16
Misterton Marshes: 20
Moira: 19, 83
Moira Junction: 90, 108, 115, 116, 118, 122, 152, 168, 194, 201, 209, 214, 219
montana (Hahnia): 25, **114**, 216, 219, 223, 229, 233
montana (Linyphia): 29, 42, **185**, 218, 225, 228, 231, 235, 237

245

montana (Tetragnatha): 25, **125**, 216, 220, 223, 226, 230, 233, 236
monticola (Pardosa): 24, 77, **104**
moorland: 21, 196, 222-225
Morcott Gullet: 112
Mountsorrel: 78
Murphy, F.M.: 62
mystaceum (Theridion): 25, 42, 53, 54, 69, **120**, 216, 223, 229, 233, 236

Nailstone Wiggs: 12, 113, 127, 136, 157, 164, 165, 193, 201, 209
Nanpantan: 19
Narborough Bog: 20, 32, 79, 112, 125, 135, 136, 144, 155, 158, 166, 201, 209, 214, 226
narrow specialists: 195
National Forest: 199
nationally notable species: 22, 50, 197
Natural Areas: 188-194
Nature Conservancy Council: 67
nava (Hahnia): 25, **115**, 194, 216, 219, 226
nebulosus (Lepthyphantes): 28, **178**, 237
Needwood Forest: 199
neglecta (Clubiona): 23, **88**, 219
neglectus (Robertus): 25, 47, **123**, 216, 233
Nellist, David: 61, 62, 68
nemoralis (Pelecopsis): 27, 77, **149**, 193, 234
NESTICIDAE : 25, 124
Nether Broughton: 86
neutral grassland: 20, 196
New Forest: 61
Newbold Verdon: 83
Newfield Colliery: 194
Newfield Heath: 83, 85, 91, 108, 110, 122, 136, 138, 150, 151, 152, 166, 188, 193, 201, 209, 214, 219
Newton Burgoland: 20, 54, 96
Newton Burgoland Marshes: 87, 150, 201, 209
Newtown Linford: 200, 201, 202, 210, 211
Nielsen, B.O.: 72
nigriceps (Pardosa): 24, **105**, 216, 222, 229, 232
nigrinus (Bathyphantes): 28, **174**, 225, 228, 231, 235
nigrita (Tetragnatha): 25, **126**, 216, 220, 226, 230, 233
nigrum (Dicymbium): 26, **140**, 217, 220, 223, 227, 230, 234
Normanton: 142, 148
Norris Hill: 77, 136, 138, 144
North Brook: 201
North Kilworth: 19
North Luffenham Quarry: 83, 85, 91, 92, 95, 112, 113, 115, 149, 190, 209
North Luffenham,: 100
north-east Leicestershire: 13, 45, 47, 61, 64, 68
north-west Leicestershire: 13, 107
nudipalpis (Walckenaeria): 26, **138**, 220, 223, 227, 230, 234, 237

Oakham: 11, 19
Oakham Museum: 30, 51
Oakley Wood: 165, 197
oblongus (Tibellus): 24, 42, **99**, 215, 219, 222, 226, 229, 232, 236
obscurus (Cnephalocotes): 27, 43, **150**, 194, 217, 220, 224, 227, 230, 234
obscurus (Lepthyphantes): 28, **180**, 197, 218, 225, 231, 235
obtusa (Tetragnatha): 25, 41, **125**, 216, 223, 230, 233
OONOPIDAE : 22, 80
opistographa (Araniella): 26, 34, 35, 37, 38, 41, 54, **133**, 216, 220, 223, 227, 230, 233
Osgathorpe: 19
Outwoods: 32, 67, 77, 91, 115, 138, 141, 145, 154, 165, 176, 184, 193, 201, 208, 214, 232
ovata (Enoplognatha): 25, 38, 41, **122**, 193, 216, 219, 223, 226, 230, 233, 236
Owen, J.: 52-53 61
Owston Wood: 16, 17, 34, 35, 36, 76, 77, 92, 93, 113, 123, 125, 127, 130, 134, 136, 140, 145, 155, 157, 174, 179, 181, 182, 190, 201, 209, 214, 232

pagana (Tegenaria): 44, 47
paganus (Asthenargus): 28, **164**, 193, 197, 224, 234
pallens (Tapinocyba): 27, **153**, 193, 224, 234

pallens (Theridion): 25, 40, 41, **122**, 216, 219, 223, 226, 229, 233, 236
pallidula (Clubiona): 23, 87, **215**, 226, 229, 232, 236
pallidum (Porrhomma): 28, **165**, 192, 224, 235
pallidus (Lepthyphantes): 29, 53, **183**, 194, 218, 221, 225, 228, 231, 235, 237
palustris (Pardosa): 24, **104**, 215, 219, 222, 226, 229
parallela (Pelecopsis): 27, **149**, 194, 217, 220
parasiticus (Thyreosthenius): 27, **154**, 217, 224, 234
parietina (Tegenaria): 25, 30, 33, 34, 35, 37, 44, 47, 48, 49, 77, **111**
parvulus (Bathyphantes): 28, **174**, 218, 221, 225, 228, 235
pasture woodland: 17
patagiatus (Larinioides): 26, 35, 36, 37, 48, **131**, 216, 230, 233
peltata (Linyphia): 29, **186**, 218, 225, 231, 235
penicillata (Moebelia): 26, 77, **141**, 193, 234
perita (Arctosa): 24, 38, 40, 47, **108**, 194, 219
permanent pasture: 12
permixtus (Diplocephalus): 27, **159**, 217, 220, 224, 227, 230, 234
phalangioides (Pholcus): 23, 52, 82, **195**, 236
Phillips, Ian: 62
PHILODROMIDAE : 24, 97-99
PHOLCIDAE : 23, 82
phragmitis (Clubiona): 23, 41, 52, **87**, 226, 229, 236
phrygianus (Pityohyphantes): 22, 29, 50, 69, 77, **184**, 195, 235
picinus (Diplocephalus): 27, 43, **159**, 217, 220, 224, 227, 230, 234
Pickard-Cambridge, O.: 38, 39, 45, 49
Pickworth Great Wood: 17, 77, 80, 81, 93, 98, 106, 134, 140, 182, 184, 190, 201, 209, 214, 232
pictum (Theridion): 25, **119**, 226, 229, 233
Pignut Spinney Marsh: 149, 177, 179, 201, 210
Pilawski, S.: 72
pinicola (Tetragnatha): 25, **125**, 197, 216, 226, 233
pioneer species: 195
Piper Wood: 86, 197
piraticus (Pirata): 24, 43, **108**, 219, 222, 226, 229, 232
PISAURIDAE : 24, 109
pitfall trapping: 64
Pleistocene: 15, 70
praecox (Tapinocyba): 27, **152**, 220, 224, 234
praedatus (Philodromus): 22, 24, 50, 69, **98**
pratense (Baryphyma): 27, 35, 37, 41, **144**, 191, 194, 196, 227, 234
praticola (Ozyptila): 24, 53, **96**, 215, 226, 232, 236
prativaga (Pardosa): 24, **105**, 216, 219, 226, 229, 232, 236
Pre-Cambrian rocks: 13, 14, 70
Prior's Coppice: 16
prominens (Cercidia): **134**, 217
prominulus (Metopobactrus): 27, **144**, 223
proxima (Agroeca): 23, **91**, 215, 219, 222, 232
prudens (Centromerus): 28, **171**, 218, 224, 235
pubescens (Sitticus): 24, **103**, 194, 215, 222
Puddledyke: 135, 162, 163, 186, 201, 210
pulcher (Oonops): 22, 43, **80**, 219, 222, 226, 229, 232, 236
pulicaria (Micaria): 23, **86**, 215, 219, 222, 226, 229, 232
pullata (Kaestneria): 28, **175**, 218, 225, 228, 231, 235
pullata (Pardosa): 24, 40, **104**, 193, 216, 219, 222, 226, 229, 232, 236
pulverulenta (Alopecosa): 24, 43, **106**, 216, 219, 222, 226, 232, 236
pumila (Pocadicnemis): 27, **146**, 220, 224, 227, 230, 234, 237
punctatum (Lophomma): 27, 35, 41, **155**, 220, 224, 227, 230, 234
pusilla (Microlinyphia): 29, 38, 41, 53, **186**, 218, 221, 225, 231, 235, 237
pusillus (Drassyllus): 23, **85**, 190, 215
pusillus (Minyriolus): 27, 77, **152**
pygmaea (Hypsosinga): 26, 60, **133**, 190, 191, 196, 217

pygmaeum (Porrhomma): 28, 54, **165**, 217, 220, 224, 227, 231, 234, 237

quadratus (Araneus): 26, 38, **130**, 216, 223, 233
Quenby Hall: 32
Queniborough: 86
Quorn: 84, 87, 137, 182, 184, 200, 203, 205
Quorn House Park: 198

RA65 recording card: 53, 68
Rainfall: 15
ramosa (Agyneta): 28, **168**, 220, 231, 235
rare spiders: 76
Ratby: 201, 209
Rattrap, D.B.F.: 32
Ravenstone: 161, 183
Ray Society: 30
reclusa (Clubiona): 23, 39, **86**, 215, 219, 222, 226, 229, 232
Red Data Book spiders: 22, 50, 197
red wood ants: 197, 198, 199
redii (Agalenatea): 26, 77, **132**, 190, 216
Redmile: 19, 78
Reference Collection: 65, 66, 69
Reservoirs: 19-20, 87, 97, 229-231
  Blackbrook: 67, 100, 115, 128, 200, 202, 214, 229
  Cropston: 20, 46, 67, 86, 95, 102, 138, 142, 148, 158, 159, 173, 175, 200, 204, 214, 229
  Eyebrook: 80, 125, 126, 129, 131, 181, 200, 205, 214, 229
  Knipton: 201, 207
  Rutland Water: 12, 18, 20, 54, 87, 118 162, 186, 194, 201, 211, 214, 229
  Saddington: 19, 35, 82, 100, 119, 160, 163, 186,187, 201, 211, 214, 229
  Staunton Harold: 20
  Swithland: 20, 67, 79, 103, 128, 131, 137, 142, 160, 168, 177, 184, 194, 197, 201, 211, 214, 229
  Thornton: 20
reticulatus (Neon): 24, **101**, 222, 232
retusus (Oedothorax): 27, 41, 54, **148**, 217, 220, 224, 227, 230, 234
reussi (Silometopus): 27, 53, **150**, 217, 227, 237
ridge and furrow: 12
riparian spiders: 196, 226-228
River Sence Marsh: 201, 210
Rivers: 17-18, 87
  Anker: 17, 18, 190
  Avon: 17, 18, 188
  Chater: 17, 18, 148, 201, 210
  Devon: 17, 18, 189
  Eye, 17, 18, 201, 210
  Eye Brook: 17, 18
  Gwash: 17, 18, 148
  Langton Brook: 17
  Mease: 17, 18, 190
  North Brook: 209
  Saddington Brook: 17
  Sence: 12, 17, 18, 87, 210
  Smite: 17, 18, 189
  Soar: 10, 11, 12, 13, 15, 17, 18, 20, 45, 55, 61, 62, 87, 119, 131, 135, 137, 142, 144, 148, 150, 160, 162, 164, 166, 189, 194, 196, 201, 210, 226
  Swift: 17, 18
  Trent: 9, 10, 11, 17, 18, 19, 20, 142, 145, 152, 162, 164, 189, 190, 194, 201, 210, 214, 226
  Tweed: 12, 17, 18
  Welland: 10, 12, 15, 17, 18, 20, 55, 90, 103, 119, 162, 190, 194, 201, 210
  Witham: 17
  Wreake: 11, 15, 17, 18, 20, 201, 210
Roberts, M. J.: 31, 45, 51, 61, 62, 63, 68, 71, 73
Robin-a-Tiptoe: 12
robusta (Trochosa): 24, 77, **107**, 190, 196, 216
robustum (Leptorhoptrum): 28, **163**, 220, 224, 227, 231, 234
Rothley Common: 120
rubellum (Gonatium): 27, **145**, 224, 227, 234
rubens (Gonatium): 27, **145**, 217, 220, 224, 227, 230, 234

rufipes (Gongylidium): 26, **142**, 217, 227, 230, 234, 237
rufus (Macrargus): 28, **173**, 225, 235
rurestris (Meioneta): 28, **169**, 217, 221, 224, 228, 231, 235, 237
ruricola (Trochosa): 35, 41, **107**, 216, 219, 222, 226, 229, 232, 236
Russell-Smith, A.: 73
rusticus (Urozelotes): 23, 46, **85**
Rutland: 10, 11, 12, 37, 40, 50, 51
Rutland Natural History Society: 30, 51
Ruzicka, Vlastimil: 73

saeva (Tegenaria): 25, 48, 54, 77, **111**
Saltby: 47, 122
Saltersford Brook: 12
SALTICIDAE : 24, 99-103
sanctuaria (Ozyptila): 24, 47, **95**, 190, 215
Sanderson, A.: 62
sandstone: 13
saxatilis (Meioneta): 28, 53, **169**, 217, 221, 224, 228, 231, 235, 237
scabricula (Ozyptila): 24, 77, **95**, 190, 196, 215
scabriculus (Troxochrus): 27, **152**, 194, 195, 217, 224, 227
scabrosa (Ceratinella): 26, **135**, 217, 227
scenicus (Salticus): 24, 35, 41, **99**, 193, 215, 222, 229, 232, 236
sclopetarius (Larinioides): 26, **131**, 226, 230, 233
Scraptoft: 43, 54, 86, 108, 120, 161, 166, 181
scrubland: 16
Seal Wood: 36
Seaton: 78
Seaton Meadow: 77, 90, 103, 104, 110, 135, 160, 190, 201, 211
Seaton Meadows: 55
SEGESTRIIDAE : 23, 82
segmentata (Metellina): 26, 35, 37, 38, 41, **127**, 193, 216, 223, 226, 230, 233, 237
Selden, P.A.: 51
senoculata (Segestria): 23, 42, **82**, 215, 222, 226, 229, 232
Shacklewell Hollow: 146, 190, 201, 211
Sharnford: 17, 205
Sharnford Meadows: 119, 201, 211
Shawell Gravel Pits: 100, 118, 119
Sheepy Wood: 197
Sheet Hedges Wood: 67, 77, 79, 94, 129, 134, 145, 201, 211
Shepshed: 78, 79, 85, 96, 133, 139, 200, 207
Sherwood Forest: 47, 49, 61, 73, 74, 101, 153, 196
signifer (Haplodrassus): 23, **83**, 215, 219, 222
silvestris (Haplodrassus): 23, 77, **84**, 193, 196, 197, 232
silvestris (Tegenaria): 25, 38, **112**, 194, 226, 233
silvicola (Cryphoeca): 25, **113**, 223, 233
simile (Theridion): 25, **120**, 216, 233
similis (Amaurobius): 22, 41, **78**, 193, 215, 222, 229, 232, 236
simoni (Psilochorus): 23, 77, **82**, 195
simulans (Achaearanea): 25, **118**, 194, 219
sisyphium (Theridion): 25, 35, 38, 41, **118**, 216, 219, 223, 229, 233
site collection: 65
site evaluation: 196
site scores: 196-197, 200
Six Hills: 79, 125, 211
Skeffington Wood: 16, 106, 127, 140, 167, 175, 181, 190, 201, 211
Slade, Lorna: 62
slate: 13
Smith, Clifford: 47, 53, 54, 62
Snarestone: 19
Soar navigation: 19
Soar Valley: 11, 15, 67, 190
socialis (Drapetisca): 28, **176**, 225, 231, 235
Somerby: 200, 204
south-west Leicestershire: 12
spider collecting activity: 64
Spider Recording Scheme: 46, 47, 53, 66, 238
spinimana (Zora): 23, 43, **92**, 215, 222, 232
SPIREC software: 8, 68, 69, 76
spirit collection: 65, 66, 68

Sproxton: 201
Squires, A.E. 57, 62
St. Mary's Allotments: 115, 149, 152, 166, 181, 194, 201, 211
stagnatilis (Clubiona): 23, 38, 39, **87**, 222
Stamford: 43
Stathern: 120
stativa (Ceratinopsis): 27, **151**, 217, 220
Stoke Dry: 122
Stonesby: 55, 100
Stonesby Quarry Nature Reserve: 55, 84, 86, 90, 103, 104, 109, 115, 122, 137, 148, 152, 157, 166, 201, 211, 214, 215
Stoneywell Wood: 115, 124, 138, 154, 164, 165, 167, 173, 182, 201, 211, 214, 232
striata (Tetragnatha): 26, 77, **126**, 194, 230
sturmi (Atea): 26, **132**, 223, 233
subaequalis (Micrargus): 27, **157**, 194, 217, 220, 224, 227, 234, 237
subitaneus (Microctenonyx): 27, **153**, 194, 224, 227, 237
subtilis (Agyneta): 28, **167**, 224, 235
subtilis (Clubiona): 23, 77, **90**, 190, 196
sulcifrons (Panamomops): 27, **160**, 217, 227
sundevalli (Maso): 27, **145**, 217, 220, 224, 227, 230, 234
sunshine: 15
Sutton Cheney: 36, 177, 201, 207
Swannington: 13, 19
Swepstone: 201, 209
Swithland: 13, 35, 78, 200, 203
Swithland Wood: 16, 40, 46, 67, 77, 80, 81, 85, 92, 96, 100, 101, 105, 115, 127, 136, 138, 145, 148, 149, 157, 160, 163, 165, 171, 178, 182, 184, 186, 196, 197, 201, 212, 214, 232
sylvaticus (Centromerus): 28, **170**, 218, 221, 224, 228, 231, 235, 237
synanthropic species: 22
Syston: 201, 212

tepidariorum (Achaearenea): 44
temperature: 15
tenebricola (Lepthyphantes): 29, **182**, 197, 225, 235
tenuis (Lepthyphantes): 28, 41, **180**, 193, 218, 221, 225, 228, 231, 235, 237
terrestris (Clubiona): 23, 43, **88**, 215, 222, 226, 232, 236
terricola (Trochosa): 24, 43, **107**, 216, 219, 222, 226, 232
TETRAGNATHIDAE : 25, 124-127
THERIDIIDAE : 25, 116-124
Thistleton Gullet: 112, 113, 133, 190, 201, 212
THOMISIDAE : 23, 93-97
thoracica (Enoplognatha): 25, **122**, 194, 216, 219, 223, 233
thoracica (Labulla): 28, 42, **177**, 225, 231, 235, 237
Thorpe Langton: 86
Thringstone Fault: 12
tibiale (Dicymbium): 26, **140**, 220, 223, 234
Tickencote: 203
Tilton-on-the-Hill: 12, 16, 32, 211
Timberwood Hill: 21, 137, 155, 193
tinctum (Theridion): 25, 41, **121**, 216, 229, 233, 236
Tinwell Marshes: 54, 152
Tixover Quarry: 110, 190
Toft, S.: 72
Trent Valley and Levels: 188-191, 194, 196
triangularis (Linyphia): 29, 38, 41, 42, **184**, 193, 218, 221, 225, 228, 231, 235, 237
Triassic rocks: 13, 14
trifrons (Baryphyma): 27, 77, **144**, 193, 224
trivialis (Clubiona): 23, **89**, 193, 222
trux (Ozyptila): 24, **96**, 215, 219, 232
Tugby Wood: 147, 190, 212
Tullgren Funnel: 64
Twenty Acre Piece: 125, 201, 212

uliginosus (Pirata): 24, **109**, 216
ulmi (Xysticus): 24, **95**, 215, 229, 232
Ulverscroft: 201, 211

Ulverscroft Nature Reserve: 46, 67, 77, 79, 91, 92, 100, 109, 124, 136, 137, 140, 144, 153, 164, 165, 184, 201, 212, 214, 222
umbratica (Nuctenea): 26, 35, 37, 41, **132**, 193, 216, 220, 223, 226, 230, 233, 237
uncatus (Drepanotylus): 28, **163**, 224, 231
uncinata (Dictyna): 22, 41, 79, 215, 226, 229, 232, 236
unicornis (Walckenaeria): 26, **139**, 217, 220, 223, 227, 230, 234
unimproved grassland: 20
Uppingham: 40
Uppingham School Museum: 30, 51

vagans (Prinerigone): 28, **163**, 194, 224, 227, 231, 234
vagans (Tiso): 27, **151**, 217, 220, 224, 227, 234, 237
varians (Theridion): 25, 41, **119**, 216, 223, 229, 233, 236
variata (Torania): 44
variegata (Poeciloneta): 28, 43, **176**, 218, 225, 235
vatia (Misumena): 23, 41, 52, 53, **93**, 232, 236
venatoria (Heteropoda): 44
viaria (Micronета): 28, 53, **170**, 218, 221, 224, 228, 235, 237
Victoria County History: 30, 36
Victoria Park: 33
vidua (Allomengea): 29, **187**, 194, 228, 231, 235
vigilax (Walckenaeria): 26, **139**, 194, 220, 223
Viking Way: 97
virescens (Cheiracanthium): 23, **90**, 194, 215, 219
virescens (Micrommata): 23, 30, 35, 36, 37, 48, 76, 77, **92**, 197
vittatus (Anelosimus): 25, **117**, 216, 219, 223, 226, 229, 233, 236
vivum (Gongylidiellum): 27, **156**, 217, 220, 224, 227, 230, 234
voucher material: 68, 69

Walden, T.A.: 31, 43, 61
Waltham Quarry: 40, 108
Waltham-on-the-Wolds: 43, 84, 106
Wanless, F.R.: 61, 62, 68, 71
Wanlip Gravel Pits: 177, 201, 212
Wardley Wood: 78, 108, 190
Warren Hill: 155
waste ground: 21
Watermead Country Park: 160, 166, 186, 188, 201, 212, 214, 226
Watsonian Vice-county 55 (VC55): 9, 10, 50
Welby Osier Beds: 123, 212
Welford: 19
Welland basin: 188
west Leicestershire: 13
wetland: 17
Whatborough Hill: 12
Whetstone: 82
Whitwell: 51
Whitwick: 11, 199, 200, 206
whymperi (Lepthyphantes): 43, 48
Wild, A.M.: 42, 51, 68, 69
Wild, Robert: 62
Wilson, Jane: 55, 62
wind: 15
Windsor Forest: 61, 73, 74
Wistow: 141, 143
Woodbrook: 67
Woodhouse: 200, 202, 203
Woodhouse Eaves: 70, 103
woodland: 16, 197, 232-235
Worthington: 150, 170, 208, 219
Wunderlich, J.: 71
Wymondham: 171
Wymondham Rough Nature Reserve: 93, 129, 190, 201, 212

x-notata (Zygiella): 26, 33, 34, 38, 39, 41, **129**, 193, 223, 226, 230, 233, 237

zimmermanni (Lepthyphantes): 28, 43, **180**, 218, 221, 225, 228, 231, 235, 237
ZORIDAE : 23, 92

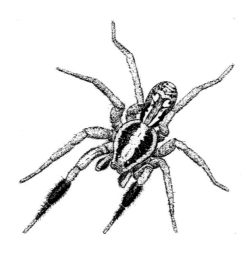

Alopecosa barbipes m. (x3.2)

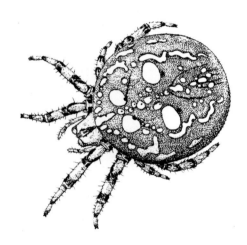

Araneus quadratus f. (x3)